Contemporary Trends and Issues in Science Education

Volume 45

SCOPE

The book series Contemporary Trends and Issues in Science Education provides a forum for innovative trends and issues connected to science education. Scholarship that focuses on advancing new visions, understanding, and is at the forefront of the field is found in this series. Accordingly, authoritative works based on empirical research and writings from disciplines external to science education, including historical, philosophical, psychological and sociological traditions, are represented here.

More information about this series at http://www.springer.com/series/6512

Ling L. Liang • Xiufeng Liu • Gavin W. Fulmer
Editors

Chinese Science Education in the 21st Century: Policy, Practice, and Research

21 世纪中国科学教育：政策、实践与研究

Springer

Editors
Ling L. Liang
La Salle University
Philadelphia, PA, USA

Gavin W. Fulmer
Department of Teaching and Learning
University of Iowa
Iowa City, IA, USA

Xiufeng Liu
Department of Learning & Instruction
Graduation School of Education University at Buffalo, SUNY
Buffalo, NY, USA

ISSN 1878-0482 ISSN 1878-0784 (electronic)
Contemporary Trends and Issues in Science Education
ISBN 978-94-017-9863-1 ISBN 978-94-017-9864-8 (eBook)
DOI 10.1007/978-94-017-9864-8

Library of Congress Control Number: 2016946348

Printed on acid-free paper

This Springer imprint is published by Springer Nature
The registered company is Springer Science+Business Media B.V. Dordrecht

Foreword

When the editors of this book, Ling Liang, Xiufeng Liu, and Gavin W. Fulmer, approached me (and Springer) with a proposal for this book, I was immediately excited and intrigued about the idea of learning about the practice of science education in a part of the globe that remains partially "hidden" from many mainstream sources of scholarship in the field. It was appropriate that our first fact-to-face meeting took place in 2012 in Indianapolis, Indiana, at NARST, where the organization was realizing its new tagline "A Worldwide Organization for Improving Science Teaching and Learning through Research." It was there that three insightful and respected scholars recognized the importance of understanding our field in a global context and using a wide lens to turn that exploration back onto itself with an examination of science education practices and policy in China. Given that China awards a number of advanced degrees in science and engineering equal to or exceeding virtually every country on the planet and is consistently ranked among the top-performing nations in math and science, an examination of the practices associated with the teaching and learning of science is both timely and warranted.

What is interesting, as the editors point out, is the interplay that exists between the unique cultural traditions and values of China and its willingness to integrate research from around the global science education community in ways that are synergistic to that heritage. This balancing act is one that is carefully crafted in a manner that captures the underlying conceptual significance of research conducted in varied settings and translates that work into the Chinese context. This is not to suggest that there is one uniform Chinese context; there are many subcultures and differential community and student needs that need to be addressed. What it does mean is that there is a concerted effort on the part of Chinese science educators to apply, in educationally meaningful ways, the work of the greater science education community. No matter what country we reside in, we need to be vigilant about not becoming insular in our work and in our thinking.

I would never presume to know what it is like to see, think, or feel through an Asian lens. But I do appreciate, in Eisner's (1998) sense of connoisseurship, Asian sensibilities and philosophy. Having studied and practiced a form of Asian martial arts, Isshin-Ryu Karate, for 35 years, and having worked professionally with international colleagues from Hong Kong and Mainland China, Okinawa, South Korea, and Taiwan, I have developed at least a superficial understanding and an unbridled respect for Asian mindsets. Isshin-Ryu provides a good example of what I find quite interesting. Translated, it means "One Heart Way." To practice that art, to achieve the level of a "master," one must realize that the mind, body, and spirit need to work as one, in harmony with the environment and those around you. This is akin to the philosophy of Confucianism that manifests itself in a variety of related philosophies found throughout Asia. At its core, it advances a type of secular humanism that permeates all aspects of Asian culture, including formal institutions, educational settings, social relationships, and the like. I find it quite interesting that the scholars contributing to this volume reflect this core principle in a multitude of ways, sometimes subtly, by concealing multiple meanings just below the surface of the written word, and sometimes overtly, by coaxing the reader to capture, synthesize, and thread connecting ideas together throughout this work. Sometimes, not looking for these relationships is the best approach to finding them, consistent with the Zen-like mindset.

So while the chapters of this book may examine topics familiar to science educators, the manner in which such topics are realized and interpreted likely engenders subtle differences from works expressing, for example, a European or Canadian perspective. The six sections that serve as an organizing theme for this book include Chinese science education reform policies, science curriculum and instruction, environmental and socioscientific issues, science assessment, informal science learning, and science teacher education. The scholars who contributed to these sections reflect perspectives from Hong Kong, Mainland China, and the USA, the latter of which have close ties to Asia.

The editors have crafted well-conceived overviews, advanced organizers if you will, that masterfully provide rich contexts for each of the six sections stated above. I would not, however, be doing justice to these portions of the book by leaving it at that. Each of these section overviews presents a frame of reference, a way of thinking about the works within those sections. Doing so not only enables the non-Asian mindset to enjoy a partial lens into the sociocultural factors that make this work unique but "see" how the writing resonates within the broader science education community. Likewise, the epilogue further embeds these works into a global context. Therefore, the reader has the pleasure of intimately discovering the nuances of Chinese science education as well as the opportunity to position those nuances squarely within the context of our global science education community.

In a recent cross-cultural study examining epistemological orientations and reasoning about socioscientific issues and beliefs about science among students from Jamaica, South Africa, Sweden, Taiwan, and the USA, my colleagues and I found consistent if not universal trends in terms of how students framed their reasoning. Such commonalities are not entirely surprising given how people reflect and contribute to the human condition. Interestingly, however, there were distinguishing characteristics between the Asian (in this case Taiwanese) frames of mind and those of students from other countries in that they scored significantly higher in viewing science as an interrelated network of highly integrated concepts and in terms of conceptualizing scientific understanding and the construction of ideas as reflective and creative acts. Even the Asian students' selection and ranking of "patients" in dealing with aspects of distributive justice in contexts reflected decisions that factored in the patients' broader impacts to others and the community where they resided, indicating a robust interconnectedness to broader societal networks. While we don't presume that Taiwanese students' conceptualizations in this study are isomorphic with all Asian cultures, it did share an underlying current of Confucianism and related Zen-like perspectives as noted above.

It is quite interesting to note that while the concept of systems thinking, coming of age in the second half of the twentieth century, became both popular and useful in thinking about the interconnectedness of themes, patterns, and trends within psychology, biology, sociology, the arts, etc. (Betalanffy 1968; Luhmann 1995; Parsons 1977), the very same idea has permeated Asian culture for hundreds, if not thousands of years, but in much more subtle ways bound up in the Confucianism mindset. The fabric that binds together those orientations forms an epistemological belief system that manifests decisions tempered by forms of benevolence and humaneness in all crafts that are practiced within a community whereby each individual's actions touches others (Shun and Wong 2004). In an educational context, teachers have a responsibility to their students, while students in turn have a responsibility to their teachers. Each is bound to the community by a web of social norms that direct human action toward an arch of dignity, respect, goodness, moral worth, and human fulfillment. True to this philosophy, Professors Liu, Fulmer, and Liang, along with their co-authors, have skillfully crafted a scholarly text that captures how other international works inform the elements of Chinese science education and how other global contexts of science education may be informed by Chinese science education. We all have much to learn from one another.

Tampa, FL, USA Dana L. Zeidler

References

Eisner, E. W. (1998). *The enlightened eye: Qualitative inquiry and the enhancement of educational practice*. Upper Saddle River: Prentice Hall.

Luhmann, N. (1995). *Social systems*. Stanford: Stanford University Press.

Parsons, T. (1977). *Social systems and the evolution of action theory*. New York: Free Press.

Shun, K.-L., & Wong, D. B. (2004). *Confucian ethics, a comparative study of self, autonomy and community*. Cambridge: Cambridge University Press.

von Bertalanffy, L. (1968). *General system theory*. New York: George Brazillier.

Zeidler, D. L., Herman, B., Ruzek, M., Linder, A., & Lin, S. S. (2013). Cross-cultural epistemological orientations to socioscientific issues. *Journal of Research in Science Teaching*, *50*(3), 251–283.

Preface

According to the most recent PISA results released in 2013, Shanghai, China, is the best-performing economy in the world. This repeats the 2009 results, when almost 15 % of students in Shanghai, China, achieved the highest levels of proficiency in all three of the assessment subjects: mathematics, science, and reading. By comparison, the OECD countries had an average of only 4.1 % of students achieving the highest levels of proficiency in all three assessment subject areas. The Shanghai students' consistent top performance in PISA has certainly generated curiosity for people outside China. As PISA seeks to assess not only basic knowledge but also application of that knowledge in situations relevant to everyday life, these results have also challenged the stereotypical image of a Chinese education system that focuses solely on rote learning and memorization. Of course, we understand that, as a municipality directly under the central government, Shanghai is the biggest economic center and one of the most developed regions in China. The science education achievement in Shanghai certainly does not represent what is happening in the entire country. Given that China has the largest science education system and maintains the world's largest supply of science and engineering talents (National Science Board 2012), people in the rest of the world have become increasingly curious about the insiders' stories related to Chinese educational policies, practices, and research. What do Chinese science educators, teachers, and students do to achieve what has been accomplished? What are Chinese students and teachers actually doing inside their classrooms? What educational policies have been helpful in promoting student learning? How do Chinese scholars perceive their successes and/or challenges in the twenty-first century? What lessons can be shared within the international science education community?

The purpose of the book is to answer some of these questions. The chapters provide an overview of science education policies, research, and practices in Mainland China, with specific examples of the most recent development in these areas. Due to differences in language and culture and various other reasons, only a small fraction of science educators and scholars from Mainland China have shared their ideas, research findings, and practices in English science education journals and books to date. Thus, there is a wealth of knowledge and insights that has been out of reach to the international community—especially the Western world. For instance, nowadays, many science educators and/or education researchers in the English world know that "Lesson Study" is an effective teacher professional development process that originated in Japan. However, few people appear to be aware that Chinese teachers have a long tradition in engaging in similar professional development practices called "Collective Lesson Preparation" (CLP or 集体备课in Chinese) in their work routines. The CLP process has provided many formal and informal opportunities for interactions among teachers. In this book, the CLP is introduced as a powerful tool in developing professional learning communities, especially in mentoring novice teachers including student teachers in Chinese schools. As a matter of fact, the CLP is only one of the four most frequently adopted in-service teacher professional development models in China.

As co-editors of the book, we have had the privilege of interacting with many authors in China and conducting collaborative research with some of them over the years. Although we work outside China, all of us have published about Chinese science education research. For example, Dr. Xiufeng Liu edited/authored a volume entitled *Mathematics and Science Curriculum Change in the People's Republic of China* in 1996. Recently, Drs. Xiufeng Liu, Ling Liang, and Enshan Liu have also co-edited a special issue on Chinese science education research in the *International Journal of Science Education* (2012). In the current book, we focus on the major science education reform efforts and changes which have occurred in Mainland China since 2001. This book is unique in the following ways. It is the first comprehensive book that addresses multiple components in the current science education system authored mainly by native Chinese speakers living with firsthand experiences. All section editors are recognized leading Chinese science educators or researchers in the particular field and currently working inside China. The first authors of individual chapters are all currently playing key roles in teaching and/or research in their respective institutions. The chapter contributors include both established and novice Chinese scholars and researchers. Some have completed their doctoral dissertations/theses recently, and some have previously published in Chinese academic journals only. Some graduate students also contributed to the chapters with their supervisors. All chapters have undergone three rounds of internal and two rounds of external reviews. During the first two rounds of internal review, each chapter was examined by a section editor and at least two book editors for content accuracy and significance. The third rounds of the internal reviews were mainly conducted by the three co-editors of the book. The co-editors also provided substantial assistance in content structure and text editing for many chapter authors. Further, Springer provided language editing service prior to the second round of external review. The final version of the entire book was read and edited by both Liang and Fulmer for consistent style and coherent language across the different contributions.

The book is not intended to be an exhaustive review of Chinese science education development, research, or practices. When soliciting chapter proposals, the editors had the following goals in mind:

- Provide a broader overview of the current status of the science education policies, practices, and research in China
- Provide sample empirical studies on some important questions for Chinese science education
- Present unique characteristics of Chinese science education research and practices

The book is primarily written for outsiders using the integrated perspectives of both insiders and outsiders. As insiders, the chapter authors all have the firsthand knowledge of Chinese science education. On the other hand, two of the three co-editors of the book (Liang and Liu) are native Chinese speakers but received doctoral degrees in science education in North America. Both have had personal experiences of K-16+ science education in China and are currently teaching in the USA, while the third co-editor (Fulmer), a native English speaker, has also been conducting some collaborative research with Chinese scholars as an "outsider." Our unique background allows us to integrate both insider and outsider perspectives in writing the section introductions or commentaries. We envision that the primary audience for this book will include science education researchers, comparative education researchers, science educators, graduate students, and state science education leaders and officers in international communities. Some chapters can also be used as supplementary readings for graduate seminars and science education research courses outside of China. Finally, the university faculty of science education inside China may also consider adopting the entire book or certain chapters in their graduate-level research courses. Both Chinese students and science education faculty will find this book helpful as they explore effective ways to share their science education stories and research with the rest of the world.

Philadelphia, PA, USA — Ling L. Liang
Buffalo, NY, USA — Xiufeng Liu
Iowa City, IA, USA — Gavin W. Fulmer

Acknowledgments

It has been our privilege and honor to work with chapter authors and section editors from different parts of China. We thank the series editor, Dana L. Zeidler, and external reviewers for their thoughtful comments, as well as Bernadette Ohmer at Springer for providing extensive copy editing service. We are also grateful for the support we have received from our respective institutions, i.e., La Salle University, USA; the State University of New York at Buffalo, USA; the National Institute of Education in Singapore; and the University of Iowa.

Contents

Introduction: Chinese Science Education in a Global Context

China has the world's biggest science education system and maintains the world's largest supply of science and engineering (S&E) talent. As revealed in the Science and Engineering Indicators (National Science Board 2014), as of 2010, more than 5.5 million first university degrees were awarded in S&E worldwide. Students in China comprised about 24 % of these degrees, compared to about 17 % in the European Union (EU), and about 10 % in the United States. In the past two decades, China's capacity for advanced S&E education has also substantially increased. In 1996, China awarded about 4000 S&E doctorates. In 2010, more than 31,000 S&E doctorates were awarded in China, compared to about 33,000 S&E doctoral degrees awarded in the United States, about 16,000 in Russia, about 12,000 in Germany, and about 11,000 in the United Kingdom. Furthermore, in the recent Program for International Student Assessment (PISA) in 2012 and 2009, Shanghai, China, was consistently ranked as the best-performing economy in the world in math, science, and reading. Without doubt, China's science education has become a point of interest in the world, and study of both its quality and quantity could have an impact on the rest of the world.

Despite this potential contribution, participation to date in exchange of knowledge in science education and research by Chinese scholars has been very limited in terms of publications in English-language journals and presentations in international science education conferences. As a result, a few inside stories about the current reforms of the Chinese science education are known to outsiders, especially to the Western world. This is certainly incompatible with the high impact of the rapid Chinese economic development, its contribution to the world's highest share of the S &E talent, and its role in the globalization of science education. The purpose of this book is to give science educators in China an opportunity to tell their own stories of a decade-long reform directly and collectively to the English-literate world. In this introduction section, we provide an overview of the current education system and some historical context of science education in China, followed by a brief description of the sections covered in the book.

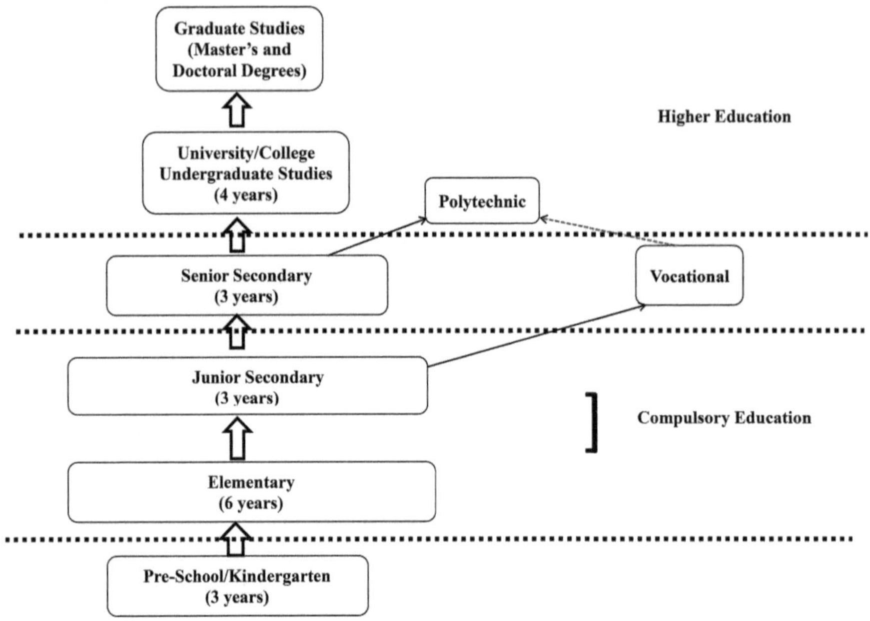

Fig. 1 Structure of the current Chinese educational system

Educational System in China

In the book entitled, *The Clash of Civilizations and the Remaking of World Order*, Huntington (1997) divides the 1990s world into nine major civilizations: Western, Latin American, African, Islamic, Sinic, Hindu, Orthodox, Buddhist, and Japanese. Civilizations as cultural entities are believed to have more enduring influence over educational practices than do political governing systems. Following this civilization-based framework, cultures of mainland China and related Chinese communities in Southeast Asia belong to the Sinic civilization, with Confucianism as a major cultural component. Confucianism, a philosophy advocating self-cultivation, humanism, collectivism, and a hierarchy of human relationships in the society, has been a guiding principle of governmental and educational systems in Chinese history for thousands of years. Chinese people and the government used to be content with a vast society being overseen by a small group of highly educated elite. With the challenges arising from the economic globalization and rapid domestic development in the past two decades, it has become apparent that human capital is the key and a mass education model must be adopted to replace of the traditional elite-oriented education system. Figure 1 illustrates the current structure of the Chinese education system, which consists of the following major categories: early childhood/kindergarten, elementary, secondary, vocational, and higher education including graduate studies. Students with physical or mental disabilities are enrolled in special education schools separately.

In mainland China, there are 31 provincial-level administrative units (including provinces, municipalities, and autonomous regions). The educational facilities and development vary greatly across the nation due to geographical, economical, and political factors. As a response to the demands of an increasingly globalized and highly competitive world, the Chinese government has strived to improve K-12 education and expand enrollment in higher education through issuing and implementing a series of reform policies, establishing new national curriculum standards, restructuring the education system, enhancing decentralization and school-based management, and aiming for quality education for all. In 2014, the kindergarten enrollment rate (for ages 3–5) across the nation reached 70.5 % (MOE 2015). The enrollment rate of school-age children in compulsory education (Grades 1–9) is approaching 100 %, while the higher education enrollment rate (for ages 18–22) was about 37.5 % in 2014. The country is moving toward massive higher education in the twenty-first century. Tables 1 and 2 present the number of students and full-time faculty in 2014 by education levels, respectively.

The latest initiative in China is a national comprehensive campaign to further improve education in the next decade, as articulated in the Outline of China's National Plan for Mid- and Long-Term Education Reform and Development (2010–2020) (MOE 2010). In this document, specific objectives for reforming both basic and higher education through 2020 have been developed and elaborated. It intends to move the nation toward an era of education system promoting quality, equity, and individuality.

Chinese Science Education

Chinese science education can be traced back to 1904 when the government of the Qing Dynasty, the last feudal empire in Chinese history, released a school regulation policy making physics, chemistry, and nature part of school curriculum (Liu et al. 2012). School science was initially imported into Chinese education system almost wholesale by the end of the nineteenth century (Wang 1997), as a result of a craze for Western learning – after the Chinese government was forced to sign multiple "unequal treaties" with foreign powers following repeated military defeats in the modern history of China. The well-known May Fourth movement initiated by college students in Beijing on May 4, 1919, followed by New Culture movement, represented part of a vast modernization movement which sought to reform China through intellectual and social means. Two terms, "Mr. Science" and "Mr. Democracy," were also coined during that time period. Science and technology have since been perceived by the Chinese society as the pathway toward a stronger, independent, and modern China.

The modern Chinese school system with primary, secondary, and tertiary levels was first built on Western models in the early 1900s. Since then, school curriculum

Table 1 Number of students in 2014 by education level

Pre-school/ kindergarten	Elementary school (1–6)	Junior secondary school (7–9)	Senior secondary (10–12)	Special education schools	Undergraduate (4-year)	Graduate	
						master's	doctoral
40,507,145	94,510,651	43,846,297	24,153,687	394,870	15,410,653	1,535,013	312,676

MOE (2015)

Table 2 Number of full-time teachers/faculty in 2014 by education level

Pre-school/ kindergarten	Elementary school (1–6)	Junior secondary school (7–9)	Senior secondary (10–12)	Special education schools	Undergraduate (4-year)
1,844,148	5,633,906	3,488,430	1,670,720	48,125	1,091,654

MOE (2015)

and organization have been influenced by the systems of multiple countries including Japan (at the beginning of the twentieth century), the United States (1920s–1940s), and the former Soviet Union (1950s). During the Cultural Revolution (1966–1976), the school system was largely dismantled during the upheavals in a chaotic society. In 1978, the Chinese government announced and adopted the "reform and opening-up policy" led by Mr. Deng Xiaoping, a top government official and advocate for promoting economic and educational reforms. Since then, many foreign textbooks and scholarly works and materials have been translated and imported to China, while government-sponsored Chinese scholars have also been sent to developed countries to study international trends in learning and instruction, and the development in curriculum and standards.

As a result of the intellectual exchanges and international influences, Chinese science education has gone through the following major stages after 1978 (Wang 2012; Wei and Thomas 2007): (1) The recovery period of the conventional school science curriculum ruined by the Cultural Revolution, from 1978 to mid-1980s. Influenced by science curriculum reforms in the USA and other Western countries in 1950s and 1960s, the main goal of school science was to prepare future scientists and engineers by placing limited resources into a small number of so-called key schools to promote the education of "elite" students. (2) Science curriculum reform in the context of compulsory education, from late 1980s to 1990s. Influenced by the call for "science for all" in the West in the 1980s, China transformed its elite-oriented education system into a mass education system. Both foundational scientific knowledge and basic science skills were emphasized. (3) Curriculum reform in pursuit of quality education, from the end of the twentieth century to the present. Influenced by the literature on the development of "scientific literacy" in the West, the traditional focus of school science curriculums has been expanded into the three-dimensional goals involving the development of scientific knowledge and skills, scientific process and methods, and scientific attitudes and values. This reform also intends to transform the exam-oriented school education into a quality-oriented one emphasizing student development. About 10 years after the release of the trial version of the national science curriculum standards in 2001, the revision of the science curriculum standards was first completed at the junior high school level and published in 2011. The revision of the elementary science standards and the senior high school science standards are still works in progress.

Although the recent development of Chinese science education has been influenced by foreign ideas, the adoption of educational approaches or ideas from foreign countries has never been a process of simple imitation. Many Chinese educators and curriculum scholars, deeply influenced by the Chinese cultural values and heritage, adopt a generally pragmatic view of treating or borrowing foreign ideas: ie, borrow or apply foreign ideas only if they are applicable to the Chinese context (Wei and Thomas 2007). The new round of science education reforms that China has implemented are likely of interest to readers outside China who wish to learn about the current status of science education in China. This book addresses that interest. In its chapters, we present research findings, reflections, and syntheses by many science educators who have played key roles in science education reforms in China over the past decade.

Sections in the Book

In this book, we cover six areas of Chinese science education: science education reform policies, science curriculum and instruction, environmental and socioscientific issues (SSI) in science education, science assessment, informal science learning, and science teacher education. While focusing on China, we also address broader implications about science education worldwide through editors' section introductions and commentaries from an international perspective.

In Part I, Chap. 1, Zhang provides an overview of recent and ongoing science education reform initiatives in China. Specifically, this section reviews major curriculum reform initiatives in elementary and secondary schools, science teacher education, science education research, science education system, science education program evaluation, and informal science education. This chapter provides background on and policy foundation for the subsequent chapters presented in the rest of the book.

Part II, including both literature reviews and empirical studies at K-12 levels, provides a status report on the reformed curriculum and classroom practices in early childhood science education, basic compulsory science education (Grades 1–9), and senior secondary science education (Grades 10–12).

In Part III, the chapter authors examine the current practices of school environmental education and discuss challenges and opportunities of environmental education toward education for sustainable development in Chinese communities. In Chap. 7, Cheng and So describe transitions in the educational policy on environmental education in Hong Kong and mainland China with a comparison to Taiwan. This chapter shows how the policies reflect changes from a focus on protecting the environment toward the concept of education for sustainable development (ESD). In Chap. 8, Lee addresses work on socioscientific issues in Hong Kong and parallel study in southeastern mainland China. Through four studies, he demonstrates how students reason about SSI and examines some of these findings' similarities with and differences from those of Western counterparts.

In Part IV, the chapter authors examine assessment in science education in China from a variety of perspectives ranging from the national to the classroom level. The chapters describe the current policies on assessment in science education within

China, examine the alignment of standardized tests with the associated national curriculum standards, and explore teachers' conceptions of assessment. The chapters will prove valuable reading for colleagues interested in understanding current policy and practice for assessment in science education in China.

In Part V, Chaps. 12, 13, 14, 15, 16 by researchers at the China Research Institute for Science Popularization provide both an overview of informal science education policies and practices and sample research studies in China. Informal science education research is still in its infancy in China; this set of chapters represents the current state of the art for research in this area in China.

Part VI is on science teacher education. First, typical pre-service science teacher preparation programs and the four most common professional development models for in-service science teachers are presented. This is followed by an empirical study on junior secondary biology teachers' misconceptions and its implication for enhancing science teacher professional development. The last chapter authors describe the use of a video case instruction approach designed for pre-service chemistry teachers. Teacher candidates' perceptions of the intervention are also reported.

Finally, in the epilogue, the co-editors of the book summarize the successes and challenges faced by Chinese science education as presented in the individual chapters. Then, we further discuss some broader implications about science education in the international community, followed by recommendations and suggestions for potential collaboration and research. We hope this book will inform science educators around the world on the current status of policy, practices, and research activities in science education in China in order to further facilitate intellectual exchanges between China and the rest of the world.

La Salle University — Ling L. Liang
Philadelphia, PA, USA
Graduate School of Education, — Xiufeng Liu
Department of Learning & Instruction
University at Buffalo, SUNY
Buffalo, NY, USA
University of Iowa — Gavin W. Fulmer
Iowa City, IA, USA

References

Huntington, S. (1997). *The clash of civilizations and the remaking of world order*. Viking: Penguin books.

Liu, X., Liang, L., & Liu, E. (2012). Science education research in China: Challenges and promises. *International Journal of Science Education*, *34*(13), 1961–1970.

Ministry of Education (MOE). (2010). *Outline of China's national plan for medium and long-term education reform and development (2010–2020)*. [国家中长期教育改革和发展规划纲要

(2010–2020)]. Retrieved from http://www.gov.cn/jrzg/2010-07/29/content_1667143.htm (in Chinese).

Ministry of Education. (2015). *Educational statistics in 2014*. (2014年教育统计数据). Retrieved from http://www.moe.gov.cn/s78/A03/moe_560/jytjsj_2014/2014_qg/ (in Chinese).

National Science Board. (2014). *Science and engineering indicators 2014*. Arlington: National Science Foundation (NSB 14-01). Retrieved from http://www.nsf.gov/statistics/seind12/c0/c0i.htm

Wang, J. (2012). Curriculum reform in mainland china, 1978–2008. *Chinese Education and Society*, *45*(1), 59–68.

Wei, B., & Thomas, G. P. (2007). The post-Mao junior secondary school chemistry curriculum in the People's Republic of China: A case study in the internationalization of science education. In B. Atweh et al. (Eds.), *Internationalisation and globalisation in mathematics and science education* (pp. 487–507). Dordrecht: Springer.

About the Contributors

Li Cai is a middle school physics teacher in Nanjing City. She earned her master's degree in teacher education at Nanjing Normal University in 2012.

Bo Chen is a lecturer in the School of Chemistry and Chemical Engineering, Guangzhou University. He received his Ph.D. in science education at the University of Macau. His research interests include scientific literacy, science curriculum materials, science teacher education, and science curriculum implementation.

Ling Chen is a professor and head of the Office of Research Administration at the China Research Institute for Science Popularization. She has a Ph.D. from the Chinese Academy of Sciences.

Xian Chen received her Ph.D. in education in 2005. She is now a professor and associate director in the Department of Subject Instruction in the Teacher College of Nanjing Normal University. She is also a member of the Chinese Society of Education.

Irene Nga-yee Cheng is an assistant professor of the Department of Science and Environmental Studies at the Hong Kong Institute of Education. She is now the external examiner of education courses in the Open University of Hong Kong and also leads a consultant team for developing the general studies curriculum in Macau in 2012–2013.

Lin Ding is an assistant professor of science education in the School of Teaching and Learning at the Ohio State University. His research work focuses on investigating and improving student conceptual understanding, problem solving, and learning attitudes in the physics domain.

Jing Du received her master's degree in science education from Nanjing Normal University in 2012 while teaching physics full-time at the Nanjing Foreign

Language School. She has been working there for 8 years and earned the honor teacher title for her teaching excellence.

Cuidian Feng is a faculty member of the School of Teacher Education of Taizhou University in Zhejiang Province. She earned her Ph.D. in curriculum and instruction from South China Normal University in 2011.

Xiaoyi Gao earned her Ph.D. in early childhood education from Beijing Normal University, and she is an associate professor of education at Beijing Normal University. Her research has focused on early childhood science education, teachers' conceptions of nature of science, science teacher education, and professional development.

Bingyuan Hu is a professor and doctoral thesis advisor at the East China Normal University. His research focuses on nanotechnology education and basic physics education. As a core designer/committee member of the Senior High School Physics Curriculum Standards, Prof. Hu is currently the chair of the Council in Advanced Physics Education Research Association, the director of Nanotechnology Education Communication Center, and the deputy editor of *College Physics* (education special issue) and *Physics Teaching*.

Xiao Huang earned her Ph.D. in science education from East China Normal University. Now she is an associate professor at Zhejiang Normal University and the director of the Science Education Institute of the Zhejiang Basic Education Research Center.

Minyan Jiang is a high school physics teacher in Suzhou of Jiangsu Province. She earned her master's degree in science education from Nanjing Normal University in 2012.

Lyna Kwan currently works as an educational consultant and postdoctoral research fellow at the National Institute of Education (NIE) of Nanyang Technological University, Singapore. She obtained her Ph.D. degree in science (physics) education in January 2009 from NIE.

Yeung Chung Lee is an associate professor at the Department of Science and Environmental Studies, Hong Kong Institute of Education. His research focuses on science-technology-society-environment education, the nature of science, and inquiry-based learning.

Mingyu Li earned her Ph. D. in Biological Science Education from Beijing Normal University. She is now working at the Admissions and Career Development Center in Capital Normal University, and mainly engaged in teacher professional development.

Xiuju Li is an assistant professor of the China Research Institute for Science Popularization. She received her doctorate in science education from Beijing Normal University in 2008. She conducts research in areas of students' scientific literacy and civic scientific literacy.

Cheng Liu earned his Ph.D. in curriculum and pedagogy from Beijing Normal University and finished his postdoctoral program in the Mathematics and Science Education Department at Illinois Institute of Technology. He is presently a faculty member of the College of Life Sciences at Beijing Normal University.

Zhen Lu is a professor and the director of the Institute of Teacher Education Research of Nanjing Normal University (NJNU). Meanwhile he has served as vice president of the Second Affiliated High School (NJNU) since 2010. His research focuses on curriculum and pedagogy, science education, teacher education, and information technology integrated with chemistry instruction.

Hongjia Ma is a professor of Nanjing Normal University. She is currently secretary-general of the Jiangsu Province Association of Chemistry Teachers, the chairman of the Jiangsu Province Committee of High School Chemistry Olympiad, and editor-in-chief of the *Chemistry Teaching and Learning* journal (in Chinese).

Fujun Ren received his M.Sc. and Ph.D. from Harbin Institute of Technology. He is currently the director-general of the China Research Institute for Science Popularization (CRISP), vice chairman of the China Science Writers' Association, vice chairman of the Beijing Association for Science and Technology, and deputy director of the Research Center of CAST and Tsinghua University. He is also a professor of the University of Science and Technology of China.

Shunke Shi is professor and secretary-general of the China Science Writers' Association and has been working with the China Research Institute for Science Popularization since 1988. His main research interest falls in the history of science popularization and the theoretical study of S&T communication development around the world.

Dongsheng Wan is an associate professor at Jiangsu Second Normal University. He received his Ph.D. degree from the Institution of Education at Nanjing University.

Jingying Wang is an associate professor of Capital Normal University in Beijing. She received a doctoral degree in comparative education from East China Normal University in 2009, a master's degree in physics curriculum and instruction in 2006, and a bachelor's degree in physics in 2003. Her research interests focus on science, technology, and engineering education, particularly the learning and assessment in both in-school and out-of-school settings.

Lihui Wang is an assistant professor at the China Research Institute for Science Popularization. She received her Ph.D. degree in philosophy from Beijing Normal University in 2008. She has been engaged in science communication theory study and informal education research.

Weining Wu is an associate professor of physics at Hubei University. He earned his MS and Ph.D. from South China Normal University. In 2012, he was a visiting scholar at the State University of New York at Buffalo. His research interests focus on physics education and educational measurement and evaluation.

Yan Yan is an assistant professor and an editor of *Science Popularization* at the China Research Institute for Science Popularization. Her research areas primarily include mass media science communication and history of science communication. She received a master's degree in Renmin University.

Jie Yuan received a Ph. D. degree in sociology at Minzu University of China and is now working at the China Research Institute for Science Popularization (CRISP) as a Postdoctoral Fellow.

Maorong Zhan received a Master's degree in Curriculum and Instruction from South China Normal University. Zhen is currently working as a learning analyzer at Jing-Han Talent Co. of Science and Technology Education in Beijing.

Baohui Zhang is a professor and dean of the School of Education, Shaanxi Normal University. He received his master's degree in chemistry education from Beijing Normal University in 1994 and Ph.D. degree in educational technology from the University of Michigan, USA, in 2003. His research has been in the intersection of educational technology, science education, and the learning sciences, which focuses on how to integrate ICT into inquiry science learning in formal and informal settings and make the intervention sustainable and scalable.

Hongshia Zhang is a professor and the head of the Institution of Education in Nanjing University. She is a standing committee member of the Chinese Association of Science Education attached to the Chinese Society of Education and a committee member of the Chinese Society of Higher Education. She is also a member of the Undergraduate Education Committee and a member of the International Students Education Committee of Nanjing University.

Huiliang Zhang is an assistant professor at the China Research Institute for Science Popularization. She has a master's degree in mass communications from the University of Leicester and a Ph.D. in education from Zhejiang University. She works primarily on science communication policy study and learning science in the informal education settings.

Xueping Zheng graduated from Biology Department of Guangzhou University. She is currently working as a science/biology advisor at the Advisory Center of Teaching and Learning in Tianhe Guangzhou.

Yan Zhou is a physics teacher in Jiangyin First Middle School of Jiangsu Province. She received a master's degree in science education from Nanjing Normal University in 2011.

Part I
Science Education Reform Policies

Xiufeng Liu

Editor's Introduction: Part I

Education is always closely tied to a country's cultural and political systems. After analyzing the mathematics and science curriculum changes from China's founding in 1949 to the mid-1980s, Liu (1996) concludes that Chinese mathematics and science curriculum changes were the direct result of the social and political changes taking place in the country and a "thermometer of the country's social and political climate" (p. 152). Liu's conclusion applies to the recent science education reforms outlined in Chap. 1. Specifically, although the country's open door to foreign investment policy and market economy started in the late 1970s, major economic reforms that have been transforming the country have only been taking place over the past 20 years or so. There were a few significant political and economic events taking place during this period. During a Chinese New Year visit to southern China in early 1992, China's "paramount leader" at the time, Deng Xiaoping, made a series of political pronouncements intended to further stimulate the country's economic reforms. The 14th National Communist Party Congress later in the year officially adopted Deng's bold vision for market economy-oriented reforms, identifying China's key task in the 1990s as creating a "socialist market economy." In November 2001, China officially joined the World Trade Organization (WTO), marking a milestone in its open-door economic reforms. With significant foreign investments and domestic economic stimuli, such as designating various special economic zones in coastal cities, China has been able to maintain an annual economic growth rate of above 8 %. In 2010, China became the second largest world economy after the USA. The rapid economic development engineered by the Chinese central government has created not only a need to reform its science education in accordance with science education reforms in developed countries but also the financial capacity to implement these reforms. As Zhang and Wan state in Chap. 1, Chinese science education reforms in K-12 curricula, teacher education, informal science education, and so on have been influenced by science education reforms in the USA and other developed countries, consistent with its open-door economic reforms.

Chinese education system is a highly centralized one, and any reforms in Chinese education must be understood within such a centralized context. Although Chinese science education reforms reviewed in Chap. 1 are aligned with science education reforms in developed countries, there are noticeable differences between China and developed countries such as the USA. One obvious difference is that Chinese science education reforms are top down. It is clear from Chap. 1 that every initiative can be attributed to a Chinese central government directive or policy document. For example, the ongoing science curriculum reforms in the elementary and secondary schools were initiated by the 2001 Chinese Ministry of Education directive entitled "Outlines of Curriculum Reform in Elementary and Secondary Education" (MOE 2001). The top-down science education reform is also characterized by systematic implementation. On the other hand, in many developed countries including the USA, science education reforms are often initiated by nongovernmental and professional organizations, indirectly supported by federal governments and followed by voluntary implementation by local governments. This difference is obviously due to the difference in education systems.

The Chinese education system can be characterized as a centralized system because:

1. The teaching program, which prescribes the aims of public education, the school system, subjects to be taught, objectives of each subject teaching, and teaching hours for each subject, is stipulated by the Ministry of Education of China.
2. The content standard and teaching syllabus for each subject, which prescribes objectives, principles of instruction, topics to be taught in the subject, and teaching hours allocated to the subject teaching units, are stipulated by the Ministry of Education of China.
3. Textbooks for each subject of each grade are developed by publishing houses approved by the Ministry of Education and reviewed by a national expert panel appointed by the Ministry of Education of China.
4. The unified high school entrance examination covering all major school subjects is conducted by local education authorities, and the unified university entrance examination covering core school subjects is conducted by a testing agency directly affiliated to the Ministry of Education in coordination with a similar testing agency directly affiliated to the provincial Departments of Education.

Comparing centralized education systems with decentralized education systems such as that in the USA, one clear advantage is that any reform initiative can be implemented nationally in a relatively short time period. Within a decentralized education system such as that in the USA, there have been numerous examples of reform initiatives that have not been fully implemented. Ravitch (2000) reviewed over a century of US school reforms, particularly curriculum reforms, and found that they can be characterized as being disconnected, fragmented, and certainly not fully implemented. Even with such federal legislation as No Child Left Behind (NCLB) in 2001, curriculum reforms remain more a slogan than substance due to a lack of truly national curricula and the alignment of many system components such as teacher education and standardized testing (Carnegie Corporate of New York 2009). In fact, the NCLB legislation is virtually dead now because its renewal is long past due at the time of this writing and most states have been granted an exemption from the law by the Obama administration.

Fully comparing centralized education systems with decentralized education systems is beyond the scope of this introduction. The one thing that seems clear is that there are both advantages and disadvantages with each type of system; it is more beneficial to seek a balance between centralization/standardization and decentralization/local autonomy. A common thread in the recent and current science education reforms in China reviewed in Chap. 1 is the relaxation of central government control in curricula, textbook development, teacher education, and assessment by allowing more diversity and local autonomy. This change may be considered as moving toward decentralization. On the other hand, it seems that the US curriculum reforms have shifted toward more national coordination and standardization in recent years. The Common Core State Standards in Language Arts and Mathematics (Common Core State Standards 2010) have been adopted by more than 46 states plus the District of Columbia at the time of the writing of this introduction, and the new standardized assessments aligned with common core standards were implemented for the first time in Summer 2014. In March 2013, the Next-Generation Science Standards (Achieve 2013), jointly developed by more than half of US states, was officially released and more and more states are in the process of officially adopting it. Standardized testing aligned with NGSS is also under development, and teacher evaluation tied to student performance is currently being considered. Of course, the above moves toward more curriculum coordination and standardization in the USA do not change the nature of the US decentralized education system, which is determined by the US constitution, specifically the Tenth Amendment.

Chapter 1 provides an overview of recent and ongoing science education reform initiatives in China. Specifically, it reviews major curriculum reform initiatives in elementary and secondary schools, science teacher education, science education research, science education systems, science education program evaluation, and informal science education. As a whole, the recent and current science education reforms in China clearly demonstrate not only a common trend with those in western countries in terms of increasing emphases on student conceptual understanding and science inquiry abilities but also an inertia due to China's long history of the centralized education system in terms of resistance to substantive changes.

References

Achieve Inc. (2013). Next generation science standards. Achieve, Inc. Retrieved from www.nextgenscience.org

Carnegie Corporate of New York. (2009). *The opportunity equation: Transforming mathematics and science education for citizenship and global economy.*

Common Core State Standards Initiative. (2010). *Common core state standards.* National Governors Association Center for Best Practices and Council of Chief State School Officers. Retrieved from: http://www.corestandards.org

Liu, X. (Ed.). (1996). *Mathematics and science curriculum change in the People's Republic of China.* Lewiston: Edwin Mellen Press.

Ministry of Education. (2001). *Outlines for basic education curriculum reform (trial version)* [基础教育课程改革纲要(试行)]. Beijing: Beijing Normal University Press (in Chinese).

Ravitch, D. (2000). *Left back: A century of battles over school reform.* New York: A Touchtone Book.

Chapter 1
Status of Chinese Science Education Reforms: Policies and Development Framework

Hongshia Zhang and Dongsheng Wan

K-12 basic education in China is largely a system of public education run by the Ministry of Education (MOE). All citizens in most provinces and districts of the country must attend school for at least 9 years, which is known as "nine-year compulsory education" including 6 years of primary education, starting at age 6, and 3 years of junior secondary education for ages 12–14. Education at the kindergarten level in China has not been fully developed, and the enrollment rate was only 56.6 % in 2010 according to the MOE (2014a). So the term "basic education" in China normally means Grades 1–12 rather than K-12. After compulsory education, students can go to either a senior secondary school for 3 years or a technical school for 2–3 years. The Ministry of Education reported an attendance rate of more than 99 % for primary schools and 80 % for primary and junior secondary schools combined. The gross enrollment ratio for higher education reached 34 % in 2013 (MOE 2014b).

In 2013, there were 213,500 primary schools with 93,605,500 pupils and 5,584,600 full-time teachers in China. The pupil-teacher ratio, including part-time teachers, was 16.76:1. At the secondary education level, there were 52,800 schools with 44,412,900 students and 3,481,000 full-time teachers. The student-teacher ratio was 13.59:1. The entrance examinations for higher education and senior secondary education as well are very competitive. The examination for higher education had been held by the Ministry of Education as a national uniform exam until 2000, when various provincial examination authorities took the place of the

H. Zhang (✉)
Institution of Education, Nanjing University, 22 Hankou Road, Nanjing, Jiangsu 210093, China
e-mail: hzhang@nju.edu.cn

D. Wan
School of Education Science, Jiangsu Second Normal University, 77 West Beijing Road, Nanjing, Jiangsu 210013, China

L.L. Liang et al. (eds.), *Chinese Science Education in the 21st Century: Policy, Practice, and Research*, Contemporary Trends and Issues in Science Education 45, DOI 10.1007/978-94-017-9864-8_1

central government (Zhen and Yang 2003). At present, a more differentiated examination scheme has been applied in the city of Shanghai and Zhejiang Province as a pilot project of the central government aiming at further nationwide promotion in the near future (State Council of China 2014).

Current policies and practices in Chinese education reform can be directly traced back to the historic National Conference of Science and Technology held in 1978. Several months before the conference, Deng Xiaoping, the "chief designer" of the great national reform, had proclaimed that ideological arguments could not help the country realize modernization; knowledge and talents were needed, and an atmosphere of respecting the knowledge and talents in the Communist Party must be built up (Deng 1977). So for preparing the historic national conference, he hosted the National Symposium of Science and Education Development, and dozens of well-known scientists and scholars, some of whom were still treated as "class enemies" or even criminals at that time, were invited by him to attend the symposium. distinguished educator, Mogeng Liu, an alumnus of Tsinghua and determinant talk to the scientists and scholars on reconstructing the national systems of sciences, technology, and education in order to enhance the quality and academic atmosphere in educational and research institutions. He declared that science and technology modernization was the foundation for modernization of all other sectors in China. This instrumentalist policy had dominated almost all sectors of science education until the eve of the new century when the phrases like "quality education" and "children centered" emerged in policy documents. This chapter will review the process of the reform and development of science education in the past few decades in six areas, covering both practical and infrastructural domains: school science education, preservice and in-service science teacher training, the national evaluation system, research in science education, the administrative structure of the science curriculum and textbook compiling, and informal science education. In particular, the underlying policies, consequences, and challenges encountered will be discussed, and suggestions for further research will be provided.

1.1 School Science Education

The development of school science education together with the overall education enterprise in China can be divided into three stages from the end of the Cultural Revolution: the Recovery stage from 1977 to 1985, the Transformation stage from 1985 to 2001, and the Fast Development stage from 2001 to 2013, the eve of the Third Plenary Session of the Eighteenth Congress of the Communist Party of China, which constitutes a milestone for a new era in the country. The transition within these stages is manifest in three historic documents: the *Decision for Educational Infrastructure Reform* issued by the Central Committee of the Communist Party of China ([CCCPC] 1985), the *Outlines for Curriculum Reform in Elementary and Secondary Education* (*Trial Version*, hereafter referred to as *Outlines*) issued by the

MOE (2001a), and the *Decision on Several Major Issues for Deepening the Reform* (CCCPC 2013), which was approved by the Third Plenary Session.

The earliest academic articles that formally applied the concept "scientific literacy" in Chinese curriculum reforms were in the 1990s (e.g., Zhong 1997). "Scientific literacy" was very often interpreted as the combination of (a) the knowledge of science, (b) the investigative nature of science, (c) science as a way of knowing, and (d) the interaction of science, technology, and society (Chiappetta et al. 1991; Bybee 1997). Since then, the objectives prescribed in the national curriculum at all levels have included three dimensions: knowledge and skills, processes and methods, and emotions, attitudes, and values. However, this new concept of science, together with its accompanying educational theories and administrative infrastructure, has encountered many challenges in the Chinese education system during the past decades at all levels and sectors.

1.1.1 Science Education in Primary Schools

The science curriculum "Nature" in primary schools was resumed in 1981 (MOE 1981) soon after the 10-year turbulence of the Cultural Revolution, during which the course "Common Sense" was taught with contents directly related to industrial and agricultural activities (e.g., how to grow rice plants). The objectives of the new science curriculum that was formulated by the Committee of Education of China (1992), the precedent of MOE, moved away from learning sheer technological skills and factual knowledge to scientific literacy, including attitudes toward observations, communication skills for scientific exploration, and the understanding of the relationships among science, technology, and society (STS, Li 1989), which was the pervading theory in Western countries. Chinese scholars and teachers worked hard to assimilate and accommodate the modern curricular ideas and teaching methods into their teaching practice. One distinguished educator, Mogeng Liu, an alumnus of Tsinghua University, had achieved great success using an instructional design that helped develop pupils' competence in observation, discovery, and investigation (Liu 1998). However, most teachers found it difficult to apply or even understand these modern theories in regard to the loose connections among STS in contemporary China. Classroom teaching remained largely focused on factual knowledge and the rote learning style aimed for passing examinations.

This preliminary reform was significantly accelerated by the authoritative document *Decisions on Deepening the Reform and Promoting the Quality Education* (*Quality Education* hereafter) issued jointly by the CCCPC and the State Council in 1999. The phrase "quality education" as a signal of a new revolutionary change in educational aims meant to take an opposite direction from the traditional "examination-oriented education," the real portrayal of Chinese education practice in the past.

As the new century approached, the nation's target for building a knowledge-based economy became more urgent especially under the pressure of the upcoming

2001 entrance into the World Trade Organization. In particular, accompanied with the ambitious call by the central government for building world-class universities at higher education level, curriculum reforms at all levels of education were required to be reconstructed in order to cultivate student's creative ability. As a result, a national expert board then was assembled by the MOE in order to execute the task of curriculum revision at the primary and the secondary education levels. The board had worked for more than 2 years to complete the *Outlines* document which was issued in 2001. The document clearly set up six goals for a new national curriculum across all school subjects: (a) paying attention to children's attitude toward learning instead of factual knowledge transmission; (b) paying more attention to the integrative nature of knowledge, rather than the current segregated structure of curricula, and designing a coherent system throughout the 9 years of the compulsory education period (i.e., years 1–9); (c) strengthening the connection between teaching content and real lives of children; (d) paying attention to the inquiry process of learning in order to cultivate competence in obtaining and retrieving information and analyzing and solving problems and in order to develop communication and cooperation skills; (e) encouraging formative evaluation in order to facilitate children's learning and teachers' professional development instead of focusing on selecting or ranking students; and, finally, (f) changing the centralized curriculum administrative structure into a "national-local-school," three-part consortium model in order to tailor the curriculum to the needs and characteristics of districts and their schools.

Under this general education reform policy, the *National Science Curriculum Standards for the Full-Time Compulsory Education* (*Grades 3–6*) (*Trial Version*) was issued almost at the same time by the MOE (2001b). The importance of this new official document lies not only in changing the title of the curriculum from "Nature" to "Science" and sketching the national curriculum standard but also in raising the status of the subject with a heightened science educational aim: "Science courses in primary schools aim to improve the scientific literacy of children" (MOE 2001b, p. 1). Moreover, this broad aim was broken down into concrete domains of objectives: "scientific inquiry process, the attitude and value towards science, and scientific knowledge" (MOE 2001b, p. 3–4). Eventually, the first *National Science Curriculum Standards for the Full-Time Compulsory Education* (*Grades 3–6*) (*Trial Version*) (MOE 2001b) was established to replace the Nature curriculum, which had been implemented for the previous 10 years.

Accordingly, a nationwide experiment for the new Science curriculum was set up in 2001 in 38 districts of 27 provinces and autonomous regions of China, or about 80 % of all provinces and autonomous regions. Four years after the experiment was initiated, all primary schoolchildren in Year 3 and above in China have had the new curriculum afterward. In 2007, organized by MOE, a meeting called "Second Session of Science Curriculum Standards Revision" was convened, and it proposed further that the Science course should be embodied in school curriculum from the very beginning of primary education, i.e., Year 1. Nowadays this desire has been fulfilled in most urban schools in China.

Generally speaking, Chinese science education at the primary school level in the new century has gone beyond factual knowledge learning. Today, modern theories of science learning have not only been written in the official documents and the national standard of science curriculum but also witnessed in some classroom activities (Wang 2010) such as inquiry learning and cooperative learning as framed by J. Dewey in his monograph entitled *Democracy and Education*. However, the inquiry activities following the prescribed steps in textbooks are often too ossified to adapt to individual students' backgrounds. As a result, classroom atmosphere tends to be either undemocratic or out of control. "Teachers talked too much," as a guest trainer from the USA commented during an in-service teacher training program carried out in Suzhou in 2010 when the first author of this article presented.

1.1.2 Science Education in Secondary Schools

Science education at the secondary level in China has a longer history than at the primary level, but with the same tradition of focusing on knowledge transmission. This inevitably resulted in most students perceiving science as too hard, boring, and irrelevant to their life (Shao 1991). This situation reached its peak during the Cultural Revolution when, for example, the physics course in senior secondary schools in an eastern coastal province even included nuclear physics.

Significant changes took place simultaneously at the secondary school level along with those at primary education level as described above. In fact, the national curriculum standard and the accompanying policy documents for primary education had their counterparts at the secondary education level. In addition to the standard for integrated science curriculum for students in Grades 7–9 (MOE 2001c), there were also a series of curriculum standards for all individual science subjects, including physics, chemistry, biology, and geography, at both junior and senior secondary school levels, which were published in the same year based on the instructional policies prescribed in the *Outlines*.

Although the policies and documents for reform were produced quite efficiently, the majority of teachers' understanding of and ability to implement these reforms took a long time. One big challenge science education reform faced at the secondary level was the traditional subject-based curriculum model. Dating back to the early twentieth century after John Dewey visited China, a kind of integrated science course was offered in junior secondary schools. But it soon returned to the single subject-based model for almost an entire century until the policy document *Outlines* was issued in 2001, encouraging a "move away from the separated nature of single subject-based curriculum through reducing the number of courses and teaching hours while adding a new integrated course to form a balanced, comprehensive curriculum, leaving more space for student choice" (MOE 2001a: p. 1). The integrated science course was resumed subsequently when the *National Science Curriculum Standard for the Full-Time Compulsory Education* (*Grades 7–9*) (*Trial*

Version) was issued in 2001 (MOE 2001c). Meanwhile, the US *National Science Education Standards* (National Research Council [NRC] 1996) and the series "Project 2061" were translated into Chinese.

Just as in primary education, the experiment for the new integrated science curriculum in secondary schools was carried out around 2001, and it encountered more challenges. The experiment was primarily carried out in four provinces (Hangzhou city in Zhejiang, Shenzhen city in Guangdong, Hubei, and Hunan) and one municipality (Shanghai)[1] but only Zhejiang has survived until today. Shenzhen and Shanghai, which are among the most developed cities in China, dropped out only a few years after the experiment commenced. The reasons were mainly poor curriculum resources, lack of qualified teachers, and poor teaching equipment in schools (Cai 2007). Moreover, many schoolteachers in the experimental districts explicitly opposed the reform (Yu 2003; Wang et al. 2007).

The astounding phenomenon of the Shanghai PISA 2009 does demonstrate great progress in Chinese education reform, especially in eastern coastal regions. In terms of the three subscales of reading literacy that were the main target of the year, access and retrieval, integration and interpretation, reflection and evaluation, Shanghai scored higher on the first subscale than the other two. In particular, in the "reflection and evaluation" subscale, which corresponds closely with scientific literacy, Shanghai scored the lowest (Shanghai Group of PISA 2013), as did some other Asian countries and districts (HKPISA Centre 2011). These findings are congruent with an earlier observation from the 2006 PISA, which had science as its focus subject, where it was found that children's self-conception of and confidence in science scores in Confucian Asian countries were lower than most other regions of the world (Ho 2009). Findings showing greater incidences in Chinese students of rote learning and surface learning and a lower development level of critical thinking and psychological well-being are numerous (Sun et al. 2010; Zhang 2010).

The most remarkable development of secondary science education in recent decades lies in the involvement of the integrative content of STS (science, technology, and society) as a component of the curriculum at both the integrated and individual subject levels. In classrooms, more hands-on activities and group work are assigned for students, although these phenomena appear mainly in some of the cities located in developed regions. The challenge resulting from the lack of qualified teachers will be elaborated more in subsequent sections.

[1] The three cities, Shanghai, Shenzhen, and Hangzhou (the capital of Zhejiang province), are in the top 10 of 2014 GDP city ranking in China. Hunan and Hubei provinces are also in the top 10 of the GDP ranking at the provincial level.

1.2 Science Teacher Education

1.2.1 Preservice Teacher Education

Most of the science teachers in primary schools in China graduated from teacher training schools with 2- to 3-year programs at the tertiary level that largely did not have a science major. This situation was changed in 2010 when all of the teacher training schools with 2-year programs have been required to upgrade to normal colleges offering 3- or 4-year programs (Gong 2000). According to a national survey conducted by the authors of this article in 2003 involving 1,737 teacher participants[2] from 21 provinces and autonomous regions, only 29.2 % of primary schoolteachers were graduates with science or math degrees; 16.8 % had Bachelor's degree or above (Zhang and Yu 2004). For longitudinal comparison, a follow-up national survey was taken in 2013 with 2,005 participants, which found that although the degree level rose enormously, in that Bachelor's or above degree holders now made up 61.7 %, the percentage of teachers in science or math degrees decreased to 21.2 % (Zhang et al. 2013).

Most teachers at secondary schools were educated at traditional normal colleges[3] for preservice teaching education where science discipline courses only constituted a part of the 4-year undergraduate program for science majors (Yan 2009). The programs typically utilized factual knowledge-centered curricula and rote learning or implanting teaching methods. In addition, science teacher education in China had been exclusively carried out in the form of separate science disciplines (Hao 2014), such as physics, chemistry, biology, and geography, until 2001 when the undergraduate major "Science Education" was established by the *Outlines* in selected normal colleges and universities (MOE 2001a, b, c). The number of such institutions is currently at least 66 (Zhang and He 2012).

The quality of the new major, however, was not as good as expected according to an investigation which was conducted in 2010 through interviewing 27 college senior students and 25 past graduates from six normal universities offering the program. The reasons identified mainly related to the quality of the educators (Zhang and He 2012). A very recent qualitative study on the undergraduates' views of the nature of science with a sample of normal university students revealed that most of the preservice students in science majors did not understand the concept of experiment, the objective nature of science, and the purpose of cooperation among scientists unless these concepts were transformed to the context of human relationships, which, as interpreted by the authors, shows the impact of the traditional Chinese philosophy of *Zhong Yong* (中庸) on the students' learning

[2] There were a total of 150,983 science teachers in primary schools in 2003, increasing to 173,505 in 2011, according to the website of CERNET (Chinese Educational Research Network): http://www.edu.cn/jcjy_9453/.2014,8,8.

[3] Today there are a total of 108 normal colleges and universities with strong teacher education commissions. Six of these are run by the Ministry of Education.

(Wan 2014). In Chinese academia, *Zhong Yong* can be generally interpreted as a school of dialectics paying emphasis on interpersonal relationship. It was also termed as “naive dialectics” by some Western scholars (Peng and Nisbett 1999).

Currently, preservice teacher education in China, including science teacher education sector, is largely conducted in the traditional normal colleges and universities where student teachers are assigned to individual science departments, where the first 2 or 3 years are devoted to subject-based learning and the last 1 or 2 years to pedagogical learning, which is called the “2 + 2” or “3 + 1” model (Yan 2009). This “normal college-based” teacher education system has been changing gradually since 1999 when the authoritative document *Quality Education* was issued. That was the first time China encouraged “institutions of all kinds of higher learning, including non-normal ones, to participate in teacher training enterprises and to set up educational departments in some of these institutions” (CCCPC and State Council of China 1999, p. 8). This policy was repeated in the *Educational Revitalization Action Plan 2003–2007*, issued by the MOE (2004), which focused primarily on developing western rural areas and the national lifelong education system. A few years later, the Ministry of Science and Technology (MST) (2007) also appealed to more institutions of higher learning and research academies to support and participate in the process of curricular reform, textbook compilation, and science teacher training through taking advantage of their own strengths. However, the implementation process for these policies was not a success. For example, in spite of the number of colleges and universities offering the integrated undergraduate major Science Education reaching 66 (MOE 2011a), only one of them, Southeast University, was a research-intensive university; its program leader was an academician and former Deputy Minister of the MOE who has made tremendous contributions to primary science education reform in the past decade.

1.2.2 Professional Development for In-Service Teachers

With the conditions in the preservice teacher education system as shown above, it is understandable that the quality of school science teachers constitutes a bottleneck in Chinese science education reform. This problem is more serious in science education than in any other subjects because the history of modern science education at any level dates back only 100 years. A recent survey of junior secondary school science teachers’ view of the nature of science was conducted in Zhejiang Province with 222 teacher participants from ten countries or cities where the integrated science course had been implemented for about 20 years. The results showed that the teachers had difficulty understanding nature of science (Wu 2011). Another study using qualitative methods in the same districts corroborated these findings and called for more investigation into teachers’ professional development programs in order to enhance their effectiveness (Wang 2010).

With regard to the serious shortage of qualified teachers of integrated science and the uneven distribution of educational resources across the vast country, a consensus was reached that the nationwide reform of science education must have nationwide scale programs for teacher training in order to support the reform (see Huang and Miao 2015). Therefore, in the *Outlines* it was proclaimed that "normal colleges and universities as well as other institutions of higher learning that participate in teacher training should adjust their programs, curriculum, and teaching methods to fit the new educational aims and targets. In-service teacher training programs should conform with the new policy, for which local governments should make a feasible and sustainable executive plan to ensure that teachers in the experimental schools are supported by the training programs" (MOE 2001a, b, c).

Consequently, the central government organized many kinds of in-service training programs. For example, a national project in 2003 for promoting the integrated science course at the junior secondary level, initiated by the MOE, targeted teachers' inquiry teaching ability. This project trained more than 500 schoolteachers and local curriculum administrators in ten sessions of the program from 2004 to 2007 (Zhang 2007). Another influential outcome of the project was the special science education website called "Xinhua Science Education" on the MOE's *Research for Teaching and Learning* website, which provides curriculum resources and a communication platform for teachers' professional development on the national scale.

Furthermore, in 2006 the State Council also called for cultivating a large number of highly qualified trainers for science teachers by issuing *Outlines of the National Scheme for Scientific Literacy (2006–2010–2020)* (State Council of China 2006). This document was the first one by the State Council specifically concerned with science education. It listed several programs entitled "Fundamental Engineering" which were to be implemented where training the trainers was to be the priority. Following this general policy, the MOE and the Ministry of Finance jointly initiated a "National Teacher Training Program" in 2010 on an unprecedented scale and effort, with a 550 million RMB annual budget from 2010 to 2012.[4] An example of the program is the immediate creation of two pioneer teacher training bases for science education at Guangxi Normal University and Southeast University.[5] Again, the latter was the sole research-intensive university that participated in this teacher training program.

However, the effectiveness of the training programs was not as great as anticipated, given that most of the trainers themselves had the same knowledge structure as their trainees (Ding 2000). In 2012, a national survey of 9,026 trainees and 298 trainers on the quality of the teacher training programs found out that the content of the training courses did not fit the trainees' needs, the teaching mode was mainly lecture, and the trainers themselves had not been adequately trained (Xue and Chen 2012).

---

[4] National Teacher Training Program. http://www.gpjh.cn/cms/gygp/index.htm. Retrieved June 24, 2012.

[5] National Teacher Training Program. http://www.gpjh.cn/cms/gygp/index.htm. Retrieved June 24, 2012.

Teacher training was largely conducted in the traditional form of "expert lecture" for a large audience rather than class discussion or individual presentation in small classes (Bo 2011). Many trainees expect and even prefer, however, to deal with their learning difficulties by themselves after the lectures. This special large audience mode has been viewed as a powerful tool for accelerating the efficiency of large-scale training programs in a country with a large population but limited resources, which definitely needs further systematic research (Zhang 2003a, b). Inquiry learning activities have appeared in a few training bases, especially after the document *Some Suggestions for Strengthening Teacher Training Programs* was issued by the MOE (2011b). In very recent years, the use of digital learning as an educational technology has been changing the situation significantly as many online teacher training programs have been emerging at all levels from the central government to local administrative departments and teacher training institutions (Wu 2012).

Generally speaking, the two "outlines" documents, *The Outlines for Curriculum Reform in Elementary and Secondary Education* and *Outlines of the National Scheme for Scientific Literacy* (*2006–2010–2020*), have played the key role in promoting science teacher training in China in the past decades. These training programs have to date made a considerable impact on filling the huge personnel shortage of teachers and trainers for school science education regardless of its quality. Another up-to-date document by the State Council of China, *Outline of China's National Plan for Medium and Long-Term Education Reform and Development* (*2010–2020*) (2010), further proclaimed "to build up an open and flexible teacher education system with normal universities and colleges as the main platform, with comprehensive universities and the whole society participating and contributing in various forms."

1.3 Evaluation System

The educational evaluation system for national quality control of basic education in general and science education in particular in China is just at its starting stage. At the beginning of this new century, the Chinese educational assessment system was largely a pen-and-paper-based test system (He and Chen 2007) which was operated at the midterm and the end of term. Evaluation activities for school quality improvement were controlled by the central government through their agents in the provincial governments with commitments at the local levels. The data source for the evaluation was confined to schools with no participation from the outside society. The results of the evaluation were kept in the hands of the government with little information open to the public. The general logic of evaluation was that high scores meant good students, good students meant good teachers, and good teachers suggested good schools, which finally led to the conclusion of a good national education system (He and Chen 2007). Furthermore, the implicit presumption before conducting the evaluation was that the system could not be wrong.

Triggered by many international assessment programs such as PISA and TIMMS, the Center for National Assessment of Education Quality[6] (CNAEQ) was established by the MOE in September 2007, at Beijing Normal University. Although the staff of the center is largely scholars from the university, it is completely financed and directed by the central government. CNAEQ functions mainly as a data warehouse to service the central government that is monitoring the reform process and making new policies. It also acts as a coordinator to mediate its branch agencies normally affiliated with universities with strong teacher education traditions or local government agencies. Each branch is responsible for designing the student assessment instrument of a particular subject for certain age groups that is used for the nationwide survey beyond its local area. The constructs of the assessment tools for student learning have been prescribed by the CNAEQ as moral and citizenship literacy, physical and health level, literature and art literacy, and practical and creative ability, all of which have been approved by the MOE. The reality, however, is that the instruments are largely focused on how much knowledge students have remembered from their textbooks. The CNAEQ itself is responsible for designing the questionnaire used to survey teachers and principals nationwide, which is mainly concerned with the issues of the qualification of teachers and the facilities and conditions of education in their schools and affiliated local districts.

The national survey, charged by the CNAEQ, was conducted by the branch institutions responsible for collecting the data of their assigned subjects and age groups by the CNAEQ. The target of the surveys was once limited to the eastern area of China, at the beginning, and now has extended gradually to the inner part of the country in recent years. In 2012, it was reported on the center's website that in the subjects of science and math, the survey sample covered 271 counties[7] with 210,000 pupils[8] from 5,290 primary and secondary schools. In addition, about 5,290 principals and 53,000 teachers[9] of these subjects participated in the survey.

The feedback mechanism of the evaluation system has not yet been well constructed. The findings from the surveys cannot be found in public media; they often appear in oral reports at official conferences normally held in the capital city of a province. In those conferences, the staff from branches of the CNAEQ would report their findings largely in the form of average scores of the districts. The attendees of the conference are mainly local officers, principals, and teacher representatives who would not be permitted to access any specific information in

---

[6] Center for National Assessment of Education Quality http://www.eachina.org.cn/eac/index.htm

[7] There are more than 2,800 counties in China.

[8] There were about 140,000,000 students in primary and secondary schools in 2012 according to the website of CERNET (Chinese Educational Research Network): http://www.edu.cn/jcjy_9453/. 2014,8,8.

[9] There were about 870,000 teachers who taught math and sciences in 2012 according to the website of CERNET (Chinese Educational Research Network): http://www.edu.cn/jcjy_9453/. 2014,8,8.

written form. Specific results of the survey are also not available on the website of the CNAEQ.

Consequently, this agenda of education quality control for K-12 is restated in the new official document *Outline of China's National Plan for Medium and Long-Term Education Reform and Development (2010–2020)* (State Council of China 2010). As one of the targets of the national education infrastructural reconstruction, a multi-criteria evaluation mode has been called for to tailor curricula to the various abilities and needs of children and to serve the developmental educational objectives rather than the traditional competitive ones for selective purposes. The multi-criteria evaluation mode also encourages all interested parties to participate in the process of evaluation, including parents, local communities, and social organizations. In terms of science education in particular, the new multi-criteria instrument has just finished its pilot stage and is going to be applied full scale in late 2014.

China has entered a new era of reform of the whole society, including the education system, which is signified by the issuing of the landmark document from the central party authority at the end of 2013: *CPC Central Committee Decision on Several Major Issues for Deepening the Reform* (CCPCC 2013). The task in the new era will be to focus on promoting the administration-execution-evaluation "three-pipe-run" mechanism and expanding provincial governments' responsibility and raising schools' overall autonomy.

1.4 Research in Science Education

The paradigm of science education research in China had been based on separate subjects, such as physics education, chemistry education, etc., before the 1980s, with little attention to the comprehensive view of science and the cross-disciplinary nature of science education among learning science, philosophy, sociology, and so on (Liu 1988). Entering the 1980s, many scholars turned to massive translation work for introducing modern theories they could access from the developed countries. For example, the monograph *Teaching Elementary Science: Through Investigation and Colloquium* by Brenda Lansdown, a Professor of Science Education at Harvard University, was translated into Chinese. Based on her teaching in person in Chinese classes, she demonstrated in this book that Chinese children, like their counterparts in the USA, could initiate their own observation without previous instruction by their teachers only if the materials and equipment provided in front of them were well structured (Lansdown et al. 1983). Later, some classroom-based observations of Chinese educational practices started to appear in domestic educational journals, although they were rather primitive by international standards (Cai and Chen 2011; Hu 2007).

The traditional mode of Chinese academic study was the same as that for writing an argumentation article full of personal opinions (Zhang 2011). Modern science, as well as many other disciplines in universities being imported into China in the late nineteenth century, is incongruent with Chinese society in many aspects, including the conceptions of what knowledge is, how to generate new knowledge, and how to assess the reliability and validity of the knowledge generation process (Zhang 2010). This has unavoidably caused it to be difficult for most researchers to adapt to the Western empiricist paradigm, so the quality of the limited number of empirical studies is generally poor. In fact, the majority of empirical studies of science education published in recent Chinese journals were conducted by Master's degree students, suggesting a bright future in China. The limitations of these empirical studies are mainly associated with poor representative sampling and little consideration of threats to validities in data interpretation (Zhang and He 2012). In addition to the cultural factor influencing the paradigm of the research, some Chinese scholars pointed to the background of the researchers themselves, who were largely graduates from the traditional colleges with little training in conducting research or inquiry learning (Ding 2000).

In 2001, *Outlines* required major normal universities to set up national research centers for elementary and secondary education practice so as to actively take part in the ongoing curriculum reform (MOE 2001a). Consequently, eight national curriculum research centers were set up across the country, as shown in Table 1.1. In addition, some provincial research centers were also set up in institutions of higher learning by local governments. These centers have played a pioneering role in guiding school practice, training hundreds of teachers and graduates and promoting research quality in science education.

A unique contribution to the development of science education research in China was the "Learning by Doing" project adopted from France, which was jointly

Table 1.1 Eight national research centers and their missions[a]

The centers	Missions and focuses
1. Curriculum Research Center at Beijing Normal University	Evaluations of teacher professional development and student learning and textbook compilation for some subjects
2. Curriculum Research Center at Central China Normal University	Standard development for both the national curriculum and the "integrated practice" curriculum at the junior secondary level
3. Curriculum Research Center at Eastern China Normal University	Evaluation of curriculum development and policies, comparative and international education, and student development and learning science
4. Curriculum Research Center at Southern China Normal University	In-service teacher training and assessment of student learning at the senior secondary level
5. Curriculum Research Center at Northwestern China Normal University	In-service teacher training and school-based curriculum practice and research
6. Curriculum Research Center at Guangxi Normal University	Teacher professional development support and research for *Science for Grades 7–9*
7. Curriculum Research Center at Southwestern China Normal University	In-service teacher training and local school experiments for the new curriculum standard
8. Curriculum Research Center at Fujian Normal University	Information technology application in teaching and learning in schools

[a]The data in this table are mainly from the websites of the institutes

initiated and sponsored in 2001 by the MOE, the Chinese Academy of Science, and the Chinese Association of Science and Technology. The project was led by academician Yu Wei, who was a Deputy Minister of the MOE during 1990s, and involved hundreds of principals and teachers who have now become leaders of science education research and practice in Beijing, Shanghai, and Jiangsu provinces (Ye 2011). The project now is expanding to more provinces such as Jilin in northeastern China. Another influential project worthwhile to mention is the experimental research on the Science and Technology for Children (STC) curriculum for application in Chinese classrooms, which was originally developed by the Center of Science Resources in the USA. The experiment was approved by the Department of Basic Education of the MOE in 2006 and was carried out for 6 years in 20 classes in ten primary schools nationwide from Year 1 to Year 6. The project has trained hundreds of high-quality trainers with the help of professional trainers sent by the center and has published a series of materials in case studies.

At the end of the twentieth century, the ideas of postmodernism spread over Chinese research journals in the social sciences including education, which was coincident with the strong domestic conservative movement of antiscientific methods while upholding Confucianism. Paradoxically, this conservative antiscientific ideology embraced John Dewey's children-centered theory by assuming that Dewey was the founder of constructivism while neglecting the fact that Dewey was a founder of pragmatism and saw "science is a name for knowledge in its most characteristic form" (Dewey 2001, p. 196). This radical constructivism in China recapitulated the progressive movement that took place in the USA in the early twentieth century (Zhang 2005). One of the consequences can be demonstrated by the survey results conducted by Liang et al. (2008) that Chinese teachers possessed higher scores in views of nature of science assessment in most of the advance dimensions than their American counterparts except the most traditional one: "observation and inference." This kind of paradoxical phenomenon in non-Western countries like China may indicate that science education should follow an evolutionary course, so different countries with different development stage of science and the corresponding society should have different science education in terms of both curriculum and teaching methods, although the territory of science itself has no boundary between countries. Considering the fact that modern science did not originate from Chinese culture, Chinese education should place emphasis on the classical elements of science, such as the importance of observation and the objectivity of data collection, before incorporating any postmodernist thoughts into the curriculum (Zhang 2003c). Some Chinese researchers, therefore, raised the issue of curriculum localization for science education and argued that the most urgent task for Chinese researchers is to create "a true, relevant, and affordable science curriculum for Chinese children" (Wei 2008; Zhang 2002).

In order to prepare high-quality researchers for the future, doctoral programs in Science Education were established 4 years ago in a few top normal universities such as South China Normal University and Southwest University, although this is obviously far from sufficient for this vast country.

Another significant event for Chinese science education research was the foundation of the Chinese National Association for Science Education in November 2009, which is affiliated to the Chinese Society of Education. The association not only serves to provide a platform for academic research but also to improve the communication between practicing teachers and researchers. Just a few months after the founding ceremony, 14 science education research projects were listed as the Special Projects in the 2010 Scheme of National Educational Research, which was the first time science education research topics were included in this scheme in this country.

Generally speaking, the most significant improvement of the research community in recent decades has taken place in the domain of organizational construction rather than research conduct itself. Thus, there have not been sufficient numbers of researchers and expertise in China for either national science curriculum standard formation or textbook compilation. In addition, most of the scientists and professors in outstanding universities are rarely interested in these jobs (Ding and Luo 2005). Therefore, there has not been a special academic journal in the science education field in China to date. Nevertheless, some specific institutions (see Table 1.1) designated by the central government to be in charge of both research conduct and teacher training, like those in Beijing Normal University, Guangxi Normal University, and Southwest University, have indeed recently played a key role in the development of science education research in China. This "Confucian model" that emphasizes the organizational function over individual or local special needs has shown some strength at present (Marginson 2011). But its limitations will emerge as soon as the organization has been established, because unlike commercial activities, academic endeavors have to be developed in a scientific and democratic environment.

1.5 Administrative Structure of Curriculum and Textbook Development

The efficiency of the well-known government control model of modernization in Confucian Asia (Marginson 2011) can be clearly demonstrated by the development process of science education in China since 1949 when the Soviet Union's education system model was adopted by the new government of the People's Republic of China. The model ran efficiently until the end of 1950s when the diplomatic relationship between the two countries was broken. The first Nature curriculum was issued in 1956 uniformly for the entire country (MOE 1956). However, the revolutionary province-based curricula issued in place of the uniform national curriculum during the Cultural Revolution from 1966 to 1976 proved to be ineffective or even disastrous. Thus the priority of the education reform which commenced in 1978 was to rebuild the common national education curriculum.

This government control model was also adopted in research and teacher training domains in addition to the curriculum domain. The themes of research projects and the funding mechanism are controlled by the government at all administrative levels. Even international cooperative projects are also largely monitored by the government because many foreign organizations nowadays have been known to take advantage of the Chinese infrastructure so as to ensure the effectiveness of their projects. In particular, the National Institute of Education Research (including its publishing house) is directly controlled by the MOE and has played a key role in the process of education reform. For example, with funding from UNESCO, the institute convened a national symposium in Jiangsu Province in 1985 on Chinese science teachers' quality, and the STS program was initiated at this conference.

This administrative model also penetrated into the textbook compilation system, since the publishing houses were responsible for training their customers. In fact, before 2001 many training programs were assigned to certain publishing houses affiliated with the MOE or local governments. Even today, after being transformed into private commercial companies, publishing houses still rely strongly on the government because the latter decides which publishing companies are qualified to participate in projects of new textbook compilation and associated teacher training programs.

The government control model began to change in 1985 when the issue was addressed in the historic document *Decision for Educational Infrastructure Reform*. From that time on, the power of the central government gradually moved to provincial governments. For example, a province-based curriculum development project was initiated and experimented with in Zhejiang Province first and then Shanghai 3 years later. In 1989, the textbook compilation scheme was further changed from the original single set of textbooks designated by the MOE to a national-local collaboration model in a half-market system with several private publishing companies competing with each other.

This national-local collaborative model was further extended to that of a national-local-school one referred to as the "three-party consortium model" after the publication of the document *Decisions on Deepening the Reform and Promoting the Quality Education* in 1999 (CCCPC and SCC 1999). Provincial governments were empowered to make decision about when and how to execute the national curriculum or even make their own local curriculum programs. Schools were also empowered to develop or adopt a curriculum to fit their own strength, traditions, and special local needs.

The government control administrative model of curriculum and textbook development changed further after entering the new century, from prescribing curriculum and textbook contents to quality control on the basis of the national standard publication. The former government-nominated board for supervising textbook writing has been replaced by the textbook examination committee. As stated in the document *Outlines*, "All textbooks, including those following the national curriculum standard as well as local products, which are intended to be distributed to other provinces should be reviewed by the national textbook examination committee. Local texts should be reviewed by the provincial committee. In addition, the writers of textbooks should not serve as members of the committee taking part in the process of review.... A competitive bidding mechanism should be in place for textbook publication."

It has been legislated that the national curriculum standard at the top level of the curriculum administration system should play the key role in ensuring the right of children to receive compulsory education and in linking the different stages of education. Local governments should pay attention to local needs while implementing the national standard. Therefore, the space for schools to creatively implement the national and local governments' requirements has been enlarged. This school-based model has special value for science education with regard to the rationale that inquiry learning for children is best presented within the local environment where they live.

Generally speaking, the Chinese centralized power structure of curriculum and textbook development has been in the process of devolution. Another significant milestone in this progress is manifested by the open and transparent process of creating the new policy for the mid- and long-term reform framework from 2010 to 2020 (State Council of China 2010). For instance, the draft of the document was put in the Internet for public consultation. Although there is little in this document directly concerning science education specifically, the general policy orientation and the scientific and democratic attempts implicated by the process definitely allude to a bright future for science education development in China.

Indeed, China has entered into a new era of reform in all sectors of its society, including education, which has been signified by the 2013 issue of landmark document from the party central committee: *CPC Central Committee Decision on Several Major Issues for Deepening the Reform* (CCPCC 2013). The tasks and purposes in the document are sharply clarified. One task is further constructing the separation mechanism for the administration, execution, and evaluation processes. For this purpose, provincial governments' rights and the schools' overall autonomy will be expanded, and the social organizations for educational evaluation will be developed. Furthermore, social resources are being encouraged in the conduct of education.

1.6 Informal Science Education

Chinese informal science education originated at the end of the 1980s when a national pilot survey of public scientific literacy, organized by the Chinese Association of Science and Technology, was conducted by use of the common inventory issued by the International Society of Science Education. The formal National Survey of Scientific Literacy of the Public has been conducted at least eight times since 1992. Based on the alarmingly negative results from the survey, many arguments for acceleration science education inside and outside schools pervaded the media. A few years later, the first National Law of Science and Technology

Popularization in China was approved by the Chinese People's Congress in 2002 (State Council of China 2002), which was also a response to the national strategy of raising the country's power and sustainability via science and technology advancement. This indicated that the top leaders of the country had recognized that it was imperative to promote public scientific literacy in order to not only benefit economic health but also social welfare. The law outlined a framework for the responsibility of the government, the duties of related social organizations, and the measures to ensure the effectiveness of public science education. It also specified the objectives, contents, and approaches of the popularization task.

In terms of the administrative duties, the law requires that governments at all levels must make a great effort to popularize science and technology by including it into their regional plans of economic and social development so as to ensure adequate support conditions and environment. By defining the popularization as a benefit for all of society, the law requires all kinds of social institutions to take part in the endeavor, including schools of all types at all levels, institutions of scientific research and development, academic associations of all disciplines including natural and social sciences, and organizations in the medical and healthcare sectors.

These authoritative arrangements have special importance for present-day China, since the attention of the country has been largely paid to the economic function of science and technology, especially focusing on prestigious organizations of scientific research and technological innovation. Chinese scientists normally look down upon and hesitate to participate in popularizing work since it is of no importance in academic and personnel promotion evaluation, especially in research-intensive universities. Furthermore, 19 of the 21 scientist interviewed did not agree to include the work of popularization into the evaluation criteria for their research projects (Wang and Li 2010). They argued that conducting research to generate new knowledge was the primary duty of scientists or professors, for which there had been already too many hard evaluative criteria; scientists' work would be interrupted negatively if a new criterion of popularization was added into the evaluation system. This reveals that several generations of scientists in this country lack both a basic awareness of the humanistic function of science and a profound view of the relationship between science and social development, a deficiency which is embedded in the long historical roots of instrumentalism in the nation (Wu 2002).

The weak point of the science education system in China, as stated earlier, is the gap between normal universities and research-intensive universities (Wan and Zhang 2011). Moreover, the segregation also exists among the MOE, the Ministry of Science and Technology, and the National Science Foundation of China, since the latter two have nothing to do with the former or even the informal education endeavor. On the other hand, the MOE and its local educational authority subbranches have paid little attention to informal science education. Science popularization in China is a job solely assigned to the Association of Science and Technology at both the national and local levels. So, unlike the National Science Foundation in the USA, for example, requiring its funding applicants to clarify their methods to popularize research findings, the Chinese counterpart, the National

Natural Science Foundation of China, addresses little of this issue. Also, unlike many famous universities in the Western world that maintain their own museums or laboratories of science and technology which are open to the public, there is none in this country so far.

The current challenge is how to encourage more participation from various organizations in the academic and social sectors through establishing an effective collaboration mechanism among them. In the recent document, *Outline of China's National Plan for Medium and Long-Term Education Reform and Development*, the task of comprehensive universities' participation has been specified as one of their institutional duties: "Science popularization should be embraced as one of the duties of social service of all institutions of higher learning in order to promote the scientific literacy of the public along with their humanistic literacy." Very recently many associations of science and technology at the university level have been newly founded at certain top Chinese universities' campuses with solemn inaugural events. However, they seem to be treated as the same kind of routine activities that would happen whenever dealing with the brainstorm commands from the central government, so the members of the associations are almost the same as those in the senate of the university. The issue of the working mechanisms of those associations has not been taken into the agenda.

As the law is configured, it is important that the tasks and approaches of popularization go beyond the form of factual transmission of knowledge and should emphasize scientific methods, reasoning, and aspiration. However, there have not been clear and consensual conceptions among the experts such as museum curators and schoolteachers about what scientific reasoning is and how it is different from pseudoscience. In fact, pseudoscience has been persistent in public media and publications, which has tragically coincided with the radical postmodernist ideology that pervaded Western countries in past decades, and could be harmful to Chinese science education practice due to the nonscientific tradition of Chinese culture (Zhang 2003c).

Although the Law of Science and Technology Popularization had definitely brought about new perspectives and a promising future for the enterprise in this country, the mechanism for monitoring practicable approaches was not developed. Therefore, the State Council of China issued the *Outlines of the National Scheme for Scientific Literacy (2006–2010–2020)* (2006) which prescribed the main targets, contents, and means in the following 5, 10, and 15 years, respectively. More specifically, a series of key programs were promulgated via the document: Scientific Literacy Activity for Youth, Scientific Literacy Activity for Farmers, Scientific Literacy Activity for Citizens, and Scientific Literacy Activity for Public Servants. Related administrative measures for these key programs were then also described. For example, the financial and personnel support projects for the programs in the next 5 years were made clear through establishing a series of "engineering projects" including Science Education and Training Engineering, Resources Development and Sharing for Science Popularization, Media Empowerment for Science and Technology, and Infrastructure Building for Science Popularization. Very recently, a new education program specially tailored for a career in informal education at the master's degree level has been created jointly by the Chinese Association of Science and Technology and a few top universities mainly located in Beijing and Shanghai. The students in these programs will work in science parks and museums after their graduation.

These policies issued by the State Council seem to have had some effectiveness. For example, the TV programs about nature, science history, and the modern development of science and technology appear more frequently nowadays especially on Chinese Central Television. Some big research projects from the Ministry of Science and Technology leave space for science popularization. Local schools, universities, and other social organizations, however, have not fully understood their duties, for it is hard to see any substantial progress in popularization activity apart from numerous associations having been funded in a rush at university campuses, as mentioned earlier. Therefore, it is imperative for China to find a more practical resolution to encourage and combine every resource in the academic and civil world to break through the segregated administrative mechanism of policy formation. Of course, it would not be an easy endeavor in China because these persistent problems are caused by the dilemma that the ultimate goal of popularization is to overcome the unscientific traditions and habits of society from which the reform leaders themselves have benefited.

1.7 Conclusion and Suggestions

Generally speaking, Chinese science education reform and development in the past decades can be divided into three stages, Recovery (1977–1985), Transformation (1985–2001), and Fast Development (2001–2012), and now it has entered a new era of reform with "internal adjustment" as its focus (Table 1.2), as designated in the report of the Third Plenary Session of the Eighteenth Central Committee of the Communist Party held in 2013. In terms of school science education, it has experienced the common sense model in the early 1980s, and the later factual knowledge-centered model, and the current model with learning theories being implemented into classroom activities through enacting the policy of national curriculum standards. The infrastructure of the endeavor has also been reformed. Many professional organizations have been established recently, including the Chinese Association of Science Education and the Center for National Assessment of Education Quality. More fundamental changes also took place in the past decades. The intensive centralized administrative system established in the 1950s has been gradually changed, especially after entering the new century. The MOE that once controlled everything including curriculum development, teacher training programs, and textbook compilation now has limited its responsibilities only to accreditation and evaluation undertakings. The research paradigm of science education has also shown apparent progress in the past decade though the dominant style remains controversial.

Table 1.2 The stages of science education development in China

	Restoration (1977–1985)	Transformation (1985–2001)	Fast development (2001–2013)	Internal adjustment (2013–future)
Curriculum	"Common Sense" course in primary schools; factual knowledge-centered learning with rote method	"Nature" replaced "Common Sense"; modern learning theories and STS ideas were imported from abroad	"Science" replaced "Nature"; integrated science course at secondary schools, the series of national curriculum standards	The national curriculum standards are in the process of revision
Teacher education	"Normal college-based" teacher education system in separated subject-based teaching mode	The old system was changing slowly with "lesson study" or "expert lecturing" dominating for in-service training	Many national programs and institutions were set up in normal colleges and universities; online training programs flourished	The system of normal university domination with participation of other universities has been established
Evaluation	"Pen-and-paper" examination of rote learning		The national central CNAEQ was set up	More social organizations will participate
Administration	The central government control model	Dilution of the central government power	National-local-school, the three-party consortium model	Administration, execution, and evaluation will be separated
Research	Introducing modern learning theories mainly through translation work	Argumentative and comparative papers on how to reform education in China	Eight research centers and the national data bases were set up; primitive empirical studies appeared	More focus will be paid on "Chinese Issue" in order to search for the "Chinese Model"
Informal education	Spontaneous activities in spare time relying on individual interest of some writers and scientists		The first law of science and technology popularization was approved and followed by some professional organizations being founded	The increasing social crisis caused by low scientific literacy will push the business into a substantial development stage

The main challenges that reform encountered in past decades were related to an autocratic administrative system that might be more effective than a democratic system for early stages of a restoration and transformation characterized by massive adoption from developed countries, but may be problematic as soon as the fast development stage is entered, because only a democratic system can guarantee the creative production of policies that will fit the variety of needs of a vast country. Thus, reform has been going into a new era of directly challenging the administrative mechanism so as to build up a more democratic system.

The fundamental challenge for school science education practice is the stereotyped and authoritative conception of knowledge by teachers and researchers that is embedded in traditional Chinese culture. Chinese epistemology perceives knowledge as the permanent and transcendent truth which is inseparable from the human mind. This belief obstructs researchers and teachers from understanding the nature of science as well as related educational theories. Rote learning and implantation teaching styles, the traditional teacher training model, and the dominant rhetoric research style are the consequences of this culture.

Thus, how to design teacher training programs to help teachers in both the preservice and in-service domains change their innate beliefs about knowledge will be a fundamental task in the future, especially for the science education sector. Science education reform in China as well as in other developing countries should follow the evolutionary law rather than the revolutionary one by putting more emphasis on classic science concepts and beliefs before introducing postmodernist ideas of science. This argument, of course, should be treated as a fundamental research topic for researchers in China and abroad.

Given this unique Chinese situation, a balance should be cautiously kept between the more fashionable postmodernist model of science teachers' professional development and that of classic conceptions of science as an objective and truth-searching endeavor (Wang 2004). In this regard, the government should act once again as a strong leader to push professional scientists and research-led universities to join this great fundamental work for their beloved science.

In conclusion, science education in the world could be divided into two models. One might be called "Education for Creative Learning" and the other "Education for Massive Learning." Chinese science education largely belongs to the latter. The unique Chinese collectivist way of running education enterprises, which has been developed over thousands of years and proven appropriate to the condition of limited resources with a large population, has also been proven very powerful in modern times. This might also imply that the government-centered model is effective for popularizing science knowledge in a developing country in its early stage. However, this model might not work so effectively when the reform gets deeper. Therefore, Chinese science education development in the future should not rely solely on transforming the teacher education system by embracing research universities' manpower, but also on the whole nation's elevation of scientific literacy of the people and the democratic system of the government, which have been addressed specifically in the emerging historic document, *Decision on Several Major Issues for Deepening the Reform* by CPC Central Committee. Of course, all these issues would constitute research topics in the near future.

Acknowledgments This research is sponsored by the Humanity and Social Sciences Foundation for universities in Jiangsu Province as one of the products of the project "Exploratory Research on Star Talents Education Model." 2010ZDIXM018.

References

Bo, L. (2011). Continuance education curriculum exploration for primary science teacher education (小学科学教师继续教育课程体系探索). *Continuance Education* (继续教育研究), *9*,122–124. (In Chinese).

Bybee, R. W. (1997). *Achieving scientific literacy: From purposes to practices*. Portsmouth: Heinemann.

Cai, T. Q. (2007). A review of the *National standard of science curriculum* for grades 7–9 (trial version) ("全日制义务教育科学(7-9)年级课程标准"述评). *Global Education* (全球教育展望), *36*(1), 88. (In Chinese).

Cai, T. Q., & Chen, L. H. (2011). A review of commentary science education research in China (我国科学教育研究述评). *Global Education (*全球教育展望), *40*(6), 74–83. (In Chinese).

CCCPC. (2013). *Decision on several major issues for deepening the reform* (中共中央关于全面深化改革若干重大问题的决定). Beijing (Official document), (In Chinese).

CCCPC., & The State Council (SC). (1999). *Decisions on deepening education reform and fully promoting the quality education* (中共中央国务院关于深化教育改革全面推进素质教育的决定). Beijing: People's Education Press. (In Chinese).

Central Committee of the Communist Party of China (CCCPC). (1985). *Decision for educational infrastructure reform* (中共中央关于教育体制改革的决定). Beijing: Ministry of Education of China. (Official document). (In Chinese).

Chiappetta, E. L., Fillman, D. A., & Sethna, G. H. (1991). A method to quantify major themes of scientific literacy in science textbooks. *Journal of Research in Science Teaching, 28*, 713–725.

Committee of Education of China. (1992). *The curriculum of "Nature" for the full-time nine-years (1–9) compulsory education (trial version)* (九年义务教育全日制小学自然教学大纲(试用)). Beijing: People's Education Publication. (In Chinese).

Deng, X. P. (1977). *Selected work of Deng Xiaoping*. Vol. 2. [EB/OL]. (邓小平文, 第2卷). Retrieved from. http://www.wxyjs.org.cn/GB/186508/186513/186675/186677/. (In Chinese).

Dewey, J. (2001). *Democracy and education. A Penn State electronic classic series publication*,. p. 196.

Ding, B. P. (2000). Science education: A new research field in China (科学教育学: 一个新兴的教育研究领域). *Studies in Foreign Education* (外国教育研究), *27*(5), 12–16. (In Chinese).

Ding, B. P., & Luo, X. K. (2005). The characteristics of American science education reform at the primary and secondary levels: some reflection of Chinese science education (美国基础科学教育改革及其主要特点----兼谈加强我国科学教育研究). *Journal of Capital Normal University (Social science edition)* (首都师范大学学报 (社会科学版)), *4*, 102–103.(In Chinese).

Gong, L. S. (2000). Several issues concerning enhancing the reform of teacher's colleges (关于加快师范专科学校改革和发展的几个问题), *Journal of Bingzhou Normal Colledge, 21*(5), 1–6. 郴州师范高等专科学校学报. (In Chinese).

Hao, Q. L. (2014). Teachers preparation for integrated science course (综合科学课教师队伍研究). *Continuance Education* (继续教育研究), *5*, 66–68. (In Chinese).

He, K., & Chen, J. (2007). Problems and suggestions concerning the construction of assessment system for basic education (基础教育评价的问题分析与对策). *Educational Review* (教育评论), *5*, 44–47. (In Chinese).

HKPISA Centre. (2011). T*he fourth HKPISA report PISA 2009: Monitoring the quality of education in Hong Kong from an international perspective*. Hong Kong: HKPISA Centre.

Ho, E. S.-c. (2009). Characteristics of East Asian learners: What we learned From PISA. *Educational Research Journal (Hong Kong), 24*(2), 327–348.

Hu,W. P. (2007). The trend and prospective of Chinese science education research (科学教育的研究趋势与展望). *The Journal of East China Normal University* (Educational issue) (华东师范大学学报 (教育科学版)), *25*(4), 44–51. (In Chinese).

Huang, J. Y., & Miao, B. Q. (2015). *A prediction for the shortage of science teachers in China between 2015–2020* (我国2015—2020年义务教育科学教师缺口量预测). doi:10.13445/j.

cnki.t.e.r.2015.04.002. Retrieved from http://www.cnki.net/kcms/detail/11.5147.G4.20150709.1007.002.html

Landsdown, B., Blackwood, P. E., Brandwein, P. F., Translated by Chen, D. Zhang, T. (1983). *Teaching elementary science: Through investigation and colloquium.*兰本达、P.E.布莱克伍德、P.F.布兰德温著,陈德彰译《小学科学教育的"探究—研讨"教学法》. Beijing: People's Education Press. (In Chinese).

Li, P. S. (1989).The educational reform of the course of Nature in primary schools: A review (小学自然教学改革的回顾与展望). *Chinese Educational Journal* (中国教育学刊), *5*, 24. (In Chinese)

Liang, L. L., Chen, S., Chen, X., Kaya, O. N., Adams, A. D., Macklin, M., & Ebenezer, J. (2008). Assessing preservice elementary teachers' views on the nature of scientific knowledge: A dual-response instrument. *Asia-Pacific Forum on Science Learning and Teaching*, *9*(1), 1. Retrieved from http://www.ied.edu.hk/apfslt/v9_issue1/liang/index.htm

Liu, X. (1988).The Crisis of and strategies for Chinese science education (论我国科学教育的危机与对策). *Educational Experiment and Research, 2*, 18–21. (In Chinese).

Liu, M. G. (1998). *Liu Mogeng's explorations for the reform of the course of nature in primary education* (刘默耕.小学自然课改革探索). Wuhan, China: Hubei Educational Publication. (In Chinese).

Liu, X., Liang, L. L., & Liu, E. (2012). Science education research in China: Challenges and promises. *International Journal of Science Education, 34*(13), 1961–1970.

Marginson, S. (2011). Higher education in east Asia and Singapore: Rise of the Confucian model. *Higher Education, 61*, 587–611.

Ministry of Science and Technology. (2007). *Suggestions for strengthening the national capability of science popularlzation* (关于加强国家科普能力建设的若干意见). Beijing: The Ministry of Science and Technology. (In Chinese).

MOE. (1956). *Syllabus of "Nature" for primary education (trial version)* (小学自然教学大纲(草案)). Beijing, China: People's Education Press. (In Chinese)

MOE. (1981). *Syllabus for fulltime five-year primary education (trial version)* 全日制五年制小学教学大纲(试行)). Beijing, China: People's Education Press. (In Chinese).

MOE. (2001a). *An outlines for the basic education curriculum reform (trial version).* Beijing: Ministry of Education of China. (Official document), No.17. (In Chinese). 教育部.基础教育课程改革纲要(试行).

MOE. (2001b). *National science curriculum standard for the full-time compulsory education (grades 3–6) (trial version)* (全日制义务教育科学(3-6年级)课程标准(实验稿)). Beijing: Beijing Normal University Press. (In Chinese).

MOE. (2001c). *National science curriculum standard for the full-time compulsory education (grades 7–9) (trial version)* (全日制义务教育科学(7–9年级)课程标准(实验稿). Beijing: Beijing Normal University Press. (In Chinese).

MOE. (2004). *Educational revitalization action plan for the 21 century 2003–2007* (面向21世纪教育振兴行动计划), Beijing: Ministry of Education of China. No. 5. (In Chinese).

MOE. (2011a). *The data set in the MOE website*: http://www.moe.edu.cn/. Retrieved 24 Feb. 2014. (In Chinese).

MOE. (2011b). *Some suggestions for strengthening teacher training programs* (关于大力加强中小学教师培训工作的意见). Beijing: Ministry of Education of China. No. 1. (In Chinese).

MOE. (2014a). *MOE: The national gross attendance rate of kinder garden in 2010*. Retrieved from http://edu.people.com.cn/GB/15104866.html. (In Chinese).

MOE. (2014b). *Ministry of Education: 2013 national education statistics report* (2013教育统计年鉴). Retrieved from http://paper.sciencenet.cn/htmlnews/2014/7/298374.shtm. (In Chinese).

National Research Council [NRC]. (1996). *National science education standard.* Washington, DC: The National Academies Press.

Peng, K., & Nisbett, R. E. (1999). Culture, dialectics, and reasoning about contradiction. *American Psychologist, 54*, 741–754.

Shanghai Group of PISA. (2013). *Quality and equity: A brief report of the program for international student assessment 2009 in Shanghai, China* (质量与公平:上海2009年国际学生评估项目(PISA)研究报告). Shanghai Education Press. (In Chinese).

Shao, Z. J. (1991). *A survey report on K-12 curriculum reform in Zhejiang Province (浙江省义务教育课程教材改革调查报告).* In: *K-12 curriculum reform in Zhejiang Province* (浙江省义务教育课程教材改革资料汇编), 227–228. Hangzhou: Zhejiang Education Publication. (In Chinese).

State Council of China. (2002). *National law of science and technology popularization of China* (中华人民共和国科学技术普及法). Beijing: the State Council of China. (In Chinese).

State Council of China. (2006). *Outlines of the national scheme for scientific literacy (2006–2010–2020).* 全民科学素质行动计划纲要(2006-2010-2020年) Beijing: The State Council of China.No.44. (In Chinese).

State Council of China. (2010). *Outline of China's national plan for medium and long-term education reform and development (2010–2020)* 国家中长期教育改革和发展规划纲要(2010-2020年)). Retrieved from http://www.gov.cn/jrzg/2010-07/29/content_1667143.htm. (In Chinese).

State Council of China. (2014). *Instructions on deepening reform of college entrance examination system* (关于深化考试招生制度改革的实施意见). Retrieved from http://news.xinhuanet.com/politics/2014-09/04/c_1112367808.htm. (In Chinese).

Sun, Z. F., Zhang, H., & Zhen, Y. (2010). The effectiveness of freshman seminar in research-intensive universities. *Qinghua University Educational Research, 31*(6), 119–124. (In Chinese).

Wan, D. S. (2014). *Research on Chinese pre-service science teachers' naïve views of the nature of science and the cultural impact.* Ph.D. thesis, Nanjing University. (In Chinese).

Wan, D. S., & Zhang, H. (2011). A Review on the *framework for science education – preliminary public trial* for K-12 in USA (美国2010《科学教育框架(草案)》述评及启示). *Comparative Education Review, 12*, 78–82. (In Chinese).

Wang, C. S. (2004). A critical reflection on the trend of "despising knowledge" in Chinese basic education (认真对待"轻视知识"的教育思潮). *Peking University Education Review* (北大教育评论), *2*(3), 5–23. (In Chinese).

Wang, J. Y. (2010). Comparing the fulfillment of the teaching objectives in their enquiry teaching between USA and Chinese teachers (科学本质观与科学探究的意义及实践——美国李德曼教授访谈录). *Global Education* (全球教育展望), *4*, 79–84. (In Chinese).

Wang, Z., & Li, D. (2010). A survey of Chinese scientists' perceptions on their willingness to do the popularization work (科学家对自身参与科学传播活动看法的调查研究). *Science Popularization Research* (科普研究), *3*, 71. (In Chinese).

Wang, X. H., Ma, Y. P., & Fan, X. Y. (2007). Are teachers ready for the integrated course of "Science". *Educational Theories and Practices, 27*(3), 45–48 (In Chinese).

Wei, B. (2008). Curriculum development of science and technology in secondary schools: A review of past three decades (中学理科课程发展:三十年的回顾与反思). *Journal of Chinese Education* (中国教育学刊). *11*, 5–8. (in Chinese).

Wu, G. (2002). *The freedom of science* (自由的科学). Fujian: Fujian Education Publication, 35. (In Chinese).

Wu, Y. Y. (2011). A survey of the secondary science teachers' view of the nature of science: A case from Zhejiang Province (初中科学教师科学本质观的调查研究---以浙江省初中科学教师为例). *Global Education* (全球教育展望), *40* (3), 82–87. (In Chinese)

Wu, X. D. (2012). *Science education in the digital age* (数字时代的科学教育). Southeastern Science and Technology University Press (华南理工大学出版社). (In Chinese).

Xue, H. N., & Chen, X. M. (2012). A national survey on the quality of present teacher training programs. *Educational Science, 28*(6), 53–57 (In Chinese).

Yan, Y. H. (2009). A century changes of Chinese science teacher training programs (中国科学教师培养课程结构百年变革). *Higher Education of Science, Math and Technology* (高等理科教育), *4*, 13. (In Chinese).

Ye, Z. N. (2011). “Learning through doing” :The experimental project for science education reform (“做中学”:科学教育改革实验项目的探索). *Curriculum for Basic Education* (基础教育课程). (1–2), 48–51.

Yu, Z. Q. (2003). The milestone of sciendary science education development in China: The curriculum standard analysis (中国中学科学教育发展的里 程碑:课程标准分析). *The Journal of Teaching Monthly* (教学月刊). *2*, 2–5. (In Chinese).

Zhang, H. (2002). Significance of scientific literacy education and its localization (科学素养教育的意义及其本土化诠释). *Qinghua University Educational Research (*清华大学教育研究), *4*, 20–26. (In Chinese)

Zhang, D. Z. (2003a). We should pay attention to special Chinese characteristics in education (我们应该关注中国教育的特殊问题). *Curriculum, Textbooks and Teaching methods, 6*, 4 (In Chinese).

Zhang, H. (2003b). *What is the thing called science (for secondary and primary teachers training) (科学究竟是什么).* Beijing: Education Science Press (In Chinese).

Zhang, H. (2003c). The contribution and limitation of the constructivism to Chinese science education (建构主义对科学教育理论的贡献与局限). *Educational Research* (教育研究), *7*, 79–84. (In Chinese).

Zhang, H. (2005). The traps and outlets of current Chinese curricular and instructional research (我国课程与教学研究的困境与出路). *Educational Development Research* (教育发展研究), *3*, 39–43. (In Chinese).

Zhang, K. (2007). X. K Luo:Experiment of the new curriculum of Science: Hope for the future with a long march (罗星凯:科学新课程实验,任重道远的希望之路). *The Era of Internet Technology* (网络科技时代), *17*, 35. (In Chinese).

Zhang, H. (2010). Higher education development in China: Reforms and challenges from globalization. In F. Maringe & N. H. Foskett (Eds.), *Globalization and internationalization in higher education: Theoretical, strategic and management perspectives* (pp. 125–138). London: Continuum International Publishing.

Zhang, H., & He, H. (2012). Effectiveness of undergraduate science education program in China: Perceptions of seniors and alumni. *International Journal of Science Education, 34*(13), 1991–2013.

Zhang, H., & Yu, B. (2004). A survey of primary science teacher’s scientific literacy (小学科学教师科学素养调查研究). *Educational Research* (教育研究), *11*, 58–67. (In Chinese).

Zhang, H., Wan, D. S., Yu, B. (2013). Ten-years change of scientific literacy of Chinese primary science teachers. *Journal of Suzhou University*, Initial issue, 72–76. (In Chinese).

Zhen, R. L., & Yang, X. D. (2003). The reform of college entrance examination: Historical and Realistic thinking (高考改革:历史与现实的思考). *Journal of Xiamen University, 1*, 108–114. (In Chinese).

Zhong, Q. Q. (1997). The “Scientific Literacy” theory abroad and its relationship with curriculum reform in science and math subjects. *Comparative Education Review, 1*, 16–21.

Part II
Science Curriculum and Instruction

Ling L. Liang

Editor's Introduction: Part II

The five chapters in this section report on the current state of K-12 school science curriculum and instruction in China, which provides either an overview or a snapshot of some notable outcomes achieved recently in various regions. In China, the compulsory education for children covers the first through ninth grade. Although preprimary programs are not part of compulsory education, Chinese kindergarten or preschool education has made impressive progress over the last decade. The current kindergarten enrollment rate (for children age 3–5) across the nation has reached 70.5 % (MOE 2015). Especially under the one-child policy, Chinese parents have been eager to invest in the early education of their children to ensure they are ready for formal school learning. In the new *Guidelines for Kindergarten Education* (Trial version, Ministry of Education 2001), science was listed as a subject in the kindergarten curriculum in China for the first time in history. The new curriculum guidelines call for a shift away from the previous knowledge-oriented, teacher-centered model to a more student-centered model. Kindergarteners are encouraged to conduct self-directed, hands-on inquiries and connect learning with daily life experiences.

The above changes are well in line with the recent developments in early childhood science education in developed countries (e.g., Copple and Bredekamp 2009; Krogh and Morehouse 2008; Worth and Grollman 2003). According to the literature cited above, the following common characteristics are shared by many high-quality early childhood science programs:

1. The learning and teaching of science is built on children's prior experiences, backgrounds, and early theories.
2. Children are encouraged to pursue their own questions and develop their own ideas.
3. Children are engaged in in-depth exploration of a topic over time, in a carefully prepared and supportive learning environment. The teacher seeks to find a

balance between children's pursuit of their own interests and the pursuit of ideas generated by the teacher herself or by peers.

4. Children are encouraged to reflect on, represent, and document their experiences and communicate their ideas with others.
5. Children learn science through an integrated curriculum and through play (e.g., dramatic play, exploratory play, and constructive play). In dramatic play, for instance, children can learn science concepts while also developing self-regulation, symbolic thinking, memory, language, and social competence.

From Kindergarten through grade 12 and beyond, constructivist views of learning have had strong influences on science education policy and classroom practices worldwide in the past two decades (Guo 2007). A fundamental assumption of constructivism is that learners construct understanding through interactions with the physical and/or social environment (Piaget 1970; Vygotsky 1978). Student minds are not blank slates prior to learning. Students come to the classroom with prior knowledge of how the world works. From a constructivist perspective, learning is an individual process that involves linking new ideas and experiences with what the learner already knows. A teacher's role is to create a learning environment and to act as a guide or facilitator to help students construct their understandings. Stemming from this constructivist viewpoint, various models of instruction have been proposed and implemented. The 5-E model is one of the widely adopted student-centered approaches that integrate a constructivist view of learning with scientific inquiry practices, built on Karplus' initial three-phase learning cycle teaching sequence of "exploration-invention-discovery" (Karplus 1977). In the 5-E model, learners are guided through a learning cycle—engage, explore, explain, elaborate, and evaluate—to develop their scientific understandings and abilities to conduct scientific inquiry. During the engagement phase, students are encouraged to ask questions about objects, organisms, or events created by the teacher. For exploration, students plan and conduct investigations to collect relevant data and then analyze the data and use scientific knowledge to generate explanations in the explanation phase. In the elaboration phase, they are expected to apply the newly learned concepts and principles to new problems and questions. Finally, evaluation should be connected with all phases of instruction, during which students demonstrate knowledge, understanding, and ability to do science through formative and summative assessment (Bass et al. 2009).

Modeling instruction is another inquiry-based and constructivist-oriented approach, also built on a learning cycle sequence and first introduced into physics teaching by Wells et al. (1995). The modeling approach is designed to engage students in developing scientific understandings by constructing and using scientific representations or models to describe, explain, and predict physical phenomena. During the instruction, students conduct investigations in small cooperative learning groups and constantly model physical objects and processes using verbal, diagrammatic, graphical, and mathematical representations. Modeling teachers are also informed by students' preconceptions of physics and guide student discourse through scaffolding and asking probing questions (e.g., Socratic dialog) to elucidate the models (Wells et al. 1995). Both 5-E and modeling instruction have been used in designing science curricula. Other models such as problem-based learning (PBL) and the project-based approach are also promoted by science educators. PBL is the learning that results from the process of working toward the understanding or resolution of a problem, whereas in a project-based approach, the curriculum units are developed with clearly defined learning goals (core ideas, crosscutting concepts, and scientific practices), as well as carefully designed learning performances and driving questions that allow students to explore the targeted concepts (Krajcik and Czerniak 2014).

In Chinese culture, "thinking" and generalized knowledge receive much more respect compared to "doing" and empirical data or evidence. Historically, while a school textbook presents a small portion of generalized knowledge, a teacher's main role has been perceived as a "knowledge dispenser"—to transmit knowledge and answer students' questions. Therefore, teacher-centered "direct teaching" focusing on explanation of knowledge has been the dominant instructional approach in all Chinese classrooms. This presents great challenges to the reform-based and inquiry-oriented science learning and teaching.

Chapter Introductions

There are five chapters in this section. In Chap. 2, Gao and Zhang describe some major developments in early childhood science education or kindergarten education since the 1980s. They also summarize some new initiatives and effective practices to promote early childhood science education, through illustrations of several exemplary cases from classrooms. In conclusion, the authors discuss challenges and directions for future development of early childhood science education in China.

In Chap. 3, Gao et al. provide a brief overview of the elementary science curriculum reform in Guangzhou and across the country since 2001. They also report the results of a survey administered to both students and teachers in the Tianhe district in Guangzhou city to show the conditions and quality of school learning as a result of the decade-long reform efforts. The study reveals some encouraging results, including generally supportive science learning and teaching environments, a reasonable level of teacher satisfaction with the new curriculum and school facilities, a high level of student motivation in science learning, an adequate level of student understanding of the nature of science, and generally positive student attitudes toward the natural environment. Students also demonstrate good understanding of the science concepts related to life science and earth/space science, but their understanding of physics-related concepts appears relatively weak. This might be attributed to the teachers' lack of professional training in both physics content and pedagogy as most elementary science teachers do not have a major in science. Moreover, the teaching survey results indicate that "experimentation" and "demonstration" are the least frequently used teaching methods compared to other teaching modes such as lecturing and class discussion, whereas the physics education research literature in the USA and elsewhere has repeatedly demonstrated the ineffectiveness of the traditional lecture method in developing student understanding of physics concepts (e.g., Hake 1998). Finally, the authors' claim regarding students' adequate understanding of the nature of science should be accepted with caution, as the five-question instrument on the nature of science may not fully capture the essence of the nature or process of the scientific knowledge development (Lederman et al. 2002). For instance, when students agree that "scientific knowledge changes with time," this does not necessarily mean that they understand how and why scientific knowledge changes—scientific knowledge may change based on new observations (new empirical evidence or data) or due to reinterpretation of existing data based on alternative scientific models or theoretical frameworks. It would be difficult for students to develop a sound understanding of the nature of science without actively being engaged in experimenting, testing, and modifying their own ideas and/or mental models in science classrooms.

Chapter 4 presents an investigation into science classroom practices in junior high schools. The students' reports of actual instruction indicate that they often or very often experience teacher-directed learning, occasionally experience cooperative learning, and seldom or rarely experience conceptual/experiential learning. This pattern is consistent across all three school types (with academic standings ranked as high, medium, and low), although the students in higher-ranked schools report significantly lower cooperative learning and conceptual/experiential learning than their counterparts in the lower-ranked schools. The researchers also surveyed students' preferred teaching methods and found that the direct teaching subscale showed the smallest discrepancy between the actual and preferred occurrence. Apparently, direct teaching is the most desired method embraced by both teachers and students. In fact, similar results have been found in other countries such as Finland (Juuti et al. 2010). As part of the culture, Chinese teachers strongly believe in the importance of knowledge and its structure and feel that they are obligated to explain the knowledge and structure to their students in teaching.

Ma's study further reveals that the students also desire to experience more cooperative and conceptual/experiential learning than they currently do. The students particularly welcome more instructional practices incorporating writing, brainstorming, and debating, as well as science field trips. On the attitudes scale, students' attitudes toward science are generally positive. There is a small but statistically significant difference among the school types, with students in the higher-ranked schools having less positive attitudes. The findings also show that students' attitudes are statistically significantly more positive when they experience instructional strategies that encourage more student cooperative inquiry and conceptual development. Traditionally, in the knowledge-oriented and exam-driven Chinese education model, direct teaching has been deemed as the most efficient instructional approach in school learning. However, in order to achieve the three-dimensional goals emphasized in the new science curriculum standards—i.e., knowledge and skills, processes and methods, and attitudes and values—the use of cooperative inquiry and conceptual/experiential instructional strategies has become increasingly important. Further research in this area in the Chinese context is definitely needed.

When reading the two chapters, the readers should be aware that the samples in both Chaps. 3 and 4 represent economically developed cities only. Rural areas surely have different stories. For instance, as summarized by Wang and Zhao in an article on basic education curriculum reform in rural China (2011), some real challenges or problems related to rural school education are identified as a serious shortage of funding, teaching resources, and information technology. Apparently, more investment, resources, and action are needed in rural education, while empirical research can be used as a tool to identify problems and measure the effectiveness of the reform-based actions.

In Chap. 5, Huang and her colleagues describe the structure and framework of the new discipline-based science curriculum in senior high school. The authors also examine the implementation of science curriculum standards and report results based on an investigation of physics curriculum standard implementation across the country. To understand teachers' and students' awareness of the curriculum goals and their classroom experiences associated with the new curriculum, the researchers surveyed over 4000 and interviewed approximately 1000 teachers and students. The findings of this study are important and necessary for informing both educators and policy makers. As revealed in the survey and interviews, at the senior high school stage, the influence of the national college entrance exams on curriculum choice and instructional practices is huge. In general, it appears that most teachers are aware of and embrace the three-dimensional goals (i.e., knowledge and skills, process and methods, and attitudes and values) of the new curriculum. However, in reality, many teachers would most likely focus on the "knowledge and skills" dimension only as it is the one to be tested in the external, high-stakes exams. Such findings are not unique among Chinese teachers. Abell's research literature review on science teacher knowledge (2007) revealed that science teacher knowledge of curricular goals has been studied around the world (e.g., in Australia, Finland, Israel, Spain, the UK, and the USA) and that, while the teachers recognize a variety of goals for science teaching, they tend to emphasize content goals over attitudinal or process goals.

In China, most science educators believe that physics is the foundation of all sciences. Students are required to take 3 years of physics courses starting in the 8th grade. Students who plan to major in science or engineering take physics for five consecutive years from 8th grade through 12th grade (usually two class periods per week). The courses are algebra based and emphasize the development of conceptual understanding and rigorous problem-solving skills. In the USA, by contrast, many high schools still adopt a 3-year course sequence of "biology-chemistry-physics" in their science programs in which only about 30 % of the students take a year-long physics course in the 11th/12th grades. As such, the physics course is traditionally reserved for the best students in the USA (AAPT 2006). In a research article published in *Science* in 2009, Bao and his colleagues compared Chinese high school graduates with their US counterparts in two domains: conceptual understanding of physics (the Force Concept Inventory and the Brief Electricity and

Magnetism Assessment) and scientific reasoning. Scientific reasoning ability is defined as domain-general reasoning skills such as the abilities to systematically explore a problem, to formulate and test hypotheses, to manipulate and isolate variables, and to observe and evaluate the consequences. As expected, after the rigorous training in physics in the middle and high school years, the Chinese students' performance on the concept/content tests is far better than that of their US counterparts. However, the research found no difference between the two groups of students on the scientific reasoning scale. This finding suggests that the traditional content-rich direct teaching may be effective in improving student concept test scores but not in the development of students' scientific reasoning abilities. Such research could also be used in teacher preparation or professional development programs to further inform preservice/in-service teachers about the necessity of reforming traditional instruction and/or curriculum design approaches.

In Chap. 6, Wei and Chen examine the embedding of the idea of scientific literacy in the new senior secondary school chemistry curriculum, on four themes: (a) the knowledge of science, (b) the investigative nature of science, (c) science as a way of knowing, and (d) the interaction of science, technology, and society. Most recently, by adding one additional dimension of metacognition and developing global citizenship, Choi et al. (2011) proposed a new framework for twenty-first-century scientific literacy which includes five aspects: scientific content knowledge, scientific habits of mind, character and values, science as a human endeavor, and metacognition and self-direction. Metacognition and self-direction are important for developing lifelong learners who seek new information, act, check, and make informed decisions on their own initiative. Although this framework was developed primarily for South Korea, it certainly adds to the dialog on scientific literacy across the globe.

Wei and Chen's analysis of the new series of chemistry textbooks shows that the themes of the "investigative nature of science" and the "interaction of science, technology, and society" have been added, but the theme of scientific knowledge remains dominant. The least represented theme in the new curriculum programs is "science as a way of knowing" or nature of science. This finding is quite revealing. Similar studies need to be conducted on other science disciplines. If Chinese educators and policy makers are serious about the development of scientific literacy for all, it is imperative for the theme of nature of science to be integrated into the science curriculum. Research has suggested that students may not be able to develop a sound understanding of the nature of science through conducting inquiry activities alone. Instead, they need to be explicitly instructed that scientific knowledge is a human construct, theory laden, tentative, based on evidence, testable, and creative while they explore and reflect on science learning (Adb-El-Khalick and Lederman 2000a, b). Other aspects of scientific literacy such as metacognition also need to be examined in future studies (Choi et al. 2011).

References

Abell, S. K. (2007). Research on science teacher knowledge. In S. K. Abell & N. G. Lederman (Eds.), *Handbook of research on science education*. Mahwah: Lawrence Erlbaum Associates, Inc., Publishers.

Adb-El-Khalick, F., & Lederman, N. G. (2000a). Improving science teachers' conceptions of nature of science: A critical review of the literature. *International Journal of Science Education, 22*, 665–701.

Adb-El-Khalick, F., & Lederman, N. G. (2000b). The influence of history of science courses on students' views of nature of science. *Journal of Research in Science Teaching, 37*, 1057–1095.

Bao, L., et al. (2009). Learning and scientific reasoning. *Science, 323*, 586–587.

Bass, J. E., Contant, T. L., & Carin, A. A. (2009). *Teaching science as inquiry* (11th ed.). Boston: Pearson Education.

Bransford, J., Brown, A., & Cocking, R. (Eds). (2000). *How people learn: Brain, mind, experience, and school* (Chapter One). Washington, DC: National Academy Press. Available on line at http://www.nap.edu/html/howpeople1/

Choi, K., Lee, H., Shin, N., Kim, S.-W., & Krajcik, J. (2011). Re-conceptualization of scientific literacy in South Korea for the 21st century. *Journal of Research in Science Teaching, 48*(6), 670–697.

Copple, C., & Bredekamp, S. (2009). *Developmentally appropriate practice in early childhood programs serving children from birth through age 8* (3rd ed.). Washington, DC: National Association for the Education of Young Children (NAEYC).

Guo, C.-J. (2007). Issues in science learning: An international perspective. In S. K. Abell & N. G. Lederman (Eds.), *Handbook of research on science education*. Mahwah: Lawrence Erlbaum Associates, Inc., Publishers.

Hake, R. R. (1998). Interactive engagement vs. traditional methods: A six-thousand student survey of mechanics test data for introductory physics courses. *American Journal of Physics, 66*(1), 64–74.

High School Committee of the American Association of Physics Teachers. (2006). *Physics first: An informational guide for teachers, school administrators, parents, scientists, and the public*. College Park: American Association of Physics Teachers.

Juuti, K., Lavonen, J., Uitto, A., Byman, R., & Meisalo, V. (2010). Science teaching methods preferred by grade 9 students in Finland. *International Journal of Science and Mathematics Education, 8*(4), 611–632.

Karplus, R. (1977). Science teaching and the development of reasoning. *Journal of Research in Science Teaching, 14*, 169–175.

Krajcik, J. S., & Czerniak, C. M. (2014). *Teaching science in elementary and middle school: A project-based approach* (4th ed.). New York: Routledge.

Krogh, S., & Morehouse, P. (2008). *The early childhood curriculum: Inquiry learning through integration*. New York: McGraw-Hill.

Lederman, N. G., Abd-El-Khalick, F., Bell, R. L., & Schwartz, R. S. (2002). Views of nature of science questionnaire: Toward valid and meaningful assessment of learners' conceptions of nature of science. *Journal of Research in Science Teaching, 39*, 497–521.

Ministry of Education. (2015). *Educational Statistics in 2014* (2014年教育统计数据). Retrieved from http://www.moe.gov.cn/s78/A03/moe_560/jytjsj_2014/2014_qg/

Ministry of Education (MOE). (2001). *Guidelines for Kindergarten education (Trail version, 幼儿园教育指导纲要[试行])*. Beijing: Beijing Normal University Press.

Piaget, J. (1970). Piaget's theory. In P. H. Mussen (Ed.), *Carmichael's manual of child psychology* (3rd ed., pp. 703–732). New York: Wiley.

Vygotsky, L. S. (1978). *Mind in society: The development of higher psychological processes* (M. Cole, Trans.). Cambridge, MA: Harvard University Press.
Wang, J., & Zhao, Z. (2011). Basic education curriculum reform in rural China: Achievements, problems, and solutions. *Chinese Education and Society, 44*(6), 36–46.
Wells, M., Hestenes, D., & Swackhamer, G. (1995). A modeling method for high school physics instruction. *American Journal of Physics, 63*, 606–619.
Worth, K., & Grollman, S. (2003). *Worms, shadow, and whirlpools: Science in the early childhood classroom.* Portsmouth: Heinemann.

Chapter 2
An Overview of Early Childhood Science Education in China

Xiaoyi Gao and Baohui Zhang

2.1 Introduction

In this era of globalization, science education is becoming increasingly important. Since 2001, China has been undergoing its eighth basic education reform, and science education has been one of the areas attracting most attention. The eighth basic education reform was initiated by the Ministry of Education (MOE), with the main purpose being to improve the overall quality of China's basic education and to update the school curriculum. The structure and content of curriculum have been adjusted or reformed, so as to meet the requirements of quality-oriented education. It is obvious that early childhood science education (ECSE) is very important for children because science education is the starting point that lays the foundation for their further studies. In this chapter we will follow the Chinese tradition of using "early childhood" and "kindergarten" interchangeably to refer to the time period when children might be educated in formal kindergarten to the time when they enter primary school (ages 3–7). According to the *Kindergarten Work Regulation* issued by the MOE in 1996, kindergarten is an institution that provides care and education for 3- to 6-year-old children. Given that kindergarten is not part of the 9-year compulsory education (Grades 1–9) in China, its duration could range from 1 to 3 years, including beginning, intermediate, and advanced levels. It could be full-time, part-time, seasonal, or boarding schools or a combination of the above forms. According to the statistical report on the development of national education in 2014 published by the MOE, there are about 229,900 kindergartens and 40.5 million

X. Gao (✉)
Faculty of Education, Beijing Normal University, Beijing 100875, China
e-mail: gaoxiaoyi@bnu.edu.cn

B. Zhang
School of Education, Shaanxi Normal University, Xian 710062, China

L.L. Liang et al. (eds.), *Chinese Science Education in the 21st Century: Policy, Practice, and Research*, Contemporary Trends and Issues in Science Education 45, DOI 10.1007/978-94-017-9864-8_2

children enrolled in kindergartens with an enrollment rate of about 70.5 % (MOE 2015).

Ever since the MOE promulgated the 2001 *Guidelines for Kindergarten Education* (trial version), the concept, research, and practices of ECSE have undergone great changes, and there have been accumulated experiences and advancements in practice. In the following sections, we will summarize the main stages from reform and opening up in the late 1980s to the present and the background and characteristics during different stages in Chinese kindergarten science education. The chapter is approximately organized in chronological order.

2.2 The Background and Characteristics of Reform and Development of Early Childhood Science Education in China

As mentioned before, the modern science education reform in China began in 1978, and ECSE is of no exception. Before the 1980s (starting from the establishment of PRC to the outbreak of the Cultural Revolution), science education for Chinese kindergarten was modeled after the former Soviet Union: it emphasized the knowledge system of different disciplines. During the 10 years of the Cultural Revolution, ECSE was almost stagnant. Before 1980, the only one subject included in the kindergarten course was understanding of the natural environment (Shi 1999), for example, seasonal changes. The teaching methods were mainly teacher-centered instruction, supplemented by children's observation, planting, and animal feeding (Yuan and Zhang 2009).

After the reform and opening up, kindergarten science education underwent three stages: the general knowledge education of the 1980s; the specific ECSE of the 1990s, which put children at the center of teaching and learning; and the establishment of ECSE during the 2000s, which put the focus on inquiry. Following the development of ECSE, related ideas, philosophy, and practices have been ever changing and evolving. In 2012, the MOE issued the *Guide for 3- to 6-Year-Old Children's Learning and Development*, setting specific goals and educational suggestions for science education for different age groups (MOE 2012).

2.2.1 General Knowledge Education and Its Basic Characteristics

Kindergarten science education was characterized by "general knowledge education" at the time of reform and opening up in the late 1980s. After that, in order to reverse the chaos caused by the "Cultural Revolution" and to ensure normal kindergarten teaching work, the MOE issued the *Educational Outline for Kindergartens* (draft) (referred to as the *Old Outline* hereafter) in October 1981, which still followed the division system of the 1950s. During that time, by modeling the former Soviet Union, ECSE paid more attention to the systemic and progressive education of knowledge. It mainly involved education on knowledge of nature based on seasonal changes, and its main method of teaching is to impart the knowledge to the kids by teachers, occasionally supplemented by the activities for kids to observe, plant, and feed animals. This value orientation of the curriculum of 1950s has a lasting influence on the later development of ECSE. The common goals in 1981 were as follows: (1) enriching children's general knowledge about society and nature and developing their worldview; (2) fostering children's interests and curiosity about nature in order to gradually develop positive attitudes toward people and the environment; and (3) developing children's attention, observation, memory, imagination, thinking, and language skills. Compared to the education system that had "understanding the natural environment" as the main content of science education in the 1950s and 1960s, "general knowledge" in science education is more extensive, including knowledge about society and nature, and is more diverse in teaching methods, such as observation, experimentation, physical activity, game playing, speech making, and poetry recitation (Liu 2008).

The essence of "general knowledge education" is to teach children about nature and human society, i.e., knowledge about society and nature. According to its mission, there were nine specific learning objectives at the beginning level, nine at the intermediate level, and 12 at the advanced level. For instance, some sample topics and requirements of the instruction for the beginning groups in the old outline include (1) knowing the student's own name, gender, and names of the family; (2) being acquainted with the kindergarten learning environment, the teacher, and the classmates; (3) recognizing three common types of vegetables and/or fruits and one or two kinds of flowers and/or trees and knowing their names, obvious characteristics, and main usage; and (4) with the teacher's help, learning how to grow one or two types of plants whose seeds are big and easy to grow (Shi 1999).

The goal of general knowledge education was also to develop students' interests, attitudes, and capacities. However, judging from the education content and implementation of the teaching process, there was still too much emphasis on imparting knowledge; in contrast, children's interests, attitudes, and abilities were not emphasized. The *Old Outline* reflected the value of knowledge-based orientation. What teachers cared most was to teach children based on the prescribed syllabus. In the end, instruction led by teachers was predetermined; children had to accept, exercise, and memorize passively. As a result, children's initiative and development of creativity was ignored.

2.2.2 The Establishment of ECSE and Its Basic Characteristics

In the late 1980s and early 1990s, with the rapid development of modern science and technology, as well as increasingly frequent international academic exchanges, Western child psychology and educational theories, such as Jean Piaget's cognitive psychology, Bruner's theory of instruction, and Gagne's cognitive learning theory and curriculum theories, were introduced to China. They had a tremendous impact on the traditional philosophy of child education in China. With the right timing, early childhood scholars drew on the successful experience of foreign science education for children and put forward the concept of "science education for preschool children." After that, "general knowledge education" evolved into "preschool science education." In the following 10 years, ECSE developed continuously in practice and exploration.

At this stage, in order to reverse the spoon-fed condition, new goals of preschool science education were proposed, which included scientific knowledge, scientific methods, and scientific attitudes. More specific objectives included "helping children access a wide range of scientific experience of the world and develop basic science concepts based on their experience; helping young children learn how to explore the world around them and how to study science, to observe, to think, to solve problems, and to use their hands; stimulating children's curiosity and interest in exploring the world around them and in learning science; and cultivating children's positive attitudes towards caring for and protecting the natural environment" (Liu 2008). In the mid-1990s, general knowledge courses were largely replaced by early childhood science curricula (Yuan and Zhang 2009; Zhang 2006).

Based on the information mentioned above, we can see an apparent shift in the goals of education in this stage. However, kindergarten science curriculum was certainly inherited from the general knowledge education model. Knowledge was still set as the top priority and attitudes were relatively neglected in the practice. At this stage, the primary characteristic was that teachers did not fully understand the nature of kindergarten science curriculum. Second, they still spoon-fed children as they did in general knowledge courses (Yuan and Zhang 2009). Third, although children had more opportunities for sensory perception and operation, the practice and operation that focused on understanding knowledge and relationships could not inspire children's interests in exploring the world (Liu 2008).

2.2.3 The Basic Characteristics of ECSE in the Twenty-First Century

In the twenty-first century, along with the new round of curriculum reforms in China and the worldwide development of science education, ECSE in China changed accordingly. In 2001, the *Guidelines for Kindergarten Education* (trial version) (MOE 2001), which was promulgated by the MOE, explicitly classified "science" as one of the five areas of teaching content in kindergarten (health, language, society, science and mathematics, and art) and refined the goals of kindergarten science education. At the same time, the MOE and the Science and Technology Association cosponsored a reform plan for kindergarten science education called "Learning by Doing" (LBD) in May 2001. Since then, the practice of ECSE started to change profoundly. As a result, kindergarten science education in China entered a new phase.

Compared to prior stages, the major change in this phase was that scientific knowledge was no longer set as the top priority. Instead, fostering children's creativity, curiosity, and inquiry abilities was emphasized (Yuan and Zhang 2009), which are all essential to the scientific spirit and scientific literacy. "Independent inquiry" and "returning to life" became the keynotes of science education in kindergarten (Zhang 2006).

There were many changes in the practice of kindergarten science education in this phase. Teachers began to pay attention to children's interests and daily life experiences and select issues that could lead children to explore. Taking the subject of flora and fauna as an example, teaching content was no longer confined to the "factual knowledge" about flora and fauna, but instead, teachers were expected to emphasize conceptual knowledge that would stimulate children's interests in exploration, for example, "surveying the growth of plants during their life spans and getting to know the diverse species of plants are the primary goals for children" (Beijing "Learning by Doing" Project Team 2003). Science education in this stage made a great effort to guide children to experience and do science. In terms of teaching strategies, teachers were also expected to emphasize the development of children's exploration skills, such as how to effectively "question," "speculate," "verify," "record," and "discuss."

Despite the new developments in philosophy and practice, there were still some problems in the practice of kindergarten science education. For instance, children's explorations were to some extent still controlled by teachers, and the so-called explorative activities did not actually reflect the essence of scientific inquiry. These issues reflected teachers' lack of understanding of the nature of science and scientific knowledge development and a lack of ability to carry out the reform-oriented science education practices. The Early Childhood Teacher Education programs certainly need to be strengthened in China. Table 2.1 summarizes the main characteristics of each ECSE development stage since 1980s.

Table 2.1 Characteristics of ECSE in periods of 1980s, 1990s, and twenty-first

Time	Stage	Main characteristics	Orientation of values
1980s	General knowledge	Transmit the factual knowledge of nature and society to children according to the syllabus	Preparation for elementary schools
1990s	The formation of ECSE	Attach great importance to the children's perceptual experience and stress scientific methods such as observation, classification, and measurement	Preparation for elementary schools
Twenty-first century	Focus on scientific inquiry	Inquiry is not only an important goal but also an important method in science learning, and knowledge is acquired in the process of inquiry and problem solving	Preparation for life-long learning and development

2.3 Current Early Childhood Science Education Reform and New Development in China

2.3.1 New Changes in Policies in Preschool Science Education

Since 2000, positive changes have been seen in ECSE theory and practice in China. The *Guidelines for Kindergarten Education* (trial version) issued in 2001 highlighted the change in the value orientation of kindergarten science education. The new *Guidelines* pointed out that the content for kindergarten education should be comprehensive and enlightening and that it can be roughly divided into five areas: health, language, society, science, and art (MOE 2001). Thus, the *Guidelines* proposed a new set of objectives, new content and requirements. Table 2.2 shows the detailed content and requirements of kindergarten science education (MOE 2001).

In 2012, in order to promote early childhood education and children's comprehensive development, the MOE issued the *Guide for 3- to 6-Year-Old Children's Learning and Development*. As mentioned above, serving as the general guidance to the reform and development of preschool education in China, the *Guidelines for Kindergarten Education* (trial version) issued in 2001 points out the direction, the principles, and the requirements of the reform and development. In this sense, the *Guide* (2012) can be seen as a bridge that helps transform the general *Guidelines for Kindergarten Education* (2001) into educational practice.

Specifically, the *Guide* (2012) provides the goals and the educational suggestions for children's learning and development in the areas of fitness, language, social studies, science, and art. The overarching goals are the integrity of children's learning and development, respecting individual differences in learning, understanding children's learning styles and characteristics, and attaching great importance to the qualities and characters involved in children's learning. As a leading document to guide ECSE reform and development, the *Guide* pointed out some new

Table 2.2 Content and requirements in the field of science education in the *guidelines for kindergarten education* (trial version), MOE (2001)

Content and requirements	Guiding keys
1. Stimulate children's interest and desire to explore the surrounding world and search for patterns	1. Children's science education in essence is an elementary and enlightening education, which emphasizes the development of children's interests and desire to explore
2. Create a nurturing environment for children's exploration activities that engage learners, support and encourage them to ask bold questions, express independent thoughts, and learn to respect others' opinions and experiences	2. Try to create opportunities for children to actually participate in inquiry activities in order to help them understand the process and methods of scientific exploration and experience the pleasure of discovery
3. Provide a variety of manipulative activities for every child to be engaged in with multiple senses and in various ways	3. Science education must be closely related to children's real life, using objects, events, and phenomena around them for scientific exploration
4. Through methods such as group discussion and exploratory activities, develop children's cooperative awareness and capability, and help them learn to express, communicate, and share exploratory processes and outcomes in various ways	
5. Cultivate children's interests in phenomena related to number, volume, shape, time, space, and so on in the surrounding environment in order to construct basic quantitative concepts and learn to solve some basic problems in life and games with simple mathematical methods	
6. Lead children to appreciate the impact of technologies in our lives and develop their interests in science and admiration for scientists, starting from familiar scientific and technological achievements in life or the media	
7. Help children understand the relationship among nature, the environment, and human society based on their living experiences, and develop their preliminary awareness and behavior for environmental protection starting from small things around them	

directions: (1) Children's science learning is the process of finding out commonalities, differences, and associations among things during exploration and problem solving. (2) The heart of science learning is to spark interest, to experience the process of exploration, and to develop the basic abilities of inquiry. (3) As children thinking concretely, science learning should be carried on through direct perception, experience, and operation, while any kind of knowledge transmission and/or intensive training for the mastering of knowledge and skills is discouraged or prohibited. In the *Guide* (2012), the goals of ECSE are defined in detail for each age group (please see Table 2.3).

Table 2.3 Goals and criterion of ECSE

Objective 1: To get close to nature and to be interested in inquiry		
3–4 years of age	**4–5 years of age**	**5–6 years of age**
(1) Like getting in touch with nature and to be interested in lots of surrounding things and phenomena.	(1) Like getting in touch with new things and asking questions related to new things frequently.	(1) Like to dig the meaning out by the roots about the questions they are interested in.
(2) Have the habit of asking all sorts of questions or be curious about playing with objects.	(2) Enjoy exploring objects and materials during the course of thinking and working by themselves.	(2) Often find the answer to the questions during the course of thinking and working by themselves.
		(3) Be excited and satisfied when making a discovery in inquiry.
This core objective embodies the great importance attached to children's curiosity and interest, which is the primary goal of science inquiry at the early childhood stage.		
3–4 years of age	**4–5 years of age**	**5–6 years of age**
(1) Observe carefully and find the distinguishing features of things of interest.	(1) Observe and compare things or phenomena to find their similarities and differences.	(1) Find and describe the characteristics of different kinds of objects or the change of things through observation, comparison and analysis.
(2) Explore objects through sensory-motor activities, and be sensitive to the consequences of movements.	(2) Raise a question and speculate answers boldly based on observations.	(2) Use methods to verify speculations.
	(3) Gather information through simple survey.	(3) Develop a simple research plan and execute it with the help of an adult.
	(4) Use pictures or other symbols to record.	(4) Use numbers, pictures, charts, or other symbols to record. Cooperate and communicate with others in inquiry.
This objective includes having children experience the process of inquiry and acquire the ability of conducting inquiry, which are two interrelated aspects in ECSE.		
3–4 years of age	**4–5 years of age**	**5–6 years of age**
(1) Know common animals and plants and realize and discover that animals and plants are varied.	(1) Be aware of the dynamic growth and changes of animals and plants and the basic conditions supporting them.	(1) Be aware of that the appearances and habits of animals and plants are adapted to their living environment.
(2) Observe and discover the characteristics of materials, such as softness or hardness, smoothness or roughness.	(2) Observe and discover the dissolubility or heat transmissibility of common materials and their uses.	(2) Find out the relationship between the structure and function of common objects.

(continued)

Table 2.3 (continued)

Objective 1: To get close to nature and to be interested in inquiry		
(3) Experience and perceive the impact of weather on life and activities.	(3) Observe and discover simple physical phenomena such as change of form or position of objects.	(3) Explore and discover the conditions or influential factors of common physical phenomena such as shadows, sinking and floating.
(4) Have a preliminary understanding of the relationship between animals, plants, and humans.	(4) Observe and discover the characteristics of different seasons and feel the effects of season on people, animals, and plants.	(4) Observe and understand the periodicity of the change of seasons and know the order of change.
	(5) Have a preliminary appreciation of the relationship between common technological products and their own lives and know that technological products have both advantages and disadvantages.	(5) Have a preliminary understanding of the relationship between people's lives and the natural environment, have an awareness of respecting and cherishing life and protecting the environment.

MOE (2012)

Children's understanding of the natural world and phenomena described under this objective is acquired naturally during the process of observing, experiencing, inquiring, and analyzing.

Based on these objectives, the *Guide for 3- to 6-Year-Old Children's Learning and Development* also provides some educational suggestions on how to conduct science activities. The implementation of this document will promote the future development of science education for young children in China.

2.3.2 New Philosophy of Preschool Science Education

Under the guidance of the national education policies, new changes in philosophy of ECSE took place in the new century as described below.

Firstly, "scientific inquiry" has been an important direction for ECSE. The focus and stress on "scientific inquiry" is a prominent characteristic for children's scientific and educational development in ECSE. It is the essence of the reform and development in kindergarten science education. At present, the emphasis is to encourage children to explore things with their own hands. There are four goals for science education: (1) Cultivate children's passion to explore things. (2) Encourage them to experience the scientific process. (3) Develop their scientific inquiry competence. (4) Help them obtain perceptual knowledge through exploring how to solve problems. Therefore, the new policies advocate cultivation of children's ability and appreciation of exploration in practice, by posing questions, conjectures, and assumptions and verifying assumptions through investigation, recording and analyzing data, making conclusions, and discussing their ideas with others.

Secondly, more attention has been paid to key concepts and critical experiences. Traditional science education in China paid close attention to "scientific facts" that were around the core ideas of factual knowledge and themes (Gao 2009b). However, the current development of science education has changed its focus to "key concepts." The reforms advocate that ECSE should not be limited to specific factual knowledge and themes, but should stress the foundational concepts that are very important for children, closely related to their early life experience, and worthy of children's exploration. They advocate that scientific inquiry should focus on "key concepts" and that activities should connect to key concepts in science teaching (Liu 2006). Taking plants and animals as an example, scientific activities to explore begin with the clues about "diversity" and "life cycle," "the basic needs of plants and animals," and "their relationship to the environment" and should have multiple exploration activities designed and organized around one theme.

Finally, ECSE should stay connected to real life. "Returning to life" is a philosophy which is particularly stressed by the *Guidelines* (2001). Science education should pay attention to the children's living world and not the knowledge world (Zhang 2006). The living philosophy means an orientation toward the specific scientific issues in children's lives. It begins from the scientific problems that children encounter in their lives and uses the children's existing routine scientific concepts or experience and helps children to learn science, to understand science, and gradually to construct scientific knowledge for the meaning of life by addressing specific issues in life (Wang 2009).

In addition, development of children's autonomy in science education and concern about children's interest, initiative, and science learning and development of scientific concepts are crucial aspects in the current ECSE.

2.3.3 The Development of Practice in Early Childhood Science Education: "LBD" Projects in Chinese Kindergartens

The LBD project, which was sponsored by ECSE and the China Association for Science and Technology in 2000, has greatly influenced ECSE in China. It is an important plan promoted by Yu Wei, former Deputy Minister of the MOE and an academician of the Chinese Academy of Science (CAS), and is a part of education reform to promote science education in kindergartens and primary schools. The LBD project studies the latest developments in science education abroad, including educational philosophy and instructional methods. In addition, it is developed based on the prior experiences and problems of ECSE in China. This plan is participated in, promoted by, and practiced by scientists who are keen on improving Chinese science education, educators who are devoted to science education research and reform, and teachers who are at the front lines of science education.

From piloting to implementation at a larger scale, LBD has drawn widespread attention to the whole preschool education field in China. Specifically, by observing, questioning, imagining, testing, expressing, and communicating, the aims of "the LBD" project of science education are to let all children have chances to explore the mysteries of nature, enjoy the processes of scientific discovery, gain basic scientific knowledge, acquire the primary ability of scientific exploration, and develop scientific attitudes, spirit, and thinking skills. All of these attempts will help children establish a scientific worldview and lay a solid foundation for their all-around development and ensure future citizens with good scientific literacy. Moreover, the project espoused the following criteria: developing a project for every child and respecting individual differences; laying foundations for children's lifelong learning and abilities to learn; deriving teaching cases from real life and surroundings; guiding children to learn actively and take part in exploration; encouraging teachers to be positive supporters and guides; encouraging educators and scientific researchers to work jointly as supporters and coaches; encouraging communities, family members, and student volunteers to support science education; and promoting domestic and international communication and cooperation through the Internet and other modern methods (Liu et al. 2003).

LBD demonstrates an important way to implement education for children's all-around development and it is also an important part of the new curriculum reform. The implementation of this program aims to improve teachers' ability and increase their experience in science education. By means of imitation and by testing on innovation, teachers obtain a deeper understanding of the nature of science, children and their nature of learning in a scientific way, and teachers' role as guides. After 10 years' endeavor and quest, LBD has injected new energy to children's science education in China. Many great teachers have been trained; a set of LBD science education books and multiple sets of hands-on science activities for children have been published; the theoretical framework and practical strategies of local science education have been made available; and a better atmosphere for the development of science education and management patterns has formed.

The following case (Table 2.4) titled "snail" is a typical unit reflecting the practical exploration in kindergarten led by LBD and is one of the serial cases of the LBD project.

Raising small animals is a common education activity in kindergarten, but the case of "snail" is special. It was part of the science education experimental project "LBD" in the kindergarten in China, which learns and draws experience from the scientific education method "LBD" in French. This case of "snail" was formed finally after multiple rounds of discussion, designing, and continuous experimentations in practice.

In contrast to the traditional knowledge education, this case takes the problems in which children are interested during their observation of the snail as the medium, guides children to carry out exploratory and confirmative studies, and inspires children's deep understanding of life. The inquiry activities carried out by children

Table 2.4 LBD for kindergarten children: a unit on *snail*

Activities	Questions that children focus on	Children's activities	Children's attitudes	Knowledge and skills	Communication and expression
First activity	What does a snail look like?	Please observe the configurational characteristics and body structure of snails using your eyes and your magnifying lens to discover the main body parts and then discuss your findings with your partner	They have strong interest in natural creatures and like to throw out questions during observation	Learn how to observe animals, and understand configurational characteristics of snails	Tell what a snail is called and its characteristics and then draw a picture
Second activity	Where do snails like to live?	Imagine several living environments which snails may like to live in and then to discover which kind of environment snails stay in the longest	They have great passion to explore the outside world and to design experiments. They have a matter-of-fact scientific attitude	Explore the main characteristics of snails' living environment. Learn to obtain information by paired observations	Describe snails' living environment and draw a picture
Third activity	What do snails like to eat?	Children can bring all kinds of food to the snails according to their own ideas and then observe which kind of food snails do or do not eat	They persist and focus on observation and also dare to express their own discovery	Record, summarize, and understand the feeding habits of snails by continuous observation	Discuss and form a snail diet record collectively

Fourth activity	Is the color of feces which snails evacuate same with the color of what they eat?	Design an experiment to answer the questions and try to discover the reasons for the answers	Children are diligent in thinking and are good at observing in the experiment. They are also willing to communicate with others about the meaningfulness of their experience	Do the experiment according to the proposed scheme. Understand the relationship between a snail's excretion and its food; learn to solve problems through books or Internet	Describe the experimental scheme by using drawings, paintings, photographs, and language. Form a collective record
Fifth activity	What are the white pieces?	Try to answer questions and then accurately report findings by observing	Love life; love nature	Understand the propagation characteristics of snails (oviparous)	Describe how snails give birth to babies by using drawings, photographs, and words

Data source: Science Education Experiment Project Expert Group (2004). "Learning by Doing (LBD)" ?in China: Case of Science Education in Kindergarten and Primary School. Beijing, China: Educational Science Publishing House, Page 4

lasted a few months; through a series of observations and experiments, they got an in-depth understanding of the shape characteristics, living environment, eating habit, food preference, the relationship between excretion and eating, reproduction of the snail, etc. Inquiries of children were guided by the problems they were interested in and kept becoming deeper. They experienced a whole series of scientific inquiry processes of speculation, experimentation, recording, sharing and communication, and drawing conclusion. Therefore, during the whole process of inquiry, children demonstrated unbelievable initiative for observation, the enthusiasm and persistence of inquiry, and the creativity in problem solving. "Snail" is a typical case from science education experimental project "LBD" in China, reflecting the basic idea of "LBD" which is discussed above.

Although the LBD project promotes the development of Chinese ECSE, there are still some challenges in the process of implementing this program (Gu 2011). Whereas the teacher professional development of LBD has been ongoing for more than 10 years, more work needs to be done in assessing and evaluating the implementation and effectiveness of the program.

2.4 Reflections on Development and Practices in Chinese Children's Science Education

Under the impetus brought from the development of theories of ECSE, research, and experimental projects, Chinese ECSE's practices have developed gradually and made many positive changes:

1. Teachers' concepts of science education have changed (Zou 2006), shifting from purely knowledge oriented to more inquiry oriented.
2. There has been increased emphasis on children's active engagement during the process of science education and greater attention to children's acquisition of more direct experience.
3. The selections of science education content have been getting richer and broader (Liang 2011), focusing on stimulating children's interest, with the fact that the designs of science education activities are growing gradually richer and more in-depth.
4. The designs of science education activities have begun to turn toward children's meaningful key experiences (key concepts) by designing specific activities around key concepts and highlighting children's exploration experiences during the process.

Take the scientific inquiry activity "tadpoles and the frogs," for example. In this case, based on the fact that children of different ages have different preliminary experiences, the overall goals and sub-objectives for the beginning group, intermediate group, and advanced group are designed differently, reflecting consideration of children's age characteristics and key experiences in a more logical way. At the

Table 2.5 Instructional goals and sub-objectives of the "tadpoles and the frogs" unit by group level

Goals	Sub-objectives
1. To foster children's respect for life and love for animals.	*The beginning group:*
2. To know about the eating habits of tadpoles.	1. Know that a tadpole has a mouth and can eat.
3. To know the growth process from tadpole to frogs.	2. Observe the growth process of tadpoles and know that tadpoles can turn to frogs.
4. To know that frogs are oviparous and amphibious.	*The intermediate group:*
	1. Explore what food tadpoles like to eat and know about their eating habits.
	2. Find the right growth pattern from the tadpoles to frogs by observation.
	The advanced group:
	1. Know that the frogs are oviparous and know about the growth process from spawn to tadpoles.
	2. Know about the changing pattern of tadpoles' various forms.
	3. Be able to solve problems by collecting information and collaborative teamwork.
	4. Understand frogs' amphibious living habits.

same time, a unity of multidimensional targets is achieved in the inquiry process. Table 2.5 is a case from preschool education in 2008 (Guo and Xu 2008).

The case of "tadpoles and frogs" illustrates an in-depth practice of carrying out scientific inquiry activities in kindergarten. One special aspect about this activity is that the teachers elicited the prior experience of children of different ages through different ways of investigation, and based on which, they developed the overall objective for all age groups and the specific goals for each age, thus making sure that each activity was attractive to all children and was something they were willing to inquire.

We have summarized and presented what we have learned about ECSE in China. In the following sections, we discuss some issues or problems in the current Chinese preschool science education. Although ECSE has made great progress, the current situation is not very optimistic. In the process of absorbing advanced international educational ideas, there are many discrepancies between the beliefs of reform and practical outcomes. This is not only because the basis of the ECSE in China is relatively thin and our teachers' scientific understanding and abilities are relatively limited, but also because the opportunities for science education in kindergartens are relatively limited.

Compared to other contents in the five major areas, ECSE has received relatively little attention. It can be said that ECSE is still a field that is relatively marginalized. According to the revised *Kindergarten Regulations* by MOE (2013b), a kindergarten can have full-day, half-day, boarding, or other schedule. A survey of preschool education quality that covers 11 different provinces and cities showed that science activities account for only 5.3 % in half-day reading, writing, and arithmetic activities. In other general education and teaching activities, science accounts for a very small proportion (only 2.8 %); in the design of half-day activities, there is little room for science (Liu 2011). That is to say, the practice in kindergarten science curriculum is far from meeting the requirements of the *Guidelines* (2001). In the following sections, we discuss some of the major problems in Chinese ECSE.

2.4.1 There Is a Mismatch Between the General Goals and the Specific Objectives

As we have discovered, teachers in general have a superficial understanding of the essence of science, which leads to the following issues in their practice of science education. First, the objectives of their lesson plans are too general to be operationalized, the overall goals of activities are not consistent with the specific objectives and instructional content, and the specific objectives are not well connected to each other. Second, the objectives are typically oriented toward factual knowledge, but other aspects of learning objectives, such as scientific methods, ability, emotion, and attitudes, are often ignored (Gao and Yang 2009). Consequently, the actual practice of preschool science education does not in fact reflect the true values that the LBD project advocates.

2.4.2 There Are Issues in the Selection of the Content of Preschool Science Education

There is no standard science curriculum for kindergarteners. Teachers are expected to choose appropriate content based on the new *Guidelines*. This has proven to be difficult for most teachers. "Among the kindergarten curriculum, there is no other area that is so wide and extensive as the one of science, making it a bottleneck for practitioners to select the content of science education in a scientific way" (Wang 2010). Teachers end up choosing content that is not related to the development of scientific understanding or to children's lives and interest. Many teachers are unable to select content or topics that are appropriate for students' development levels (Liang 2011). They still tend to focus on specific factual knowledge and themes, leading to the lack of explorative content for scientific education (Gao and Yang 2009). In addition, teachers select the contents for science education in a relatively casual way, resulting in unbalanced coverage of topics and omission of topics unfamiliar to them (Liang 2011).

2.4.3 The Separation of Science Education and Mathematics Education

According to the *Outline*, mathematics should be included in the science curriculum and be integrated with science. However, in the kindergarten practice, teachers pay more attention to mathematics, especially numeracy abilities. Thus, mathematics has taken a relatively high proportion of daily activities. In addition, as a survey has shown, science education is barely incorporated in these mathematical activities or other kinds of activities, and teachers seldom set up a scientific environment (Liu 2011). Therefore, mathematics education and science education are detached in the implementation of science curriculum.

2.4.4 Real Scientific Inquiry Activities Are Not Common

In the current practice of ECSE, real scientific inquiry activities are not common compared to other teaching practices. Research has found that the main learning methods for the children are listening and reading (53.4 %), followed by manual operation (21.5 %) and exploration practice (10.2 %). Because of a lack of understanding of the essence of scientific exploration, teachers often equate "inquiry activities" with easy craft-making activities, rendering inquiry a mere formality and doing exploration for the sake of exploration (Gao 2009a).

2.4.5 There Is a Lack of children's Initiative in Classrooms Dominated by Teachers

Researchers have pointed out that teachers' dominance in selecting or conducting science education activities suppresses children's autonomy (Peng and Ma 2011). It has also been found that teachers are in the status of "authority" in scientific activities. It is the teachers who select and design the activities (form, duration, etc.), provide the materials, decorate and arrange the setting, provide feedback, and evaluate the outcomes of the activities (Zhang 2004). The children have no autonomy in the process of choosing topics for research, determining the direction of exploration, and problem solving.

2.4.6 Scientific Exploration Learning Materials Are Insufficient

In the practice of ECSE, the shortage of scientific materials is a very serious problem. According to a recent survey study (Liu 2011), 38–86 % of classes do not have any exploring scientific materials and do not use such materials at all. Except for natural materials, such as live plants, magnets, and rocks, more than 60 % of classes are short on other materials. Although most of the classes grow plants or keep seeds and specimens of plants, the substances, materials, and basic tools that support children to do scientific exploration are still in extremely short supply. As a consequence, the quantity and quality of scientific inquiry activities that children can carry out cannot be guaranteed. At the same time, the gap between urban and rural areas is tremendous, with cities having much better condition than rural areas. For example, 38.5 % of the rural kindergartens do not have living plants as science activity materials, while the percentage is 21.3 % in urban kindergartens (Liu 2011).

The problems existing in the present practice of ECSE in China reflect many deep-rooted issues, ranging from factors related to the macroscopic social culture and the process of science education reform to those related to the *Guidelines* itself, and from the kindergarten curriculum orientation, the content, and the learning materials to teachers' educational beliefs and their teaching competencies.

The further development of ECSE calls for more research and guidance. These should be provided at the national and policy-making levels, including refining the content standards of science education, constructing curriculum resources, and providing a monitoring and regulating system to ensure the implementation of scientific practices in classrooms. At the same time, training of preschool teachers should also be strengthened in order to elevate teachers' scientific literacy, enhance teachers' appreciation and understanding of science, and improve teachers' competence to carry out the science education practice.

In summary, in this chapter we have presented Chinese ECSE as part of China's education system. Although we have witnessed some progress when domestic development was combined with information from overseas, especially after China's open-door policy, some problems and challenges persist. Further improvements are needed in many areas, such as curriculum goal setting, content selection, teaching methods, teachers' professional development, and research and evaluation, in order to make Chinese ECSE more effective and keep up with international trends.

References

Beijing "Learning by Doing"Project Team. (2003). *Collection of exemplar activities of children's science education by way of learning by doing*. ("做中学"幼儿科学教育活动案例集) (pp. 136–144). Beijing, China: Beijing Normal University Press.

Gao, X. (2009a). From process to inquiry: Reconsider the current goals of science education. (从过程到探究:对当前科学教育目标的再思考). *Educational Science Research, 20*(2), 57–60.

Gao, X. (2009b). Teaching for Conceptual Understanding in Science Education. (试论促进概念性理解的科学课教学). *Curriculum, Teaching Material and Method, 29*(4), 70–74.

Gao, X., & Yang, Y. (2009). On the goals of science education in early childhood exploring activities (科学探究活动目标确立中的问题及解决对策). *Studies in Preschool Education, 23*(9), 26–28.

Gu, C. (2011). Problems and related analysis in the implementation of "Learning by Doing" project (科学教育"做中学"项目实施中面临的问题及分析). *Journal of Research in Teaching, 34*(4), 84–87.

Guo, X., & Xu, X. (2008). Tadpoles and the frogs. (蝌蚪与青蛙) *Preschool Education, 25*(4), 8–11.

Liang, Y. (2011). On the contents of scientific education in kindergarten in the past 10 years. (《纲要》颁布10年来幼儿园科学教育内容改革探析). *Journal of Early Childhood Education (Educational Sciences), 30*(9), 12–15.

Liu, Z. (2006). The enlightenment of science education in Canada (加拿大小学科学教育对我们的启示). *Curriculum, Teaching Material and Method, 26*(12), 85–89.

Liu, Z. (2008). *Science education for preschool children.* (学前儿童科学教育). Beijing, China: Beijing Normal University Press.

Liu, Z. (2011). *Evaluation of the Chinese kindergarten education quality: Survey on kindergarten education quality of eleven provinces and cities.* (中国幼儿园教育质量评价——十一省市幼儿园教育质量调查). Beijing, China: Educational Science Publishing House.

Liu, S., Bai, Y., Zhu, Z. (2003). Brief Introduction about "Learning by Doing" Project. ("做中学"科学教育项目简介). *Studies in Preschool Education, 17*(3), 37–38.

Ministry of Education (MOE). (2001). *Guidelines for kindergarten education (trail version)* (幼儿园教育指导纲要(试行)). Beijing: Beijing Normal University Press.

Ministry of Education (MOE). (2012). *Guide to learning and development for 3–6 year-old children* (3-6岁儿童学习与发展指南). Beijing: Capital Normal University Press.

Ministry of Education (MOE). (2013b). *Kindergarten working regulation (revison) (draft).* (《幼儿园工作规程(修订稿)》(征求意见稿)). Retrieved 22 Mar 2013, at http://www.moe.gov.cn/publicfiles/business/htmlfiles/moe/s248/201303/149113.html.

Ministry of Education (MOE). (2015). *Educational statistics in 2014.* (2014年全国教育事业发展统计公报). Retrieved 2 Oct 2015, at http://www.moe.edu.cn/jyb_xwfb/gzdt_gzdt/s5987/201507/t20150730_196698.html.

Peng, H., & Ma, F. (2011). Problems in the science education in kindergartens and the coping strategies (当前幼儿园科学教育中存在的问题及其对策). *Journal of NeiJiang Normal University, 26*(9), 119–122.

Shi, Y. (1999). *Science education for preschool children.* (学前儿童科学教育). Shanghai, China: East China Normal University press.

Wang, C. (2009). What kind of science children need:-Thoughts on the value orientation of children scientific education. (儿童需要什么样的科学——对我国儿童科学教育价值取向的思考). *Studies in Preschool Education, 23*(5), 23–27.

Wang, D. (2010). Some suggestions for selecting content of science education in preschool. (对当前幼儿园科学教育内容选择的几点建议). *Journal of Education Guide, 28*(1), 36–38.

Yuan, A., & Zhang, S. (2009). Analysis of thirty years's preschool curriculum evolution-the reform in kindergarten scientific education curriculum (Part I). (三十年学前课程嬗变面面观之二——幼儿园科学教育课程变革(上)). *Journal of Education Guide, 27*(5), 4–7.

Zhang, Y. (2004). Understand and handle three relations in preschool science education in a right way (正解认识与处理幼儿科学教育中的三个关系) *Studies in Preschool Education, 18*(10), 25–26.

Zhang, J. (2006). Review of research on preschool science education in recent ten years. (近十年幼儿园科学领域教育研究综述). *Early Education, 24*(8), 4–5.

Zou, B. (2006). On kindergarten teachers' science education. (幼儿园科学探究活动现状探析). *Studies in Preschool Education, 20*(2), 55–57.

Chapter 3
Elementary Science Education Reform in Guangzhou: Expectations and Changes

Lingbiao Gao, Cuidian Feng, Maorong Zhan, and Xueping Zheng

3.1 Background: The National Curriculum Reform

Before 2001, the elementary science course in China was called "nature," emphasizing its focus on "general knowledge about nature" (MOE 1993). It was criticized as being elitist, knowledge centered, exam oriented, too complex and difficult, and disconnected from students' life (National Curriculum Development Committee – Elementary Science 2002). As a result, more than 70 % of the students were nervous and felt afraid of the course. They learned passively and achieved poorly in terms of scientific literacy, although they might score well on tests (MOE Project Team 2002). The course went in a direction different from the world trends. It focused only on scientific knowledge rather than students' personal development and scientific literacy. Similar problems were found in almost all school subjects. This led to the national curriculum reform in 2001. This reform signified a major shift in the general school curriculum objectives from knowledge delivery to student development in three areas: knowledge and skills; processes and methods; and emotion, attitudes, and values. The changes involved both school curriculum structure and subjects. For instance, at the compulsory stage, two new subjects, ethics and life and practical activities in society, were added into the school curriculum and the course nature was renamed science. The previous National School Curriculum Syllabi were replaced with the National Curriculum Standards. The new policy also allowed

L. Gao (✉)
South China Normal University, 55 Zhongshan Dadao, Tianhe, Guangzhou 510631, China
e-mail: gaolb@scnu.edu.cn

C. Feng • M. Zhan
School of Teachers Education, Taizhou University, Linhai, Zhejiang 317000, China

X. Zheng
Advisory Center of Teaching and Learning, Tianhe, Guangzhou, China

L.L. Liang et al. (eds.), *Chinese Science Education in the 21st Century: Policy, Practice, and Research*, Contemporary Trends and Issues in Science Education 45, DOI 10.1007/978-94-017-9864-8_3

multiple publishers to publish a variety of school textbooks. It encouraged changes in learning and teaching toward a more student-centered and interactive orientation. It also promoted a new system of student assessment which aimed at students' all-round development rather than knowledge accumulation. As a top-down, government-led movement, this reform soon spread over the entire country. Guangzhou, as the capital city of Guangdong – one of the most developed provinces in China – has been engaged in this reform since 2001.

As mentioned above, the traditional course nature was changed to science to highlight the fact that the central focus of the new course has been moved from knowledge delivery to students' scientific literacy. According to the *National Science Curriculum Standards for the Full-Time Compulsory Education* (Grades 3–6) *(trial version)* (MOE 2001a), the development of scientific literacy covers broad areas including (1) scientific knowledge and its application to daily life; (2) processes and methods of scientific inquiry and the ability to identify questions and solve problems; (3) scientific ways of thinking and exploring; (4) interest and curiosity about the world, learning, and inquiring; (5) understanding of the importance of evidence; (6) creativity; (7) environment and sustainable development; and (8) development of science and technology. With this focus, elementary science was set as an "enlightening" course and aimed at students' all-round development in three aspects: scientific knowledge and skills; scientific processes and methods; and emotion, attitude, and values. The objectives of the course are to help students to understand the simple and essential scientific knowledge in their everyday life, to help students to understand the scientific process and methods of scientific inquiry, and, further, to facilitate students to think and behave as scientists and enjoy scientific inquiry. Its aim was to direct all efforts to arouse students' curiosity about the natural phenomena around them and to attract students' interest in science so as to gradually develop a positive attitude toward science as well as an understanding of the value of science. The course introduces the very basic ideas and problem-solving techniques of daily life and helps students to understand the relationship between science and technology. Problems of resource limitation, environment pollution, and the conflicts caused by resource limitation and human development are also included in order to help students to understand the importance of science, technology, society, and environment (STSE) issues and to gradually build ideas about sustainable development.

This new elementary science course has reduced the required content knowledge by about 40 % (the number of topics/concepts was reduced from 184 to 111) to make room for inquiry learning. In the meantime, three major areas of science were still covered: the "Living World," the "Physical World," and "Earth and Space." Teachers were expected to pay more attention to the quality of students' learning processes and ability development rather than exam scores. They were also expected to change their instructional approaches from knowledge transmission oriented to more hands-on learning oriented and interactive.

As the media between the national curriculum standards and classroom instruction, a series of new textbooks were published to embody the philosophy of the new curriculum. The features of these new textbooks could be summarized as: (a) Inquiry learning has become the dominant philosophy of textbook editing. A large number of learning activities have been designed to create a proper atmosphere for exploring the world around students. (b) Scientific knowledge is still the basic content of the new textbooks. (c) Carefully designed strategies of learning and teaching have also been included to benefit both students and teachers in learning and teaching. (d) STSE issues, ethics, and moral issues that arise in the process of scientific development and issues relating to the appreciation of the beauty of the nature world, etc., have also been included in these new textbooks to cultivate the scientific literacy of students (Gao 2011).

A new evaluation system was developed to assess the process and quality of student learning. A number of formative assessment techniques were introduced, for example, all students were asked to keep a personal "Progress Record" – a kind of learning portfolio to allow students to review their learning performance and progress, to diagnose their problems, to assess their own learning, and to interact with peers, teachers, and parents.

3.2 What Really Happened: Questions

The new national curriculum drew a beautiful picture of elementary science and expected great improvements of the scientific literacy of elementary students. However, as Provus (1971) points out, a discrepancy always exists in the curriculum implementation process. This might be due to the gap between the educational experts and teachers in their values and beliefs about education and schooling due to the mismatch between the needs of the curriculum in terms of teachers' professional quality and reality of that quality and due to the level of student development and the conditions of schools and their environments. This seems to be the situation in China.

According to Gao and Watkins (2001, 2002), Chinese science teachers perceive teaching in two different ways. The first perspective views the teaching learning process as a one-way delivery of knowledge or a process of training students to achieve academic standards. Students are viewed as passive acceptors and trainees. Expected learning outcomes relate to knowledge accumulation or higher marks in examinations. The second perspective views the teaching-learning process as interactions among students, teachers, and teaching contents and environment. Students are viewed as active learners who explore, problem-solve, cooperate, discuss, construct, etc. What they learn is well beyond scientific knowledge. The ability to learn and solve problems and scientific attitudes and values become the central focus of learning. The above research also suggests that almost all teachers hold complex or even conflicting views in different times and while facing different tasks. The National Curriculum Experimental Regions, a survey conducted by Gao and his team in 2001–2003 to assess teachers' conceptions about teaching, suggested that at the beginning of the curriculum reform, teachers were oriented to the "interaction/development" point of view, which was consistent with the teaching philosophy of the new curriculum. However, when teachers encountered difficulties in practice, they switched back to the "knowledge delivery and exam-based" teaching orientation described above (Gao 2004). This result seems to suggest that teachers were not well prepared for implementing the new curriculum. They seemed to accept the philosophy of the new curriculum at the beginning but failed to translate these new ideas into classroom practices. This, in turn, pushed teachers to go back to the traditional philosophy of teaching.

Another survey conducted by the MOE project team (2010) reported that about two-thirds of elementary science teachers did not major in science in their preservice training. In addition, their main teaching responsibilities may be in art, music, Chinese, or other subjects. They teach science temporarily in order to cover the necessary teaching load as a full-time teacher. They all had less than 3 years' experience teaching science. Zhong and Gao (2007) visited several groups of these teachers and found that they had very limited amount of scientific knowledge. They did not seem to understand the nature of science or scientific inquiry and were not used to thinking in scientific ways. More importantly, they did not pay much attention to and invested little energy in this temporary and part-time teaching task. This becomes another obstacle to the new curriculum implementation.

The educational environment also seems not to support the implementation of the new curriculum (Zhong and Gao 2007). Most of the elementary schools in China do not have sufficient resources to support the new teaching-learning strategies. Many elementary schools lack the laboratory or activity rooms necessary to conduct inquiry learning activities. Outside schools, scientific museums, or activity centers are not popular in most parts of China. The large class-size system in Chinese schools allows about 40–50 students in one classroom in elementary schools. It is difficult to organize student activities in such large classes. This seems to be another obstruction to the new curriculum.

Moreover, the Ministry of Education decided to integrate the science and life and moral education courses into a new course named "ethics and life" in the first 2 years of elementary school (MOE 2001b). Science was no longer an independent subject in the first and the second grade school curriculum. The underlying philosophy focuses on encouraging integrative learning, which is the world's trend in the past decades. However, it is too far away from the Chinese teachers' perceptions of schooling. In their mind, this decision means a decrease in the importance of elementary science.

One more action which might be perceived as further lowering the status of elementary science was the abolishment of summative examinations in science at the end of elementary school, while three other subjects (Chinese, mathematics, and foreign language) were still kept on the examination list (MOE 2001a). The curriculum developers might have intended to try to create a more relaxed and friendly atmosphere for student learning in elementary schools. However, in a highly exam-oriented society, this becomes a disaster for elementary science education. As many people know, examination scores at the end of elementary school are extremely important for those students who wish to enroll in high-ranking middle schools. So, in most of the elementary schools, science was no longer treated as a core course after implementing the new curriculum in 2001. Not only school principals and teachers but also parents and society viewed elementary science as not essential. Less resource was put into elementary science (Zhong and Gao 2007). In many schools in the rural areas, students were given the textbooks and asked to learn by themselves in order to save class time for the core subjects: Chinese, mathematics, and English. Elementary science actually disappeared in these areas (Cai et al. 2009; Chen et al. 2011; Li 2007; Liu 2010; Lu 2011).

The new national curriculum in 2001 presented a beautiful dream and a high expectation of elementary science; however, the quality of science teachers, the school conditions, and the social environment in China seemed not very supportive of the implementation of the new curriculum. Moreover, some of the new designs were too far from the reality of the school teachers, educators, and society and caused negative effects on elementary science education. What then is the result of the new national curriculum? How do the students and teachers perceive elementary science under the new curriculum? What kind of change has happened in learning elementary science? How do the students achieve in learning science under the new curriculum? Do the quality of elementary science teachers and the school conditions meet the needs of the new curriculum? Have the objectives of the new national curriculum been achieved? Or, to what extent the objectives are achieved? After implementing this new curriculum for more than 10 years, it is significant to learn about the real situation of elementary science teaching and learning under the context of the new national curriculum.

3.3 A Survey in Guangzhou

To answer the above questions and to understand the conditions and quality of science learning, a survey of elementary students and their teachers in an urban district of Guangzhou City was conducted in 2009, adopting a convenience sampling technique. China is a huge country with great differences among regions. It is very difficult to cover the entire country or even the most populated areas with one survey. Therefore, the results of this survey cannot be used to interpret the situation of the whole country. However, Guangzhou can serve as a good representative sample of the coastal cities in China. Its rapid economic development and population expansion, accompanied with a variety of problems in education and society, is an example of all the coastal cities today and will be emulated by other cities in the central and the western part of China. Similarity might be found in other coastal cities, which shows the value of this survey. Given that this is the first time we tried to assess the conditions, situations, and quality of elementary science learning in China, we put more energy into building the research model and developing the instruments. In this sense, this study can be viewed as a pioneer study. We look forward to other educators applying our survey instruments to other parts of the country.

3.3.1 Theoretical Framework of the Survey

To define the scope and purpose of the survey, it is necessary to clarify what "learning conditions and quality" means. Learning in schools is influenced by many factors which form a complex system. Biggs and Telfer (1987) suggested a 3P model of learning in the school context. This model suggests that the school learning process has three phases: presage, process, and product. The presage phase includes key factors such as the learner and the teaching background, which further includes the subject, the school, and the teacher. Key factors in the process phase include the learning tasks and learning methods. Learning outcomes, including the content and structure of the learning outcomes, are the key factors in the product phase (see Fig. 3.1). This model gives a comprehensive description of the learning process.

Based on the 3P model, the conditions and quality of elementary student learning can be assessed from three aspects. The first aspect refers to the students themselves. It is believed that learning and the learner are inseparable in the process of learning (Marton 1983). Because different learners have different motivations, learners adopt different methods in the learning process. This leads to different learning outcomes. The interrelation of motivation and learning methods shows individual characteristics in the learning process, determines the possible learning outcomes, and thus reflects an individual's learning quality. In order to understand the quality of student learning, it is necessary to examine students' readiness, motivation, adopted methods, and outcomes of learning. The outcomes of learning can be indicated by the learners' understanding of knowledge, their attitudes toward science, attitudes toward nature, and understanding of the nature of science.

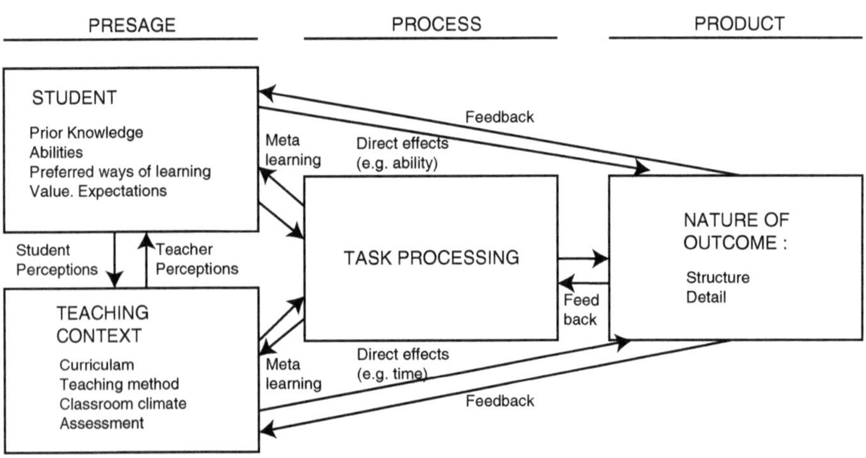

From Biggs: Why and How do Hong Kong students Learn? -- Using the Learning and Study Process Questionnaire; Education Papers 14, University of Hong Kong p.6

Fig. 3.1 The 3P model of classroom learning

The second aspect refers to the teachers, the curriculum, and the process of teaching. Teachers are the key factors affecting student learning. Their subject knowledge, beliefs or personal values of education, professional skills, and attitudes to teaching determine to a large extent the quality of teaching, and in turn influence student learning. The national curriculum is another factor important to teaching and learning since it provides the philosophy for teaching and learning, sets up the objectives and the contents of the courses, suggests a set of teaching and learning strategies, and decides the approaches of assessment. However, what really happens in the classroom, i.e., the real process of teaching and learning, is a factor not to be neglected.

The third aspect refers to the learning environment and atmosphere, including the conditions of classroom settings, laboratories, equipment and instruments, and the learning context and atmosphere in the class, the school, the family, and the community.

The abovementioned three aspects defined the scope and content of the current survey. This is similar to a number of international studies which focus on students' learning quality. For example, to assess students' learning quality in science, the PISA project tested students' understanding of science (including values, knowledge, and attitude to science) and the environment. In addition, it investigated students' understanding of the role of science in their future career, length of schooling, and the conditions of learning and teaching of science subjects in their schools. It also collected information about the students' families and the communities in which they lived (OECD 2005, 2006). The TIMSS 2007 project of IEA (2005, 2007) also took the students' personal information and the conditions of schools, life inside and outside schools, and amount and difficulty of homework into account while assessing student learning. Similarly, the NAEP 2009 project (NAGB 2008, 2009) in the United States investigated aspects such as ethnicity, languages and cultures at home, study interests and attitudes, teaching methods, assessment, and learning outcomes.

To sum up, the conditions and quality of learning can be tested from the following aspects: (1) the readiness and conditions of learners, learning approaches, and learning outcomes; (2) teachers' academic preparation and teaching methods; and (3) the learning environment and atmosphere. This is the framework for designing the questionnaires for this survey.

3.3.2 *The Instruments*

According to the above definition of situation and quality of student learning, two questionnaires were developed to collect responses from students and teachers. The student questionnaire consisted of three parts with 11 questions and 97 items. Part one focused on background information of the students, including personal information, family information, school environment, and social environment. Part two aimed to find out the learning approaches of students, including their interest in

Part	Issues	Topics
(1) Background information	Personal information family, school and social environment	Learning condition, context and atmosphere in and out of the schools, economic condition and cultural background of the families and the community
(2) Learning approach	Learning methods in class	Observation of nature, observation of demonstration, design activities, hands-on activities, group projects
	Personal habits	Out-of-class reading, memorized knowledge, comprehended knowledge, practice, educational programs watched
	Motivationand attitude	Learning motivation and interest, attitudes to science classes
(3) Learning outcomes	Core scientific concepts	Bio-diversity, metabolism, inheritance, growth and development, evolution, three phases of matter, mechanical movement, light transmission, heat and temperature, sound transmission, electrical current, orbiting of Earth, the universe, the solar system
	Understanding of science	Understanding of the essence of science, attitude to science, attitude to nature

Fig. 3.2 Structure of the questionnaire for students

Table 3.1 Distribution of topics in the student questionnaire

	Biology	Physics	Earth and the universe
N1	13	13	11
Themes selected	Biodiversity, metabolism, inheritance, growth and development, evolution	The three forms of matter, mechanical movement, light transmission, heat and temperature, sound transmission, electric current	Orbiting of Earth, the universe, the solar system
N2	5	6	3
%	38	45	27

Note: *N1* means number of themes in national curriculum standard, *N2* means number of themes selected for assessment

science and motivation for science study and their learning methods in and out of the classroom. Part three tested the students' understanding of the core scientific concepts and their construction process, students' understanding of the nature of science, and their attitudes to science and nature. Figure 3.2 summarizes the structure of the questionnaire.

To ensure a proper coverage of relevant scientific knowledge, 14 themes were selected from 37 themes in the curriculum standards. Questions were set to test students' understanding of the core science concepts of the themes (see Table 3.1).

To avoid only testing rote memorization of textbooks, three questions were designed to test students' understanding of each concept. The first question asked

Part	Issues	Topics
(1) Background of teachers	Personal information, subject knowledge, attitude toward education	Gender, age, years of teaching, qualification, major, nature of position, responsibilities, grade/class taught, school, attitude to the teaching profession, work ethics, attitude to students, views on professional development
(2) Curriculum and textbook	Appropriateness of objectives and content, match between the course and students	Coverage, difficulty, up-to-date/connection with life, compatibility with current conditions
(3) Teaching	Teaching philosophy and methods, assessment	Knowledge transmission, skills development, in-class activities, after-class activities
(4) Resources	Conditions and atmosphere of learning, resources and their usage	Learning atmosphere, family environment, equipment, facilities and other resources and their usage

Fig. 3.3 Structure of the questionnaire for teachers

directly if students had learned or heard about the concept. The second and third questions invited students to give judgments on two statements related to the concept in real life. For example, to test students' understanding of biodiversity, students were first asked if they knew or had heard about animal protection. Then they were invited to make judgments on two questions related to the concept of biodiversity: (a) mice are disgusting and should be exterminated; (b) we should not capture and kill wild animals arbitrarily. The two statements are at different levels of understanding. One is consistent with the textbook knowledge, while the other is more or less related to students' misconceptions. So, a different pass rate on items of the same concept showed the level of understanding of students on that concept. The true-false format was adopted.

The teacher questionnaire consisted of four parts with 29 questions and 80 items. The questionnaire aimed to find out (1) professional backgrounds of teachers and attitudes to teaching, (2) the teachers' views on the curriculum and the textbooks, (3) the teachers' views and styles of teaching, and (4) teachers' resources for teaching, including experiment equipment and instruments and other facilities. Figure 3.3 summarizes the questionnaire.

3.3.3 *Procedure, Samples, and Reliability of the Survey*

A team of five postgraduate students and three experienced elementary science teachers conducted the survey. They held five team meetings to construct, edit, and modify the two abovementioned instruments. Two pretests and trial interviews were conducted to examine both the student and teacher questionnaires. The pretest samples of students were 79 and 129. The samples of elementary science teachers involved in the two pretests were 30 and 29. Interviews were conducted after the test. Modifications and adjustments were made accordingly. In order to make sure

Table 3.2 Student samples by grade level and by gender

	School 1		School 2		School 3		School 4		School 5		School 6		Total		
Grade level	**4**	**5**	**4**	**5**	**4**	**5**	**4**	**5**	**4**	**5**	**4**	**5**		**4**	5
N.	44	38	40	42	35	28	37	45	36	34	40	33	**N**	234	218
													%	51.8	48.2
Gender	**M**	**F**	**M**	**F**	**M**	**F**	**M**	**F**	**M**	**F**	**M**	**F**		**M**	**F**
N.	43	39	36	46	28	35	47	35	36	34	38	35	**N**	228	224
													%	50.4	49.6
Total N.	82		82		63		82		70		73		452		
%	18.1		18.1		13.9		18.1		15.5		16.2		100		

that the instruments do focus on what we declared to investigate and have good coverage related to students' learning environment, learning approaches, and outcomes, a number of questions such as "Can you think of any other things that are important in your learning environment?" were asked in the trial interviews. Three groups of experts in Beijing, Shanghai, and Hong Kong were also invited to read the questionnaires before they were put into practice to identify the internal validity of the instruments. They were also invited to check the data collected and the interpretation of the data. Responses of all the experts confirmed that the two instruments are well designed and the results are reasonable and understandable.

The survey was conducted between September and October 2009 after the pretests and trial interviews. There are 63 public elementary schools in Tianhe District of Guangzhou City. According to their geographical locations, these schools are divided into three groups: original residential schools, new residential schools, and urban-country schools. Two schools were randomly selected from each group and a total of six schools were involved in the student survey. Five hundred and six questionnaires were distributed, with 452 returned. Table 3.2 shows the demographic data of the student sample.

Given that most of the public schools had only one science teacher and the total number of schools was 63, the project team tried to cover as many science teachers in the district as possible. The teacher questionnaires were distributed to all attendees in a staff development workshop for science teachers in the district. Forty-three questionnaires were distributed with 38 returned, i.e., about 50 % of science teachers in the district participated in the survey. Seven teachers with different backgrounds were selected as the focus group in trial interviews. Table 3.3 summarizes the teacher backgrounds of the sample.

The internal consistency index (α) was calculated and used as reliability of each of the subscales related to attitudes, understanding, and views of the students and teachers. The values of α were between 0.73 and 0.93, which showed that the survey results were reliable (see Table 3.4).

Table 3.3 Background of the teachers surveyed (sample size 38)

	Gender		Age			Final qualification		Major			Nature of post			Full-time		Years of teaching				Years as sci. teacher		
	M	F	21–30	31–45	>45	G.D.	B.A.	Science	Math	Others	Permanent	Temporary	Others	*Yes*	*No*	≤ 3	4–10	11–20	≥ 21	≤ 3	4–10	11–20
N	**7**	**31**	**13**	**20**	**5**	**8**	**30**	**17**	**5**	**16**	**34**	**2**	**2**	**25**	**13**	**6**	**10**	**14**	**8**	**20**	**12**	**6**
%	**18**	**82**	**34**	**53**	**13**	**21**	**79**	**45**	**13**	**42**	**90**	**5**	**5**	**66**	**34**	**16**	**26**	**37**	**21**	**52**	**32**	**16**

Note: *G.D.* graduation diploma, *B.A.* bachelor's degree

Table 3.4 Reliability of the survey

	Student questionnaire (452 samples)					Teacher questionnaire (38 samples)		
Subscale	Science knowledge	Learning methods	Learning interest	Understanding of science	Attitude to nature	Work attitude	Teaching perception	Teaching environment
(α)	0.85	0.88	0.79	0.73	0.78	0.93	0.76	0.81

3.4 Findings of the Survey

3.4.1 The Learning Environment

3.4.1.1 Community and Family Context

Tianhe is a new district of Guangzhou City founded in the mid-1980s. Before that, it was a large piece of farmland on the outskirts of Guangzhou. Most of its residents were farmers. With the expansion of the city, it is now a newly developed business, manufacturing, and high-tech center. A number of tertiary institutions and scientific research centers are located in this area. The majority of its population is not native residents. They come from other parts of the country. There are a high number of well-educated people among them. Public services related to science, education, and culture are provided at a higher level than they are in the rest of the country. Nevertheless, the survey shows that although libraries and science exhibition centers were available, 23 % of the students did not think they were conveniently located. Although most families could access broadband Internet, 42 % of the parents did not allow their children to go online after school due to the fear of potential distraction to academic learning.

The financial situations of most students' families were fairly good by Chinese standards. For example, 95 % of the families owned their own real estate, usually an apartment with two to four or more rooms. Eighty-four percent of families were equipped with at least one set of home electronics facilities, such as home theaters, and one car. More than 95 % of families had one or more telephones, computers with Internet access, a designated desk for studying, and more than 20 books.

According to the student responses, 60 % of the parents had tertiary diplomas or academic degrees. The educational level of 80 % of the parents was higher than secondary. Sixty-one percent of the parents cared about their children's achievement in learning science. Seventy-nine percent of the parents were able to provide a variety of support for their children's study. Eighty-three percent of the parents were willing to support or participate in science activities organized by schools. Twenty-three percent of the parents actively communicated with their children on science-related issues. Thirty percent of the parents hoped their children would go into a science-related career in the future.

Teachers were also asked for their views on parental participation and support by using a 5-point scale (5 being the highest). The average score of parental support was 3.79, 3.63 for parents' willingness to participate and cooperate with science teaching, 3.42 for caring about their children's science learning, and 3.87 for communicating about science-related topics with their children. These results are very similar to reports from the students.

The above results show that Tianhe District is well equipped with facilities and resources needed for science education. The economic, living, and study conditions of the students are fairly good, and their parents have a fairly high education level. However, parents only provide an average level of support for students' studies and have limited interaction on science-related issues with their children. The majority of the parents do not expect their children to work in science-related fields.

3.4.1.2 School Conditions and Resources

Information about the school conditions and resources was gathered from the teachers. This included information about the scale of schools and size of classes and how well they were equipped with computers, multimedia and Internet facilities, library resources, labs, and places for activities.

All the schools included in this survey were of medium scale: 60 % of them had 18 classes or less and 18 % had 24 classes or more. Compared to the government standard of 50 students per class, the class size in these schools was smaller: 90 % of them had 31–50 students per class. The schools were well equipped with modern technique equipment. Seventy-five percent of the schools equipped all or most of their classrooms with computers and multimedia equipment, but only a few of the classrooms were equipped with a digital display (e.g., projector). Fifty-five percent of the classrooms were equipped with televisions. All of the schools surveyed had Internet access. Ninety-five percent had Internet access to most or all of the offices. Seventy-eight percent had Internet access in most classrooms. When asked about science reading resources for students, the percentage of teachers who thought that they were not enough, enough, and hard to say were 40 %, 20 %, and 40 %, respectively. The follow-up interviews found that the teachers were not sure how to judge the sufficiency of library resources.

With regard to science labs, all but one teacher indicated that their schools were not equipped with science labs. Though there were several rooms for student activities in each school, they were neither specially designed nor able to fit the needs of science experiments. Teachers in the trial interviews explained that science experiments and activities can be conducted in the classroom at the elementary level. They did not think that special labs were necessary. About places specially designed for science activities, such as science corners or science gardens, 9 teachers gave no answer while 13 said yes; 16 teachers were not sure how to answer the question since their schools did have some places for all kinds of activities but none was specially designed for science. Only six teachers responded that their schools could provide all the materials needed for science experiments. The other 32 teachers thought that it was hard to say. The trial interviews revealed that the schools could provide only the chemical reagents or physical gear. However, materials used in most of the observational or experimental activities at the elementary level are daily products or foods, such as eggs or fish, which students had to bring from home.

The above results clearly portray the community and family contexts and school teaching-learning conditions in the district: average school scale and smaller class size (by Chinese standards), well equipped with computers and multimedia facilities, available labs and places for science experiments and activities but with most not specially designated, insufficient provision of materials for science experiments, and perhaps insufficient library resources.

3.4.1.3 Teachers and Teaching Environment

The majority of the elementary science teachers in Tianhe District were female. They were mostly young with qualifications in tertiary education, but most did not major in science. For a little more than half of science teachers, science was the only subject they taught. For other teachers, they taught at least one other subject while also acting as science teachers. Over 80 % of the teachers had 4 or more years of teaching experience, but about half had taught science for less than 3 years.

Results of teachers' professional attitude, including attitudes to the teaching profession, attitudes to work, attitudes to students, and attitudes to their own professional development, are shown in Table 3.5.

The results in Table 3.5 show that the teachers had a reasonably high self-identity in their profession. This is supported by the survey of the students. To the question "I like my science teacher," 60 % of the students responded they liked their science teachers, and 38 % were not sure. Only 12 % of students responded that they did not like their science teachers.

Table 3.6 shows the results of the survey relating to teachers' views on the curriculum, including its objectives, the amount and difficulty of its content, the relation between the content and students' lives, and the adaptation between the content and the learning environment. A mean score of 0.65 suggests that the teachers' level of satisfaction with the new curriculum was moderate.

A 5-point scale was used to test the teachers' teaching ideology toward two typical extremes: "knowledge transmission" and "ability development." The average scores of the teachers' responses were 3.82 and 3.91, respectively. These scores on both subscales are slightly higher than the medium value of 3, suggesting that the teachers' views of teaching were complex. They did not orientate to either extreme.

Eight kinds of the most common teaching methods and seven types of after-class activities were listed to elicit the teachers' teaching methods. Table 3.7 presents a

Table 3.5 Professional attitudes of science teacher in Tianhe District (sample size 38)

Profession	Work	Students	Development	Overall
77.8	83.9	74.3	77.9	78.5

Note: The maximum score for each column is 100

Table 3.6 Teachers' satisfaction of the curriculum (sample size 38)

	Objectives	Amount of content	Difficulty	Society relation	Life relation	Resources	Adaption	Mean
Degree of satisfaction	0.72	0.82	0.81	0.66	0.65	0.45	0.56	0.65

Note: The maximum score for each column is 1

Table 3.7 Frequencies of teaching methods and after-class activities (sample size 38)

Teaching methods	**Lecture**	**Challenge**	**Discuss**	**Demonstrate**	**Experiment**	**Read**	**Present**	**Practice**
F.	3.61	3.05	4.03	2.45	2.21	3.08	3.32	2.89
After-class activities	**Visiting exhibitions**	**Mini projects**	**Planting or feeding**	**Transmitting knowledge**	**Outdoor observing**	**DIY**	**Science competition**	
F.	3.39	3.18	3.03	2.97	3.95	3.63	3.79	

Note: *F* in this table means frequency; the frequency scales are 1 (never), 2 (seldom), 3 (sometimes), 4 (often), and 5 (very often)

summary of the results. It shows that classroom discussion was the most popular teaching method, followed by lecturing, oral presentation from students, reading textbooks, and challenging students. Teacher demonstration and student experiments were less popular in the class, though they are regarded as the most important methods in science teaching. Moreover, the frequencies of organizing after-class activities were mostly sometimes and often. Among them, outdoor observation was more popular than others.

With regard to assessment, since science was taken off the list of summative examinations at the end of each term and year in Tianhe District, a final grade of "pass" or "fail" is given to students as a record of learning. Only one teacher (about 3 %) continued using the process assessment technique and evaluated her students accordingly. Seventeen teachers (45 %) insisted in giving students in-class tests and evaluated the students on both their test results and classroom performances, while 15 teachers said they did not assess their students at all and 5 did not respond to the question. The project team members found this quite strange, so they followed up with all five teachers individually by phone or e-mails. They explained that they did not assess their students because it was practically impossible to give a student a "fail" in the final record and thus assessment was meaningless.

Summing up briefly, teachers involved in the survey were fairly satisfied with the current curriculum. Relatively, they were more satisfied with the content and degree of difficulty and less satisfied with the resources and flexibility of the course. Teachers tried to vary their teaching methods rather than lecturing all the time; however, the most popular methods adopted were still teacher-centered. Two important methods, demonstration and experiment, were not very popular. After-class activities were normally organized, especially outdoor observations. The situation of assessment was confusing and worrisome because 40 % of the science teachers said they did not assess their students. Instead they passed them all.

3.4.2 *Quality of Student Learning in Science*

3.4.2.1 The Students' Learning Approaches

As mentioned above, research shows that learning motivation and learning strategies are closely related, which reflects the learning approaches and determines the learning outcomes. In this survey, both the student and teacher questionnaires included questions on student motivation. In the student questionnaire, eight questions focused on students' motivation in learning science. The average score was 2.53 in a 3-point scale, which indicates that students' motivation in learning science was high. The views of teachers were very different from their students. Four questions in the teacher questionnaire referred to the teachers' views on students' learning motivation. The average score was 2.75 on a 5-point scale, which suggests that teachers had a slightly negative view on student motivation. The trial interviews showed that students and teachers had different judgments on motivation.

Table 3.8 Usage of the common learning approaches (sample size 452)

Record and observe	Watch exp.	Design exp.	Do exp.	Group project	Read for interest	Read textbook	Discuss	Homework	Watch TV
2.48	2.11	2.49	2.25	2.31	1.99	1.77	2.07	2.09	2.03

Note: The numbers in the column were calculated based on students' choices of 1, once a week; 2, once a month; 3, several times a year; 4, never

Students based their answers on their interest. However, teachers paid more attention to student performance. If students are involved actively and perform well in science, they will be seen as highly motivated. If students are interested in classroom activities but do not perform well in learning scientific knowledge, the teacher will not see them as having high motivation.

The student questionnaire listed some common learning approaches. A 4-point scale was used to examine how often these methods were used. Table 3.8 sums up the results. It shows that the most commonly used method was reading textbooks and books of interest. On average, students read more than once a month. The frequency of other learning activities ranged from less than once a month to about several times a year. This is consistent with the teacher-centered teaching style as reported by the teachers. Further interviews also revealed that the lack of science homework was due to science not being an exam subject.

In summary, students' learning motivation was slightly higher than the medium value, and all learning approaches were used with reading textbooks being the most popular.

3.4.2.2 Learning Outcomes

As mentioned earlier, 14 themes were selected from the 37 themes in the national curriculum to test how well the students had learned the core contents. Twenty-eight true or false items (two items per topic) related to students' daily life were set for this aim. Table 3.9 shows the details of the items and the results of the test.

The first column of Table 3.9 lists 14 scientific terms selected from the 37 themes in the national curriculum. The students were asked if they knew about these terms and where they learned about the terms. The second column lists the percentage of the students who declared that they knew the term. The third, fourth, and fifth columns show where they learned the terms. The results show that on average, the students knew 93 % of the terms. Among them, 32 % were learned from science class, 18 % were learned in everyday life, and 43 % from both. The term "metabolism" was the least known. The sixth column lists all the items testing students' understanding of the terms and the seventh column lists the pass rate. The average pass rate was 0.64. Items relating to mechanical movement, electric current, and light transmission had very low pass rates. These results may reflect students'

Table 3.9 Results of the test of students' knowledge (sample size 452)

Themes	Correct response %	Locations of learning (%)			Understanding of knowledge	
		In class	Out of class	Both	Test items	Pass rate
(1) Animal protection	96	17	15	65	Mice are disgusting, so they should be exterminated	0.52
					We should not capture and kill wild animals arbitrarily	0.95
(2) Metabolism	73	18	19	36	People can be immortal one day once science is extremely well developed	0.88
					Leaves falling in autumn is related to metabolism	0.57
(3) Genetics	92	15	31	46	Xiaoming looks almost the same as his father. This is due to genetics	0.80
					All trees have leaves – this is related to genetics	0.68
(4) Growth and development	94	25	22	47	Butterflies are not transformed from caterpillars	0.87
					Our body changes a lot as we grow	0.92
(5) Evolution	94	36	17	40	Humans do not have tails, which is a result of evolution	0.76
					Cats catching mice is a result of evolution	0.51
(6) Gas, liquid, and solid	96	45	10	42	Water can turn into ice and it can turn into steam as well	0.88
					The fact that boats can move on water means that solids can flow	0.52
(7) Mechanical movement	89	29	23	38	The movement of a car is a kind of mechanical movement. A man walking is not mechanical movement	0.11
					A nonmoving passenger sitting in a car can also be in mechanical movement	0.16
(8) Light transmission	97	45	12	39	The fact that a mirror can change the direction of light shows that light does not travel in a straight line	0.44
					Trees can block sunlight. This shows that light travels in straight line	0.58
(9) Temperature	98	44	11	43	The colder the thing is, the lower its temperature	0.76
					A padded coat can keep you warm because it can generate heat	0.65

(continued)

Table 3.9 (continued)

Themes	Correct response %	Locations of learning (%) In class	Out of class	Both	Understanding of knowledge Test items	Pass rate
(10) Sound transmission	97	42	15	40	Sound can travel in air	0.77
					Sound cannot travel in water	0.57
(11) Electric current	97	46	10	40	A lit bulb indicates that electricity is flowing through the wire	0.77
					Electric current flowing in a wire is similar to water flowing in a pipe	0.21
(12) The movement of Earth	95	35	18	41	The changing of the four seasons is related to the orbiting of Earth	0.80
					The movement of Earth has nothing to do with daybreak	0.62
(13) The universe	94	27	23	44	Earth looks like a blue planet from space	0.61
					We see the same starry sky every night	0.69
(14) The solar system	93	24	24	45	There are eight planets in the solar system.	0.69
					Earth is not a part of the solar system	0.73
Average	93	32	18	43		0.64

Note: Pass rate means the ratio of the number of students who gave the correct answer divided by the total number of students

Table 3.10 Students' responses to the nature of science ($N=452$)

	5 correct answers	4 correct answers	3 correct answers	2 correct answers	1 correct answers
N	189	110	84	38	12
%	41.8	24.3	18.6	8.4	2.7

misconceptions and that Year 4 and 5 students would not have been taught the content yet.

Another aspect of learning relates to students' understanding of the nature of science. The student questionnaire focused on five related themes and asked students if they agreed that (1) science means lots and lots of knowledge; (2) science is about experiments and investigation; (3) science can promote societal progress; (4) science can lead to a better life; and (5) scientific knowledge changes with time. Table 3.10 summarizes the students' responses and shows that about 89 % students got three or more correct answers. This means that they had a good understanding of the nature of science.

Table 3.11 Students' behavior in the natural environment ($N=452$)

Questions	Mean scores
(1) I don't litter on the street	3.54
(2) I don't throw rubbish into the river	3.54
(3) I don't mistreat pets and animals	3.60
(4) I turn off the lights when I leave the room	3.63
(5) I don't damage trees and flowers	3.46
(6) I dislike eating leftover food	2.44
Total mean score	3.35

Note: The questionnaire adopts a 4-point scale: 1= strongly disagree; 2= disagree; 3= agree; 4= strongly agree

Students' behaviors (or their attitudes toward behaviors) toward nature and the environment are another aspect of learning outcomes. The questionnaire listed six types of behaviors and asked the students to report if they behaved in the same way. Students' responses were based on a 4-point scale. As shown in Table 3.11, except for Question 6, the average scores of the questions were high. This implies that students behaved well or had a good attitude to nature and the environment.

3.5 Discussion

3.5.1 *About the General Learning Environment*

The survey shows that all the communities in Tianhe District have libraries and window displays of scientific knowledge. The Science and Technology Center of Guangdong Province is located nearby. The economic conditions of the students' families were mostly sound, and a large number of the parents had a tertiary education background. Most of the parents cared about the students' performance in science and were willing to provide possible resources to support their learning. They were also willing to support or participate in science activities organized by the schools. Moreover, about 30 % of the parents wished their children would choose a science-related career in the future. All these show that the general context supports student learning.

The conditions of the schools in Tianhe are generally good. The moderate sizes of the schools and class sizes that were not extremely large were beneficial to the application of student-centered strategies in teaching science. The schools were well equipped with computers, multimedia equipment, and wideband Internet access. Lab equipment and teaching facilities were good enough to meet the needs of the curriculum standards. All these suggest that the conditions of schools are adequate.

The elementary science teachers in Tianhe District were capable of meeting the needs of science teaching. Two-thirds of the science teachers specialized in science, which meant that science was the only subject they taught. The percentage of the so-called specialized teachers is much higher than in other parts of the country. Although many of those teachers did not major in science in their preservice training, they are experienced and learn from their work. And, most of the science teachers were young or middle aged. Seventy-nine percent of them had bachelor's degrees, while the rest were also graduates of a teachers' college[1], which meant they were qualified in terms of the government standards for elementary teachers. The professional attitudes of these teachers were reasonably high (averaging 78.5 out of 100). Sixty percent of the teachers were well received and respected by students.

However, problems existed in the environmental support to science teaching and learning. Twenty-three percent of the elementary students indicated that visiting the provincial science center was not convenient and was too expensive. The hardware and facilities were not effectively used due to the shortage of multimedia software and materials as a result of poor school management. For example, according to the general rules of school management, teachers were not allowed to use school fund to buy any food. So, if teachers want to buy eggs or fish for student observation, they need to get special permission from the principal to use the school fund; otherwise, they have to pay themselves or ask students to bring in these items. As teachers cannot always get the permission, this becomes an obstacle of student activities. Another example is that all multimedia software is supplied by companies appointed by education authorities. Their supplies were often insufficient and out of date. Teachers had to download software from the Internet themselves at their own cost.

One more serious problem is that 42 % of the elementary science teachers were trained for teaching language, social science, arts, physical education, or other subjects. They did not have a sound understanding of scientific knowledge and methods in their preservice training, and in-service training for science teachers was not sufficient. Further, the schools often switched their teaching subjects and, as a result, 52 % of them had less than 3 years of experience in teaching science. This in turn has led to problems in meeting the requirements of the new curriculum.

Generally speaking, from the above results, we found that the learning environment in Tianhe seems fairly good and supportive. This is different from what we stated in the second part of this chapter, where the learning environment was recognized as unsupportive. The reason might be due to unbalanced development in China. Guangzhou is a more developed city and, as mentioned in Sect. 3.4.1.1, Tianhe is a newly developed business, manufacturing, and high-tech center in Guangzhou City. The social and school environments in Tianhe are far better than those in the less developed regions in China. However, one might find similar results as reported here in a number of cities of the eastern part of China, such as Beijing, Shanghai, Tianjin, Nanjing, and Hangzhou.

[1] A kind of tertiary education with lower level to university

3.5.2 About the Conditions of Elementary Science Teaching

This survey found that teachers had varying degrees of satisfaction toward the new curriculum and relevant teaching materials. This suggests that the agreement among the teachers, the curriculum, and the teaching resources varies with teachers, the curriculum modules, and different kinds of resources. In general, most teachers were happy with the curriculum objectives, the coverage of content, and degree of difficulty. They were also happy that the contents of science are related closely to children's lives and the society. The teachers' personal philosophy was complex: a mixture of knowledge delivery and student development orientations. Transmission of scientific knowledge and development of scientific inquiry capacity were of equal importance in their minds. They tried to use a variety of teaching methods, and, in fact, science teaching in elementary classroom is no longer one-way knowledge transmission. This is inspiring and shows that elementary science is changing gradually toward a more student development orientation.

However, there are problems in classroom teaching. Firstly, many teachers were not happy with the flexibility of the textbooks and adaptability of other teaching resources. This might be due to the fact that these teachers did not have a science background and lacked experience in teaching science, so they wished the textbooks and teaching resources were friendlier so that they could follow step by step in class. Secondly, two important teaching strategies, experiments and demonstrations, were not widely used by teachers. As discussed above, this might be due to the teachers being ill-prepared and irrational rules in school management. Thirdly, maybe the most serious problem is that a large number of the teachers (53 %) saw little value in process assessment and lost their way in conducting processing assessment. They did not know how to practice the process assessment technique and still preferred written tests. They strongly indicated their wish to reestablish term tests and annual exams for elementary science. Changing teachers' conceptions of assessment and training them with process-assessing skills seem important for further implementation of the curriculum reform.

3.5.3 About the Quality of Elementary Students' Learning in Science

The students in Tianhe District were highly interested and motivated in learning science. Among the ten prelisted learning approaches, reading textbooks and other science-related popular readings was the most frequent way of learning. In accordance with their frequency of use, the other nine approaches are as follows: watching science-related TV programs or films, talking or discussing science-related issues with others, doing homework, observing the teacher's demonstrations, performing experiments, conducting group projects, observing and recording natural phenomena, and planning experiments. It is notable that the first four approaches are related to learning after class. This result and the result of the survey on the locations where students learned about the scientific concepts support the idea that learning after class is at least as important as learning in the classroom (Gao 2002). It might be another reason why students in Tianhe achieved well. Teachers should not overemphasize the role of knowledge delivery through the class. They should make more effort to create a proper environment for students to learn and explore by themselves and to facilitate students' understanding of the methods of exploring.

The fact that students knew 93 % of the concepts in the 14 themes chosen and that the average pass rate of the 28 test items was 0.64 suggests that students in Tianhe achieved quite well, especially considering that the students involved in the survey were in Grades 4 and 5, and many of the concepts in the survey had not been taught in science class yet. Moreover, 88.6 % of the students surveyed had an adequate understanding of the nature of science. The score of self-reported behaviors toward nature and the environment was 0.90. This shows that students knew how to behave properly to protect nature and the environment. One notable point is the low pass rate (between 0.1 and 0.2) for the three items focusing on students' misconceptions. This again suggests the importance of addressing students' misconceptions in science.

One conflicting aspect has emerged in the two questionnaires. The teachers did not think the students were motivated enough or studied hard enough, while the students reported that they were highly interested in learning science. Further interviews revealed the different criteria used by the teachers and students in judging learning motivation and attitude. The students responded to the questionnaires based on their interest, while the teachers focused more on the devotion of the students. From the teachers' point of view, many students may like science because it is interesting. They might involve actively in science activities just because they are fun. However, they might not work hard enough on acquiring the content knowledge, which was viewed as the priority by the teachers. The teachers did not consider that as a right attitude toward learning. This raises the question of whether science activities and experiments in elementary schools should focus on knowledge acquisition and understanding. We will argue that, as a starting course in science, it is more important to maintain students' interest and curiosity in science and learning. This will encourage students to learn by themselves after class and actually obtain more scientific knowledge. Some of them might even devote their whole life to science. Overemphasis on knowledge acquisition will have a negative effect on student learning, especially when their abstract thinking ability is not yet well developed.

3.6 Conclusion

In 2001, the Chinese government boosted a new round of curriculum reform. The elementary science course was reconstructed. The reform emphasized that elementary science should be an "enlightening" course and focus on students' interest in science as well as their scientific literacy. It reduced knowledge contents by a large percentage and tried to change science teaching from teacher-centered and knowledge-oriented toward a more student-centered and ability development-oriented direction. The reform abolished summative examination for elementary science and instead built a new assessment system mainly consisting of formative assessment and students' self-assessment and peer assessment in their process of learning. This caused a huge shake-up in the traditional science teaching and learning in China's elementary schools, which led to a number of questions and challenges. The new curriculum has been criticized for lowering the status of science in the elementary curriculum and undermining students' development in science. Some people also worry that the quality of science teachers is not adequate and that the teaching-learning environment does not support this new curriculum.

The survey reported in this article aimed to examine the effect of the new curriculum reform by investigating the general context of students' families and the communities; the conditions of schools, teachers, and the resources for science teaching and learning; the styles and attitudes of student learning; and results of learning. The results of the survey are encouraging: the supportive family and general environment of Tianhe District; the generally sufficient conditions of schools, teachers, and teaching resources, despite some problems; good results of students' learning; students' good motivation to learn; good learning strategies; good knowledge of science and the nature of science; and good attitude toward nature and our living environment. One thing to note is that Tianhe is located in the capital city of Guangdong Province, which is one of the most economically developed cities in China. Thus, the Tianhe case study cannot be representative of the effects of the elementary science curriculum reform in the entire country, but it may be a representative of the eastern regions of China.

References

Biggs, J. B., & Telfer, R. (1987). *The process of learning* (2nd ed.). Melbourne: Prentice-Hall of Australia.

Cai, W., Wupuer, A., Cheng, H. Liu, K. (2009). Research on the situation and countermeasures of implementation of the science curriculum in primary school of Xinjiang (新疆小学科学课程实施的现状与对策研究), *Journal of Chongqing University of Arts and Science (Natural science edition)*, No. 6, pp. 65–68.

Chen, M., He, B. Zheng, M. (2011). The implementation and improvement of science curriculum in Rural Primary School (农村小学科学课程实施现状及改进措施), *Journal of Anqing Teachers College*, No. 2, pp. 114–117.

Gao, L. (2002). Features of enquiring learning – studies of a foreign case (探究性学习的特点 – 一个国外案例的分析), *Curriculum, Teaching Materials and Method*. 2002–5, pp. 16–21.

Gao, L. (2004). A survey on school teachers' conceptions of teaching under the New National Curriculum (新课程背景下教师教学观初探). *Journal of South China Normal University (Social Science Edition)*, No.1, pp. 116–121.

Gao, L. (Ed.). (2011). *Collections of excellent texts of the elementary science textbooks of the new curriculum (新课程实验教材小学科学优秀案例选)*. Beijing: People's Education Press.

Gao, L., & Watkins, D. A. (2001). Identifying and assessing the conceptions of teaching of secondary school physics teachers in China. *British Journal of Educational Psychology, 71*, 443–469.
Gao, L., & Watkins, D. A. (2002). Conceptions of teaching held by school science teachers in P.R. China: Identification and cross-cultural comparisons. *International Journal of Science Education, 24*(1), 61–79.
IEA. (2005). *TIMSS 2007 assessment frameworks*. TIMSS & PIRLS International Study Center, Lynch School of Education, Boston College, 2005.
IEA. (2007). *TIMSS student questionnaire 2007(grade 4)*, Retrieved at http://www.iea.org/
Li, Y. (2007). Elementary science curriculum reform in the rural areas: The lack of promoting (农村小学科学课程改革亟待加强). *Educational Science Research, 7*, 25–28.
Liu, Y. (2010). Thinking on the current status of rural primary school science teaching (农村小学科学教学的思考), *Journal of South-west Guizhou Teachers College for Nationalities*, No. 1, pp. 116–118.
Lu, Chun. (2011). The situation and improvement of science education in Rural Primary School (我国农村小学科学教育的现状与改善), *Teaching and Management*, No. 1, pp. 86–87.
Marton, F. (1983). Beyond individual differences. *Educational Psychology, 3*, 189–303.
Ministry of Education (MOE). (1993). *Syllabuses of nine-year compulsory education: Elementary nature (九年制义务教育小学自然教学大纲)*. Beijing: People's Education Press.
Ministry of Education (MOE). (2001a). *National science curriculum standards for the full-time compulsory education (Grades 3–6). (Trial Version). (义务教育小学3-6年级科学课程标准(实验版))*. Beijing: Beijing Normal University Press.
Ministry of Education (MOE). (2001b). *National curriculum scheme for compulsory education (trial version)*, (义务教育课程(实验)计划), unpublished government Document, Beijing: Ministry of Education, the People's Republic of China.
MOE Project Team (a). (2002). *A survey on school curriculum and teaching* (关于课程与教学现状的调查), Unpublished report.
MOE Project Team (b). (2010). *A survey on the quality of elementary science textbooks and the conditions of elementary science teaching learning environment* (关于小学科学教科书与教学环境的调查). Unpublished report.
NAGB. (2008). *Science framework for the 2009 National Assessment of Educational Progress*. Washington: NAGB.
NAGB. (2009). *Student background questionnaire, science. (grade4)*. Retrieved at http://www.nagb.org
OECD. (2005). *Student questionnaire for PISA 2006*. Retrieved at http://www.oecd.org/pisa
OECD. (2006). *Assessing scientific, reading, and mathematical literacy: A framework for PISA 2006*. Paris: OECD.
Provus, M. M. (1971). *Discrepancy evaluation for educational program improvement and assessment*. Berkeley: McCutchan.
Team of National Curriculum Development – Elementary Science. (2002). *An interpretation of the national curriculum standards of compulsory education: Science, year 3-6, (trial version)*(义务教育小学3-6年级科学课程标准(实验版)解读). Beijing: Beijing Normal University Press.
Zhong, M., & Gao, L. (2007). *Issues in elementary science curriculum reform* (小学科学课程改革中的问题与分析). Beijing: Curriculum, Teaching Materials and Methods. 2007 (6). pp. 77–81.

Chapter 4
Science Teaching Practices in Junior Secondary Schools

Hongjia Ma, Gavin W. Fulmer, and Ling L. Liang

4.1 Introduction

One of the most important Chinese science education reform efforts in the last decade has been the introduction of scientific inquiry and active learning into science teaching. In learning, the reform-oriented approach represents a shift in the focus of school science from a goal of "mastery of scientific knowledge" to the "development of student scientific literacy," while in teaching, reform-minded educators have been advocating a shift from the traditional teacher-centered approach to a more learner-centered, inquiry-oriented classroom (MOE 2001a). The curriculum standards issued by the Ministry of Education (MOE 2001b, c, d, 2011) clearly state that the success of a science program should produce student learning outcomes in three dimensions: scientific knowledge and skills, scientific processes and methods, and attitudes and values. However, as many people are aware, at secondary school levels, student learning is often driven by high-stakes standardized exams at gateways, and creative thinking and inquiry skills are not typically emphasized in these exams (e.g., Chen et al. 2010; Liang and Yuan 2008). Consequently, learner-centered and inquiry-oriented science instruction may not be

H. Ma (✉)
College of Teacher Education, Nanjing Normal University, 122 Ninghai Road, Nanjing, Jiangsu 210097, China
e-mail: mahongjia@njnu.edu.cn

G.W. Fulmer
Department of Teaching and Learning, University of Iowa, 289 Lindquist Center North, Iowa City 52242, IA, USA

L.L. Liang
La Salle University, Philadelphia, PA, USA

L.L. Liang et al. (eds.), *Chinese Science Education in the 21st Century: Policy, Practice, and Research*, Contemporary Trends and Issues in Science Education 45, DOI 10.1007/978-94-017-9864-8_4

valued or implemented in classrooms by teachers despite the mandates of the national curriculum standards.

In this chapter, we first provide a brief overview of the reformed science curriculum in junior secondary schools. Then we describe an investigation into science classroom practices and report the findings of the study. Finally, we discuss the implications and make suggestions for future research.

4.1.1 Reformed Science Curriculum in Junior Secondary Schools

Between the establishment of the People's Republic of China in 1949 and the late 1980s, the Chinese basic and higher education systems, including the science curriculum and instruction, were largely influenced by the former Soviet Union model. In junior secondary schools, discipline-based science programs, including biology, physics, and chemistry, were offered. The release of curriculum standards (trial versions) on multiple subjects in 2001 signaled the beginning of the most recent systemic reform in basic science education in China (e.g., MOE 2001b, c, d). These curriculum standards were further revised in 2011.

As part of the reform effort, the required science courses in the junior secondary schools can be offered through either a discipline-based or integrated approach. Multiple textbook series have been published after they were approved by the National School Textbook Review Board and deemed as aligned with the respective curriculum standards. District science coordinators or curriculum leaders may choose either an integrated or separated, discipline-based textbook set for the schools under their oversight. In reality, however, the experiment involving integrated science programs has encountered serious challenges. Presently, Zhejiang province and Shanghai are the only places where integrated science programs are offered in junior high schools. School administrators, teachers, and parents fear that an integrated science program at junior secondary level may not be able to fully prepare the students for their discipline-based science courses in senior high schools and may negatively impact students' performance on college entrance exams. Other factors preventing the implementation of the integrated sciences may include a shortage of resources and quality science teacher training programs (Zhang and He 2012). For instance, in a study conducted by Zhang and He (2012), 27 preservice teacher candidates and 25 recently graduated in-service teachers majoring in science education from six normal universities were interviewed. It was found that the science teacher education curricula did not reflect the intended integration objectives of the major and that the science teacher educators were not competent in teaching integrated courses. The teacher candidates and recent graduates were least satisfied with their learning experiences related to the nature of science, scientific inquiry, and integration of subject matters.

Consequently, the majority of the junior secondary schools across the country continue offering science as separated, discipline-based subjects. Most frequently, students take biology in seventh grade, physics in eighth and ninth grades, and

Table 4.1 Content of physics curriculum in junior secondary school

Scientific inquiry	Theme/ topic	Subtheme/topic
Asking questions	Matter	Physical forms and changes; physical properties of matter; structure of matter, scale, and development; and application of new materials
Making hypotheses	Motion and interaction	Variety forms of motion, force and mechanical motion, sound and light, electricity and magnetism
Planning and designing investigations	Energy	Types of energy, the conversion and transfer of energy, the law of conservation of energy, sources of energy and sustainable development
Carrying out investigations and collecting data		
Analyzing data and engaging in argument from evidence		
Evaluating		
Communicating and collaboration		

Table 4.2 Content of chemistry curriculum in junior secondary school

Theme	Subtheme
Scientific inquiry	A better understanding of scientific inquiry, improving the ability of scientific inquiry, learning about the basic skills and ability to conduct experiments
Chemical substances around us	The atmosphere around the earth, water and common solution, metals and metallic minerals, common compounds in daily lives
Matter composition	Diversity of matter, particulate nature of matter, chemical elements, symbols of substance compositions
Chemical changes	Characteristics and classification of chemical changes, the law of conservation of mass
Chemistry and societal development	The relation between chemistry and energy resources, the usage of natural resources, common synthetic materials, chemicals and human health, environment protection

chemistry in ninth grade. Tables 4.1 and 4.2 present the content of physics and chemistry curriculum, as two examples (MOE 2001c, d, 2011). Three main themes addressed in the physics curriculum include matter, motion and interaction, and energy. In the chemistry curriculum, the main themes include chemical substances around us, matter composition, chemical changes, and chemistry and societal development. In both programs, "scientific inquiry" has been emphasized as both "content" and pedagogy.

Compared to the traditional teacher-centered instruction, the reformed science courses place more emphasis on learner-centered, hands-on/minds-on, inquiry-based, and constructivist-oriented practices. Inside the classrooms, students are encouraged to work in cooperative learning groups and communicate with the teacher and other peers orally and in writing, using diagrams, drawings, graphs, and mathematical equations. Teaching materials are designed with considerations of student diversity in development and learning styles. In general, instruction is expected to stimulate student interest and motivation and be built on student everyday life experiences. The development of student understanding is expected to connect with applications of the science content in the real world. Teachers are also expected to incorporate mass media and community resources (e.g., museums, parks, and libraries) in science learning and teaching.

4.1.2 The Need for Examining Classroom Practices and Their Potential Connections to Student Affective Learning Outcomes

To date, there have been only a few empirical studies that examined the degree of classroom implementation of inquiry-oriented learning based on teachers' or students' self-reports in schools that have been published inside China. Of these, one study reported middle school physics teachers' perceptions about inquiry and classroom implementation of inquiry teaching based on a teacher survey (Xie 2012). Xie found that the majority of the teacher participants did not implement the inquiry approach as required by the curriculum standards for various reasons, e.g., the exam-oriented education system, the teachers' limited understanding of the nature of scientific inquiry, and limited lab space and equipment. Another study used a teacher survey to investigate the characteristics and patterns of inquiry teaching practices in high school chemistry classrooms (Guo et al. 2011). According to the student reports, the majority of teachers implemented inquiry teaching less than one to two times per month. The inquiry teaching aspect was mainly implemented during designated chemistry experiment periods, and many of the experiments were confirmative in nature. In science classrooms, the majority of the teachers occasionally or rarely helped students develop scientific ways of thinking in teaching. Many teachers would implement inquiry when their classes were visited or observed by others. Overall, about 41 % of student participants perceived the inquiry approach as effective, while 31 % of students believed that inquiry teaching was a mere formality with little meaning. It was noted that both studies reported the findings based on individual survey items without including reliability and validity of instrumentation.

To further inform science teaching and policy-related decision making, it is important to obtain a more accurate picture of classroom practices and the relationship between a reform-oriented instructional approach and desirable learning outcomes. This study reports findings from an investigation into instructional practices in physics and chemistry classes from student perspectives based on a previously validated survey instrument (Juuti et al. 2010). It is among the first research studies that examine both the classroom practices and their potential connections to student affective learning outcomes in China. It certainly contributes to the overall body of knowledge about science learning and teaching in Chinese secondary schools, particularly from the student perspective.

4.2 Literature Related to Our Investigation

4.2.1 Theoretical Framework

This study draws upon the theoretical framework of the classroom learning environment, which emphasizes not only the role of teachers' actions but also of students' perceptions and interpretations of these actions (Shuell 1996). Such attention to the learning environment has proven valuable in science education as well (e.g., Fisher et al. 2011; Fraser 1998). The learning environment, and students' perceptions of their teachers' instruction within this environment, has been found to be an important predictor of student outcomes (den Brok et al. 2004). Drawing upon this theoretical framework highlights the importance of understanding teachers' actions from the students' perspective and, through this experience, the influence on students' attitudes and other learning outcomes.

4.2.2 Measurement of Science Classroom Teaching Practices

According to the science education research literature, there are multiple ways to examine teachers' instruction, including the use of classroom observations, teachers' self-reports, interviews, and student surveys (e.g., Adamson et al. 2003; Lawrenz et al. 2009; Fulmer 2008; Venville et al. 2008; Waight and Abd-el-Khalick 2007). While classroom observation methods have been found to be effective in understanding teachers' application of reformed instruction and relating this to students' outcomes (Judson and Lawson 2007; Park et al. 2011; Sawada et al. 2002), Chinese researchers have tended to avoid using these type of methods because training observers and repeated rating of classroom observations are usually costly and time consuming. Alternatively, survey methods using teachers' or students' self-reports have been frequently adopted by researchers in China. Regarding using teachers' self-reports for classroom practices research, however, some evidence has shown that teachers may give themselves higher marks on general surveys than should be expected (Centra 1973). More reliable results could be achieved by using instruments consisting of a larger number of action-specific items (Koziol and Burns 1986; Porter 2002). A good example of such instruments is the Survey of Enacted Curriculum (SEC) which includes a number of action-specific items and typically requires between 45 and 90 min to complete

(Porter 2002). As a result, teacher self-report surveys can be limited either by the participants' own bias or by the demanding time commitment of the participants.

Fulmer and Liang (2013) advocate the use of student survey as a measure of effective teaching and as direct evidence of student learning experiences in science classes and demonstrated a method for examining teachers' instruction by surveying US high school students. The main advantages of this type of approach include cost-efficiency and potentially reduced likelihood of self-confirmation bias, as the students may be less likely to over-report desirable actions than their teachers would. Furthermore, the large volume of data from students' responses allows the exploration of relationships in the data.

4.2.3 Student Attitudes Toward School Science

Student attitudes toward school science are affective in nature and involve student feelings, beliefs, and values with regard to school science (Osborne et al. 2003). Studies have revealed that there is a consistent interrelationship among the *cognitive* and *affective* factors (Beaton et al. 1996; Shrigley 1990; Simpson and Oliver 1990). Many studies have reported modest positive correlations between science attitude and science achievement (Schibeci and Riley 1986; Keeves and Morgenstern 1992; Weinburgh 1995). In a study conducted by Oliver and Simpson (1988), it was found that student achievement was strongly correlated with affective behaviors in the science classroom, particularly achievement motivation and student self-concept about their own ability in science. Research also reported that activity-based and/or issue-oriented instructional approaches had enhanced the development of positive attitudes toward science (Freedman 1997; McComas 1993; Siegel and Ranney 2003).

Whereas increasing student interest and retention in science is part of the goals of school science, research has repeatedly shown a decline in positive attitudes toward science as students move from lower grades to high school levels (Yager and Penick 1986; Catsambis 1995; Piburn and Baker 1993). Recent international studies also revealed that many students did not think that science was related to them personally, and a relatively small portion of students wanted to pursue a career involving science (OECD 2007). In a literature review of research on attitudes toward science, Osborne et al. (2003) suggested that ignoring the affective components of science education has led to the problem that many countries are currently experiencing: an alienation of youth from science. They therefore called for more research on the context of science classrooms to identify the nature and style of instruction and activities that engage students in learning science.

In order to develop student interest in science, Hidi and Renninger (2006) suggest that teaching should provide opportunities for interaction, cooperative group work, and challenges that lead to knowledge building and conceptual development. Researchers also argue that the use of student-preferred teaching methods in the science classroom supports positive social and emotional effects (Juuti et al. 2010). Thus, it is important to examine the interconnections among student interest in and attitude toward school science, enacted science teaching methods, and student-preferred instructional activities or practices.

4.2.4 Research Questions

In the present study, we attempt to investigate the following questions: (1) To what extent are students experiencing inquiry-oriented or direct teaching methods in physics or chemistry classrooms? (2) To what extent would students like to see changes of teaching practices in their physics or chemistry classrooms? (3) What are students' attitudes toward their learning of physics and chemistry in school? (4) What are the relative effects of the students' experiences of science classroom instruction on students' attitudes toward school science? (5) What are the relative effects of the discrepancy between students' *desired* and the *actual* science classroom experiences on the students' attitudes toward school science?

4.3 Methodology

This study examines students' attitudes toward science and their experience of classroom teaching practices in junior high school physics and chemistry classes. It is among the first studies that systematically explore interrelationships among student affective variables, perceived and desired classroom experiences, and other demographic characteristics. The following sections describe the instrument, the research sample and setting, and the analyses.

4.3.1 Instrument

The questionnaire used in this study was adapted from a study of science instruction in Finland (Juuti et al. 2010), first by translating the questionnaire into Chinese and then by modifying certain items to reflect the Chinese context. Both validity and reliability of the instrument were reexamined by the researchers of the study. Given the purpose of the study, we wanted to make certain that the survey items were understandable to Chinese students and reflected the Chinese context. We chose not to adopt the popular "back translation" procedure as described in other science education research literature. The questionnaire consisted of four parts: students' perceptions of classroom teaching methods and preferred instructional methods, students' attitudes about school science and science and technology in general, students' socioeconomic background information, and students' academic performance in physics and chemistry. In the teaching method section, student participants were asked to estimate the frequency (number of times over the course) that certain teaching methods actually occurred in classroom and the frequency they preferred or wished. There are 23 items in this section, using a 5-point Likert-type scale (1 = "never," 2 = "rarely, 1–5 times," 3 = "seldom, 6–10 times," 4 = "often, at least every other class," 5="very often, almost every class"). In the attitude section, students were asked to rate their perceptions and attitudes about school science on a 4-point Likert scale (1 = "disagree," "2 = mostly disagree," 3 = "mostly agree," and 4 = "agree").

4.3.2 Participants and Setting

The study involves 16 junior high schools of different levels of academic reputation or prestige in a well-developed city in Jiangsu province. The schools are of three types, A, B, and C, with the A schools being highest in academic standing and C schools being lowest in academic standing. A total of 1,334 questionnaires were distributed and collected by teachers appointed by school principals, with a survey return rate of 100 %. The study was conducted during December 2010 while the student participants (679 male and 655 female students) were taking physics and chemistry courses in 9th grade. During the survey, all participants were allowed to ask clarification questions and instructed to leave an item blank if they did not understand the question or were unsure about the answer. Five of the 1,334 returned questionnaires were fully blank, leaving 1,329 questionnaires with some data. Of these, five students completed the demographic questionnaire but did not respond to any of the other items, giving a final sample of 1,324 students (a nonresponse rate of approximately 1 %). Of these 1,324, the vast majority answered all items (almost 87 %), and there was no apparent pattern in the missing data (i.e., most items had a few students who skipped them and no item had nonresponse greater than 2 %).

4.3.3 Data Analyses

Analyses were conducted on the students' scores on measures of students' attitudes toward science and their actual versus preferred instruction. For attitude toward science, a score was calculated based on the eight attitude items, with a Cronbach's alpha of 0.68. Each student was assigned a score equal to the average of their response to all questions. For the instruction questionnaire, a factor analysis was conducted to identify subscales of items within the questionnaire based on students' responses about their teachers' instruction. Based on analysis of eigenvalues and a scree plot, a three-factor solution was found to fit the data well. The authors reviewed the content of the included items to identify any pattern or pedagogical principle connecting the practices addressed by the items and named the factor based on the form of teaching that reflected the items. Items that did not load to the

Table 4.3 Sample instructional practices survey items, by subscale

Item no.	Survey statement
Teaching with a cooperative orientation	
Q09	Students solve problems or complete tasks in small groups
Q10	Students conduct projects in small groups
Q15	Students learn cooperatively
Teaching with a constructivist orientation	
Q16	Students learn by writing essays, summaries, or stories about science
Q19	Students imagine or create various ideas (brainstorming)
Q22	Students hold a debate during the lesson
Traditional direct teaching	
Q06	Teacher leads discussion about difficult concepts and/or problems
Q12	Students learn by reading the textbook
Q14	Teacher describes phenomena and processes associated with experiments (instead of performing/demonstrating the experiments)

three main factors were removed. The first factor consisted of five items about *cooperative teaching*, such as having students work in pairs or small groups to conduct experiments or to discuss difficult problems. The second factor consisted of five items about *constructivist teaching*, such as having students draw concept maps, write stories about science, or hold a debate. The third factor consisted of eight items about *direct teaching*, such as writing content on the blackboard, having students read the textbook, or describing an experiment (rather than performing or demonstrating the experiment). The factors were examined separately for the actual versus preferred responses. The observed Cronbach's alpha values for cooperative teaching were 0.86 for actual and 0.85 for preferred; for conceptual teaching, 0.84 for actual and 0.85 for preferred; and, for direct teaching, 0.80 for actual and 0.87 for preferred. These values all indicate that the subscales had good internal consistency. For each subscale, a subscale score was computed based on the average response to the items in that factor. Table 4.3 presents a list of sample items within each subscale.

Next, to enable comparison of students' responses on the preferred and actual factors on the instruction questionnaire, a *discrepancy* score for each factor is calculated by subtracting the value for preferred from the actual score, similar to the method described by Juuti and colleagues (2010). Thus, positive values indicate that students experienced a form of instruction more than they preferred, and negative values indicate that students experienced a form of instruction less than they preferred.

After preparation into scales for attitude and for actual and preferred instruction, the data were analyzed in two ways to answer the five research questions posed. First, an analysis of variance was conducted to examine differences by school type for the attitudes, actual instruction, and discrepancies between actual and preferred instruction. Second, regression analyses were conducted to compare the relative effects of the instructional subscales on students' attitudes toward science.

4.4 Results and Discussion

Descriptive statistics for the students' attitude measures and their actual and preferred instruction are shown overall and by school type in Table 4.4. For attitudes, a higher number indicates more positive attitudes toward science. The average attitude score was near 3.0, indicating an overall positive attitude toward science (on a scale of 1–4). Because school types reflect different levels of previous academic performance and standing, we break down the findings according to school type where possible. There was a statistically significant difference in attitude among the school types (F [2,1321] = 10.23, $p < 0.01$), with students in the higher-ranked schools having lower attitude. However, this difference has an eta-squared value of 0.015, which is quite small from a practical standpoint.

The students' reports of actual instruction indicate that they often or very often experienced direct teaching, occasionally experienced cooperative teaching, and seldom or rarely experienced constructivist teaching. This pattern shown in Table 4.4 is consistent across all school types. The students in higher-ranked schools reported significantly lower cooperative teaching (F [2,1326] = 6.17, $p < 0.01$) and constructivist teaching (F [2,1326] = 16.70, $p < 0.01$) than did counterparts in the lower-ranked schools. However, while statistically significant, such school differences are small from a practical standpoint, with effect sizes (as eta-squared values) below 0.03 for all reports of actual instruction.

Table 4.5 presents the descriptive statistics on the differences between students' actual experience of instruction and preferred instruction. As can be seen, students

Table 4.4 Descriptive statistics for student attitude and instructional subscales

			Actual			Preferred		
School type	N	Attitude	Coop	Construct	Direct	Coop	Construct	Direct
Total	1,324	3.01	3.57	3.08	4.33	4.35	3.99	4.58
A	399	2.92	3.44	2.92	4.29	4.24	3.77	4.52
B	581	3.01	3.59	3.05	4.40	4.37	3.98	4.62
C	344	3.10	3.68	3.34	4.27	4.45	4.25	4.58

Coop cooperative teaching, *Construct* constructivist teaching, *Direct* direct teaching

Table 4.5 Descriptive statistics for the discrepancies between actual and preferred instructional subscales

	Discrepancies		
School type	Coop	Construct	Direct
Total	−0.78	−0.91	−0.25
A	−0.80	−0.85	−0.23
B	−0.78	−0.94	−0.22
C	−0.76	−0.92	−0.31

Coop cooperative teaching, *Construct* constructivist teaching, *Direct* direct teaching. Negative values indicate that students experienced a form of instruction less than preferred

in all schools reported experiencing less of each type of instruction than was preferred. In particular, the constructivist teaching shows the greatest discrepancy between actual and preferred, across all school types. In addition, there are no statistically significant differences in the discrepancies among schools in the students' ratings for either cooperative teaching ($F\ [2,1326] = 0.20$, $p > 0.50$) or constructivist teaching ($F\ [2,1326] = 1.21$, $p > 0.25$). Only type C schools have slightly higher discrepancy for direct teaching ($F\ [2,1326] = 4.98$, $p < 0.05$), although the direction is consistent with the higher-ranked B and A schools.

To look deeper at the differences between scales, discrepancies on sample items are informative. As shown in Fig. 4.1, the sample items demonstrate that there are consistent discrepancies across all items, regardless of school type. Additionally, it is clear that some items have particularly large discrepancies for both cooperative (e.g., Q10) and constructivist teaching (e.g., Q16 and Q22).

Next, regression analyses were conducted to examine the relationship among the instructional variables and students' attitudes toward science. The first regression analysis focused on the relationship of attitude with students' actual instructional experience. Results indicate that there are small, positive relationships between students' attitudes and their experiences of cooperative teaching ($b = 0.206$, $p < 0.001$) and of constructivist teaching ($b = 0.14$, $p < 0.001$). By contrast, there is a small, negative relationship between attitude and the experience of direct teaching ($b = -0.088$, $p < 0.01$).

The second regression analysis examined the relationships between students' attitudes and any discrepancies between actual and preferred instruction. Because earlier ANOVA results indicated that students differed in actual versus preferred instruction by school type, analyses were conducted separately for each school type. Results indicate that across all school types there was a consistent, positive relationship for cooperative teaching (for type A, $b = 0.19$, $p < 0.01$; for type B, $b = 0.206$, $p < 0.01$; for type C, $b = 0.23$, $p < 0.01$). That is, in all school types, students had higher attitudes toward science when the difference between actual and preferred instruction was positive. On the other hand, there are no significant relationships between attitude and the discrepancies for constructivist teaching or direct teaching for any of the school types.

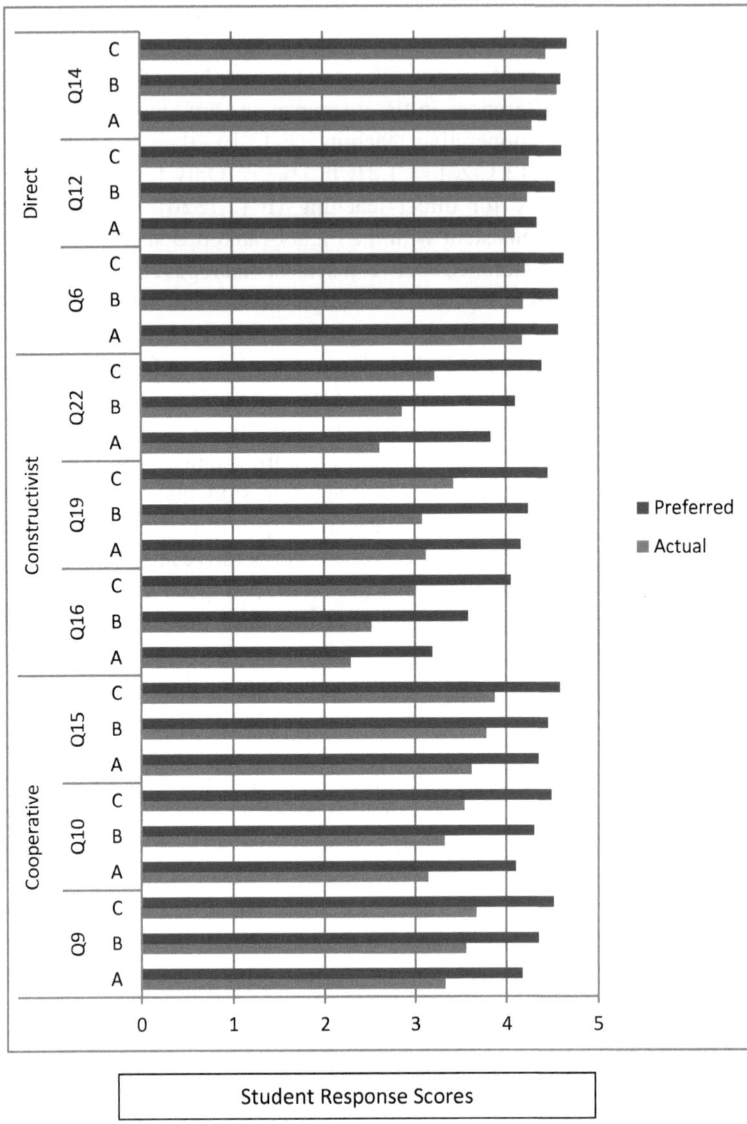

Fig. 4.1 Student responses to sample instructional practice survey items, by school type. Note that *A*, *B*, and *C* represent three types of participating schools: *A* = schools with the highest academic standing; *C* = schools with the lowest academic standing; *B* = schools with academic standing between *A* and *C*. The selected survey items are presented in Table 4.4

4.5 Conclusions

In the present study, we examined the relationships among students' attitudes toward science and their actual versus preferred forms of instruction. The findings revealed that students experience cooperative teaching and constructivist teaching occasionally (between seldom and often), whereas direct teaching occurs very often. This reflects prior work carried out in other countries (e.g., Juuti et al. 2010). Additionally, findings indicate that students would prefer to experience much more constructivist teaching than they currently do. Students would particularly welcome more instructional practices incorporating writing, brainstorming, debating, and science field trips.

We also examined the students' attitudes toward science according to their school background and the types of instruction they experience. Our finding that students from better-performing schools (i.e., "A" schools) have lower attitudes toward science was unexpected. In this education system, admission to the higher-ranked schools is competitive, and there is greater pressure for academic success and examination performance in such schools. So, it is possible that students in the higher-ranked schools have lower attitudes toward science because of the greater academic pressure they have in this subject. On a separate point, our findings regarding forms of instruction showed that students have more positive attitudes toward science when they report more frequently experiencing cooperative teaching and constructivist teaching and less frequently experiencing direct teaching. Furthermore, there is a consistent, positive relationship between students' attitudes toward science and their preference for instruction—when students' actual experience of cooperative teaching is closer to their preferred amount, they have more positive attitudes.

The findings reveal that students in the Jiangsu province report experiencing cooperative and constructivist teaching only occasionally and at a lower rate than desired. Furthermore, these students' attitudes toward science are positively related to the incidence of both cooperative and constructivist teaching, suggesting the potential importance of using cooperative and constructivist teaching strategies to raise students' attitudes toward science. Our results are consistent with the findings of Juuti and colleagues (2010), in that the students reported experiencing very high rates of direct teaching and would prefer to experience more cooperative and conceptual instruction. Furthermore, the results underscore the value of cooperative and constructivist teaching for supporting students' engagement with the class (e.g., Treagust 2007), thus supporting students' satisfaction with the classroom experience and potentially improving their performance in the subject (Nolen 2003).

4.5.1 Implications

The findings have implications for future research and for the understanding and planning of teaching. Firstly, the findings demonstrate that there are meaningful ways to measure students' experiences of instruction through a student survey of their instructional experiences and to use this information in the evaluation and planning of instruction. Much as Fulmer and Liang (2013) showed in their study of US students' experiences of science instruction, drawing on students' reports of their classroom experience can be influential in understanding instruction and may be informative for teachers and curriculum developers in planning for instruction. Such findings have been supported in other topic areas as well (e.g., Juuti et al. 2010; Nolen 2003).

Secondly, the findings show that students' attitudes are significantly higher when they experience instructional strategies that encourage student cooperation and conceptual development, as has been recommended for science education in the West (e.g., Treagust 2007). Drawing on these findings, if increased attitudes toward science are a desirable outcome of science teaching in China, then one possible avenue is to increase the use of cooperative and conceptual instructional strategies in class. However, the present study did not experimentally manipulate the extent of cooperative, conceptual, or direct teaching. Thus, additional research is needed to test such a conjecture about the potential mechanism through which cooperative and constructivist teaching can raise students' attitudes toward science.

Acknowledgments The authors thank Dr. Xiufeng Liu, Dr. Bing Wei, and two anonymous reviewers for their insightful comments.

References

Adamson, S. L., Banks, D., Burtch, M., Cox, F., III, Judson, E., Turley, J. B., Benford, R., & Lawson, A. E. (2003). Reformed undergraduate instruction and its subsequent impact on secondary school teaching practice and student achievement. *Journal of Research in Science Teaching, 40*(10), 939–957.

Beaton, A., Martin, M. O., Mullis, I., Gonzalez, E. J., Smith, T. A., & Kelley, D. L. (1996). *Science achievement in the middle school years: IEA's Third International Mathematics and Science Study*. Chestnut Hill: Boston College.

Catsambis, S. (1995). Gender, race, ethnicity, and science education in the middle grades. *Journal of Research in Science Teaching, 32*(3), 243–257.

Centra, J. A. (1973). Self-ratings of college teachers: A comparison with student ratings. *Journal of Educational Measurement, 10*(4), 287–295.

Chen, X., Jia, C. C., & Chen, N. (2010). Alignment between the physics content standard and the exam (物理内容标准与考试之间的一致性研究). *Curriculum, Teaching Material and Method (课程.教材.教法), 30*(7), 67–71. (In Chinese).

den Brok, P., Brekelmans, M., & Wubbels, T. (2004). Interpersonal teacher behaviour and student outcomes. *School Effectiveness & School Improvement, 15*(3/4), 407–442.

Fisher, D., den Brok, P., Waldrip, B., & Dorman, J. (2011). Interpersonal behaviour styles of primary education teachers during science lessons. *Learning Environments Research, 14*(3), 187–204. doi:10.1007/s10984-011-9093-z.

Fraser, B. J. (1998). Science learning environments: Assessment, effects and determinants. In B. J. Fraser & K. G. Tobin (Eds.), *International handbook of science education* (Vol. 1, pp. 527–564). London: Kluwer.

Freedman, M. P. (1997). Relationship among laboratory instruction, attitude toward science, and achievement in science knowledge. *Journal of Research in Science Teaching, 34*, 343–357.

Fulmer, G. W. (2008). *Successes and setbacks in collaboration for science instruction: Connecting teacher practices to contextual pressures and student learning outcomes*. Saarbru¨cken: VDM Publishing.

Fulmer, G. W., & Liang, L. L. (2013). Measuring model-based high school science instruction: Development and application of a student survey. *Journal of Science Education and Technology, 22*(1), 37–46. doi:10.1007/s10956-012-9374-z.

Guo, Y., Lu, Y. & Shi, J. (2011). Investigating inquiry teaching in high school chemistry classrooms: Status analysis and reflections (高中化学科学探究教学现状调查及常态化思考). *Teaching Chemistry in Secondary Schools* (中学化学教学参考) (4), 3–5. (In Chinese).

Hidi, S., & Renninger, K. A. (2006). The four-phase model of interest development. *Educational Psychologist, 4*(2), 111–127.

Judson, E., & Lawson, A. E. (2007). What is the role of constructivist teachers within faculty communication networks? *Journal of Research in Science Teaching, 44*(3), 490–505. doi:10.1002/tea.20117.

Juuti, K., Lavonen, J., Uitto, A., Byman, R., & Meisalo, V. (2010). Science teaching methods preferred by grade 9 students in Finland. *International Journal of Science and Mathematics Education, 8*(4), 611–632.

Keeves, J. P., & Morgenstern, C. (1992). Attitudes towards science: Measures and effects. In J. P. Keeves (Ed.), *The IEA study of science III: Changes in science education and achievement: 1970 to 1984* (pp. 122–142). New York: Pergamon.

Koziol, S. M., Jr., & Burns, P. (1986). Teachers' accuracy in self-reporting about instructional practices using a focused self-report inventory. *Journal of Educational Research, 79*(4), 205–209.

Lawrenz, F., Wood, N. B., Kirchhoff, A., Kim, N. K., & Eisenkraft, A. (2009). Variables affecting physics achievement. *Journal of Research in Science Teaching, 46*(9), 961–976.

Liang, L. L., & Yuan, H. (2008). Examining the alignment of Chinese national physics curriculum guidelines and 12th-grade exit examinations: A case study. *International Journal of Science Education, 30*(13), 1823–1835. doi:10.1080/09500690701689766.

McComas, W. F. (1993). STS education and the affective domain. In R. E. Yager (Ed.), *What research says to the science teacher, volume 7: The science, technology and society movement* (pp. 161–168). Washington, DC: National Science Teachers Association.

Ministry of Education (MOE). (2001a). An *outline for the basic education curriculum reform* (基础教育课程改革纲要). (official document). (In Chinese).

Ministry of Education (MOE). (2001b, 2011). National Science Curriculum Standards for the Full-Time Compulsory Education (Gr. 7–9). *(Trial Version, Revised Version)*. (义务教育科学课程标准). Beijing: Beijing Normal University Press. (In Chinese).

Ministry of Education (MOE). (2001c, 2011). National Physics Curriculum Standards for the Full-Time Compulsory Education (Gr. 7–9). *(Trial Version, Revised Version)*. (义务教育物理课程标准). Beijing: Beijing Normal University Press. (In Chinese).

Ministry of Education (MOE). (2001d, 2011). National Chemistry Curriculum Standards for the Full-Time Compulsory Education (Gr. 7–9). *(Trial Version, Revised Version)*. (义务教育化学课程标准). Beijing: Beijing Normal University Press. (In Chinese).

Nolen, S. B. (2003). Learning environment, motivation, and achievement in high school science. *Journal of Research in Science Teaching, 40*(4), 347–368. doi:10.1002/tea.10080.

OECD. (2007). *PISA 2006: Science competencies for tomorrow's world*. Retrieved July 8, 2008, from http://www.pisa.oecd.org/document/2/0,3343,en_32252351_32236191_39718850_1_1_1_1,00.html

Oliver, J. S., & Simpson, R. D. (1988). Influences of attitude toward science, achievement motivation, and science self concept on achievement in science: A longitudinal study. *Science Education, 72*(2), 143–155.

Osborne, J., Simon, S., & Collins, S. (2003). Attitudes towards science: A review of the literature and its implications. *International Journal of Science Education, 25*(9), 1049–1079.

Park, S., Jang, J.-Y., Chen, Y.-C., & Jung, J. (2011). Is pedagogical content knowledge (PCK) necessary for reformed science teaching? Evidence from an empirical study. *Research in Science Education, 41*(2), 245–260. doi:10.1007/s11165-009-9163-8.

Piburn, M. D., & Baker, D. R. (1993). If I were the teacher . . . Qualitative study of attitude toward science. *Science Education, 77*, 393–406.

Porter, A. C. (2002). Measuring the content of instruction: Uses in research and practice. *Educational Researcher, 31*(7), 3–14.

Sawada, D., Piburn, M., Judson, E., Turley, J., Falconer, K., Benford, R., & Bloom, I. (2002). Measuring reform practices in science and mathematics classrooms: The reformed teaching observation protocol. *School Science and Mathematics, 102*(6), 245–253.

Schibeci, R. A., & Riley, J. P., Jr. (1986). Influence of students' background and perceptions on science attitudes and achievement. *Journal of Research in Science Teaching, 23*, 177–187.

Shrigley, R. L. (1990). Attitude and behavior are correlates. *Journal of Research in Science Teaching, 27*, 97–113.

Shuell, T. J. (1996). Teaching and learning in a classroom context. In D. C. Berliner & R. C. Caffee (Eds.), *Handbook of educational psychology* (pp. 726–764). New York: Macmillan.

Siegel, M. A., & Ranney, M. A. (2003). Developing the changes in attitude about the relevance of science (CARS) questionnaire and assessing two high school science classes. *Journal of Research in Science Teaching, 40*, 757–775.

Simpson, R. D., & Oliver, J. S. (1990). A summary of the major influences on attitude toward and achievement in science among adolescent students. *Science Education, 74*, 1–18.

Treagust, D. F. (2007). General instructional methods and strategies. In S. K. Abell & N. G. Lederman (Eds.), *Handbook of research on science education*. Mahwah: Lawrence Erlbaum Associates.

Venville, G., Sheffield, R., Rennie, L. J., & Wallace, J. (2008). The writing on the wall: classroom context, curriculum implementation, and student learning in integrated, community-based science projects. *Journal of Research in Science Teaching, 45*(8), 857–880.

Waight, N., & Abd-El-Khalick, F. (2007). The impact of technology on the enactment of "inquiry" in a technology enthusiast's sixth grade science classroom. *Journal of Research in Science Teaching, 44*(1), 154–182.

Weinburgh, M. (1995). Gender differences in student attitudes toward science: A meta-analysis of the literature from 1970 to 1991. *Journal of Research in Science Teaching, 32*, 387–398.

Xie, S. (2012). Investigating inquiry teaching in middle school physics classrooms within minority ethnic groups (民族地区初中物理教师开展科学探究教学情况的调查), *Research in Continuing Education* (继续教育研究), Issue 4, 86–89. (In Chinese).

Yager, R. E., & Penick, J. E. (1986). Perception of four age groups toward science classes, teachers, and the value of science. *Science and Education, 70*, 355–363.

Zhang, H., & He, H. (2012). Student perceptions of the integrated 'science education' major in some Chinese universities. *International Journal of Science Education, 34*, 1991–2013.

Chapter 5
Science Curriculum and Implementation in Senior Secondary School

Xiao Huang, Lin Ding, and Bingyuan Hu

5.1 Introduction

In the fall of 2001, the Ministry of Education officially launched new schemes for ordinary senior high school curriculum and curriculum standards while the new curriculum for compulsory education was in the trial phase. In April 2003, the state issued the schema and curriculum standards related to all disciplines, which were based on extensive and thorough investigation and reviewed by experts and subsequently examined and approved by the Ministry of Education.

The reform-based documents clarify the content, evaluation, goals, structure, and teaching methods. Under the guidance of the curriculum reform for basic education, the underlying principles of science curriculum in senior high school (including physics, chemistry, and biology) were clarified and defined in the associated *Curriculum Standards for Ordinary Senior High School*.

The purpose of this chapter is twofold: (1) to situate the discussion in the historical context of the most recent high school curriculum reforms and (2) to analyze and examine the current state of science curriculum implementation and

X. Huang (✉)
Department of Education, Zhejiang Normal University, 688 Yingbin Road, Jinhua, Zhejiang Province 321004, China
e-mail: Huangxiao@zjnu.cn

L. Ding
Department of Teaching and Learning, The Ohio State University, 1945 N. High St., Columbus, OH 43210-1172, USA
e-mail: ding.65@osu.edu

B. Hu
Department of Physics, East China Normal University, 3663 Zhongshan North Road, Shanghai 200062, China

L.L. Liang et al. (eds.), *Chinese Science Education in the 21st Century: Policy, Practice, and Research*, Contemporary Trends and Issues in Science Education 45, DOI 10.1007/978-94-017-9864-8_5

challenges, especially in the investigation of the *high school physics curriculum standards* implementation.

5.2 Overview of Science Curriculum Reform

The senior high school science curriculum standards are based on comparative study of international science curricula and analysis of China's historical and current implementation efforts. The standards draw upon theoretical and empirical research into psychological characteristics of senior high school science students, as well as the influence of scientific and social development contributing to the settings of science curricula.

Science education in ordinary senior high school includes physics, chemistry, and biology. The corresponding discipline-based curriculum standards provide general guiding principles for the reform. First, these standards, especially the chemistry curriculum standards, each emphasize the importance of scientific research. They stress the important role of scientific research in learning chemistry to ensure that students "experience the process of science research, stimulate their interest in learning chemistry, and enhance their consciousness of scientific research, promote the transformation of learning skills, and cultivate students' innovative spirit and practical ability" (Ministry of Education 2001b). Second, they all focus on the cultivation of senior high school students' scientific literacy, while at the same time, the chemistry and biology curriculum standards also emphasize the importance of the relationships among science, technology, and society (STS). Third, they also aim toward student development, which diversifies the evaluation systems. The reform of physics and chemistry clearly emphasizes the evaluation of curriculum implementation and students' learning performance. For example, *high school physics curriculum standards* clearly put forward recommendations on the evaluation aligned with curriculum goals, stressing the selection function, emphasizing the process of evaluation and diversification, etc.

5.2.1 Structure, Framework, and Content of Science Curriculum

There are eight main subjects in the senior high school curriculum. These include art, comprehensive practice, humanities and social sciences, language and literature, mathematics, physical education and health, science, and technology. The subjects are connected with each other and all emphasize students' life experiences. Each follows a modular course design. According to the plan of the new curriculum, a module is a basic unit of a subject, each with a specific theme Gao and Huang (2005). With a clear educational goal, the module unfolds the content around its theme to form a relatively independent learning unit. Students can also choose electives in a

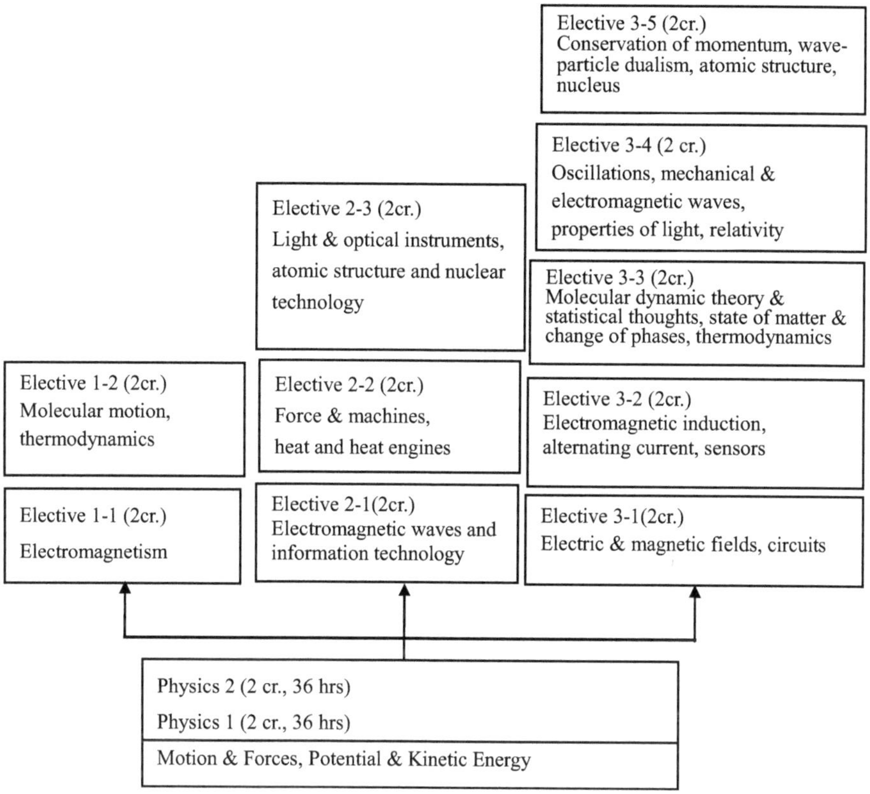

Fig. 5.1 Schematic diagram of physics curriculum

single subject or a single module in an effort to improve the flexibility and selectivity of the curriculum.

The science curriculum overall covers diverse content and presents different ways of thinking. Each science discipline contains elective modules and compulsory modules in response to different groups of students' characteristics and needs. In the physics compulsory modules, students experience some inquiry activities and are introduced to the characteristics and research methods within the discipline. The elective modules allow for student choices based on personal interests and perceived career choices. For instance, the first elective strand emphasizes the interaction between physics and society and gives prominence to the humanity of physics while highlighting the application of physics in daily life, social science, and the humanities. The second elective strand emphasizes the combination of physics and technology, stressing the applicability and practicality of physics. Finally, the third elective strand, designed for those who are interested in studying science and engineering, covers physics topics/content in more depth and shows students the application of physics in technology, economics, and society. Figure 5.1 illustrates an example of the modules, series, and strands in the physics curriculum. The physics curriculum consists of 12 modules; each

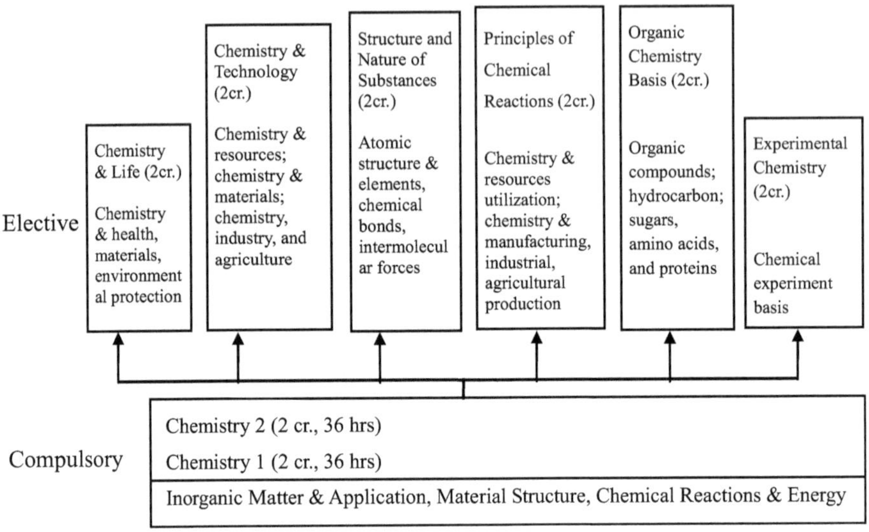

Fig. 5.2 Schematic diagram of chemistry curriculum

module means two academic credit points (Zhang et al. 2006). The first two modules, namely, Physic 1 and Physic 2, are the common required modules for each student, who must further choose a third module (2 credit points) from one of three series (series 1, series 2, or series 3) to complete six compulsory credit points of learning tasks. After completing six compulsory credit points, students can choose additional modules as electives.

Similar to the physics curriculum, chemistry gives students an opportunity to connect chemistry to daily life by building relationships between chemistry, technology, and social sciences and teaching students scientific research (Fig. 5.2). The chemistry compulsory modules contain Chemistry 1 and Chemistry 2, which help students form basic chemical concepts, improve science inquiry skills, and know ways that chemicals have an impact on the society. The elective modules consists of six modules, which involve chemistry and life, chemistry and technology, structure and nature of substances, principles of chemical reactions, the basis of organic chemistry, and experimental chemistry. Students are required to learn at least three modules (six credits) to meet the graduation requirement of chemistry curriculum. Students who are interested in chemistry are encouraged to study more modules to broaden their knowledge and improve chemistry literacy, such as 8 credits suggested for the students who will be engineers and 12 credits suggested for students who will pursue chemistry as their profession.

Biology modules are much the same, with compulsory modules and elective modules to introduce students to the interconnectedness of biology with society and technology. Figure 5.3 is the schematic diagram of the biology curriculum, which includes three compulsory modules together with three elective modules. The compulsory modules are molecules and cells, genetics and evolution, and steady state and environment. The elective modules are molecules and cells, biological science and society, and modern biotechnological special topics. Students can

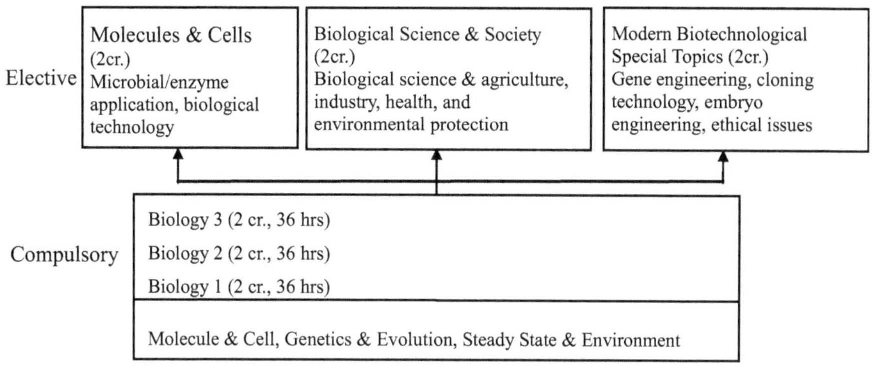

Fig. 5.3 Schematic diagram of biology curriculum

choose elective modules after they attain the 6 credits from the three compulsory modules for learning.

There are obvious commonalities between the disciplines, such as the compulsory and elective modules, as well as an emphasis on the connection between science and society. However, each discipline emphasizes its own content and strives to help students make connections between topics and students' everyday lives.

5.2.2 Science Textbook Compilation

Textbooks play a crucial role in the reform, as they are intricately tied to the process of teaching. Based on the Primary and Secondary School Textbook Diversification Policy, some senior high school science textbooks have been compiled to meet the needs of the scientific education reform and its practice. The compilation reflects not only the rationale for the basic education curriculum reform but also the understanding of curricular standards and the relationship between student learning and instruction.

In physics, there are five new versions of physics textbooks that have passed the National School Textbook Examination Committee and have been piloted in each experimental region across the country. They have been compiled, respectively, by the People's Education Press (PEP edition), the Shanghai Science and Technology Education Press (STEP edition), the Guangdong Education Press (GEP edition), Educational Science Press (ESP edition), and the Shandong Science and Technology Press (SSTP edition). In addition, three versions of chemistry and five versions of biology textbooks for senior high schools were developed and approved. All textbooks share some common features, including an emphasis on cooperative learning, problem solving, and developing scientific inquiry skills. The different versions of the textbooks in all three disciplines were developed, and the textbooks were distributed to schools based on decisions made by local educational administrative departments.

5.3 Science Curriculum Implementation

Understanding curriculum implementation and related issues is crucial to the effectiveness of any curriculum reform process (Ma 2001). Creation and implementation of a curriculum are based on the standards, curriculum concept, and the objectives of the curriculum. Below, we first summarize biology and chemistry curriculum standards implementation results based on existing research and then report our own survey study on the implementation of physics curriculum standards.

5.3.1 *Implementation of Biology and Chemistry Curriculum Standards*

Biology education researchers have investigated high school biology curriculum implementation from different angles. Research along this line is generally carried out from two approaches: (1) survey and interview studies and (2) an emphasis on curriculum research within biology education. For example, Zhang (2008) conducted a survey of the high school biology curriculum implemented over the previous year in four high schools in the Changde, Hunan Province. The purpose was to uncover the practical difficulties in implementing a new curriculum and to analyze teacher resources, textbooks, materials, and methods of evaluation. Results showed that teacher quality needs to be improved, materials need to be of higher quality, and a better system of evaluation needs to be designed. Other scholars have found that current biology education is insufficient to satisfy the needs of students, that teachers' classroom practices are inconsistent with the curriculum rationale, and that the College Entrance Examination (CEE) is mistakenly deemed as the ultimate target. The researchers' conclusions were that changes were needed in teacher professional development programs, the support system within education reform should be strengthened, and an emphasis should be placed on teaching core knowledge and skills and on implementing interactive teaching (such as establishing equality in the student-teacher relationship) (Li and Hu 2011).

There should be an increased emphasis on research into curriculum implementation during the process. However, there were only a few studies that were found to be pertinent to research regarding curriculum implementation in China, including one in which a scholar conducted and analyzed research about development, utilization, and demand for high school biology curriculum resources. This research was conducted at senior secondary schools in Jilin, Heilongjiang, and Liaoning

Provinces and found that students' participation and interest in biology can be improved when teachers use the curriculum resources to their full potential (Meng 2008). Another scholar investigated teachers' views of the biology curriculum development and clarified the shortcomings and problems in the school-based biological curriculum, highlighting that communication between different levels of schools and universities would better prepare students (Li 2006).

In high school chemistry curriculum implementation, the research mainly focuses on three areas: the status of high school chemistry curriculum implementation, the development of associated resources, and teaching and learning associated with such implementations.

First, there is a focus on the status of high school chemistry curriculum implementation. In one study, the author investigates problems and proposes recommendations for chemistry implementation based on investigation and interviews in different areas (Sun 2008). In Shandong Province, research into the high school chemistry curriculum implementation has uncovered problems and achievements in many aspects of the process. These aspects include the new curriculum concepts, reformed teaching methods, new modes of evaluation, and development of new curriculum resources (Kong et al. 2007). Second, scholars focus on the development and utilization of resources in implementing high school chemistry curricula. For example, some scholars proposed a developmental approach to high school chemistry curriculum and clarified the external conditions necessary for revision of a curriculum (Liu et al. 2009). Along with other scholars, they proposed that the development of school-based curricula should complement the national curriculum and teachers should have the ability to utilize associated resources (Liu et al. 2009).

Utilizing what has been found in the research into implementation of biology and chemistry curricula in high school, research on physics curriculum implementation should aim for several benchmarks. These include (1) understanding the difficulties and successes already identified (Wang 2012a, b), (2) ensuring that objectives are reflected in high school physics standard texts (Li 2010b; Xiao and Xing 2011; Huang 2009), (3) ensuring a connection between the physics curricula in junior and senior high schools (as well as between senior high school and college) (Li 2008; Xu 2007), and (4) selection of the modules to ensure main ideas are reflected in the curricula (Gao 2009). Still needed are studies into how the high school physics curriculum is implemented in different provinces and regions, as well as studies that examine curricular implementation in integrating three-dimensional goals in practices.

5.3.2 Methods and Instruments for Physics Curriculum Standards Implementation

In an effort to better understand new curriculum reform in experimental areas, the Ministry of Education initiated an investigation to establish a realistic basis for the revision of the high school curriculum standards. For example, in physics the investigation aims to survey students and teachers about their experiences with implementation and to uncover what works as well as the problematic areas. Additionally, this investigation analyzes the problems of the high school physics curriculum standards and offers some suggestions for revision.

5.3.2.1 Methods and Instruments

The investigation began in August 2012 and continued through October 2012. There were 72 schools involved from 8 provinces and 24 cities (see Table 5.1), and different colors in the map represent different geographical distributions (see Fig. 5.4). Areas were chosen based on their economy, culture, location, minority representation, teaching development, and students' cognitive development because they were all considered part of the Curriculum Reform Experimental Zone. Namely, the research areas cover seven regions, which include Northeast China, North China, Northwest China, Southwest China, Central China, East China, and South China. They are divided into four classes of economic and cultural development, such as Beijing is classified as the first class; Jiangsu and Guangdong belong to the second class; Ningxia, Henan, and Anhui are classified as the third class; and Yunnan belongs to the fourth class. Seven regions started new curriculum reform in different years and different batches, such as Ningxia and Guangdong as the first batch (2004), Jiangsu as the second batch (2005), Anhui as the third batch (2006), Beijing and Heilongjiang as the fourth batch (2007), Henan as the fifth batch (2008), and Yunnan as the sixth batch (2009). Also, there are some ethnic minority areas (part of Ningxia and Yunnan). In total, participants included 734 teachers and 3,494 students in the survey research (Table 5.1).

Survey questionnaires consisted of five categories, covering course design, course content, curriculum implementation, course evaluation, and curriculum demands, all of which are based on the seven aspects deemed important by the Ministry of Education (Table 5.2).

Table 5.1 Participants (students and teachers) from different provinces

Number of teachers (T) and students (S) within each participating province	Regions and municipalities within each participating province
Beijing (T, 84; S 578)	Haidian, Chaoyang, and Shunyi
Jiangsu (T, 37; S, 189)	Suzhou, Yangzhou, and Suqian
Guangdong (T, 157; S, 500)	Panyu, Huizhou, and Shaoguan
Henan (T, 124; S, 390)	Zhengzhou, Anyang, and Zhoukou
Heilongjiang (T, 47; S, 398)	Harbin, Suihua, and Yichun
Ningxia (T, 91; S, 462)	Yinchuan, Wuzhong, and Shizuishan
Anhui (T, 141; S, 619)	Hefei, Wuhu, Huaibei, and Bengbu
Yunnan (T, 53; S, 425)	Kunming, Dali, Dehong, and Mangshi

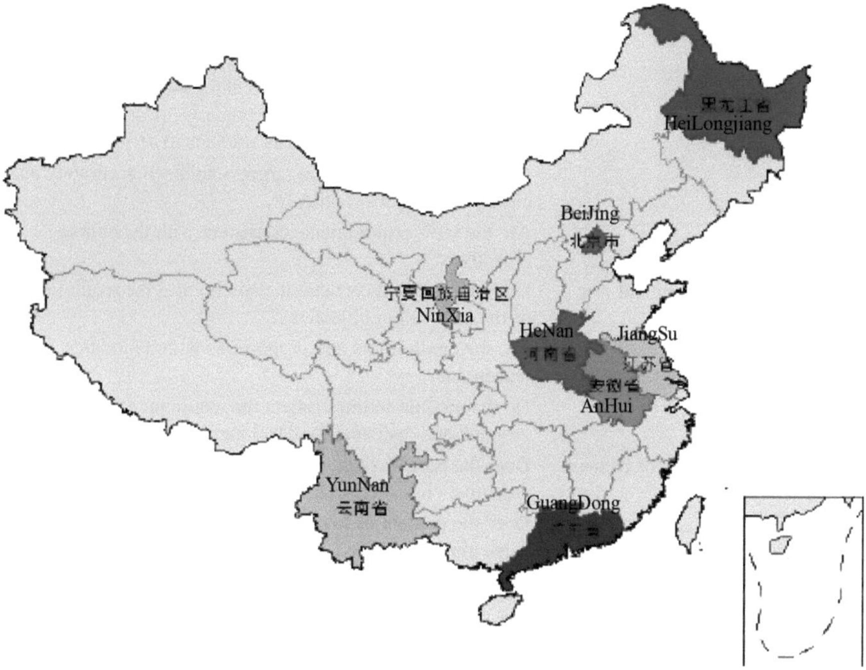

Fig. 5.4 The geographical distribution of the survey areas

In each province, the extraction method involved research objects being chosen by the principle of 3 by 3 by 3. Firstly, three districts/municipals ("high-medium-low") were extracted according to their education and economic level by the principal investigator and the person in charge of provincial education department. Secondly, three schools ("high-medium-low") were extracted according to their regional educational development by the principal investigator and the person in charge of the regional education department. Thirdly, 10th and 11th grade students were chosen according to different achievement levels by the schools' principals, and teachers were randomly selected from three schools and across different grade levels. These teachers had diverse teaching experiences, including liberal arts and science classes. Selected students ($n = 720$) and teachers ($n = 236$) also participated in interviews. However, due to limited space, we primarily report the teachers' interview responses in this chapter. All participants were interviewed about their understanding of and experiences with the curriculum standards, as well as what they perceive to be the relevant problems and solutions (see Table 5.3).

The interview protocol (qualitative and quantitative aspects) was finalized with input from a cognitive researcher, physics education researchers, and a staff member in a provincial research and education department. In order to ensure the accuracy and validity of the questionnaires and interviews, physics education researchers were consulted during the entire process. Based on the research and

Table 5.2 Ministry of education survey: aspects and associated questions

Aspects	Questions
Goals and values	What is the curriculum position and value orientation that experts expected?
	What are the values and ideas students experienced?
	Are the three-dimensional targets explained accurately and clearly for physics teachers?
	Are the CEE requirements consistent with the course objectives?
Rationality of content and structure	Does the selected curriculum content meet the requirements of the curriculum objectives?
	Are the credit hours and proportions within modules reasonable?
	Do the module settings reflect the requirement of "common ground" and "personality development"?
The moderate degree of content	Does the breadth of the curriculum standards meet students' cognitive characteristics?
	Does the content of materials meet the requirement of the curriculum standards?
Content difficulty	Does the content difficulty meet students' cognitive levels?
	Are there difficult, complex, and partial phenomena shown in the content standards?
Coherence of the content	Is there reasonable coherence between different segments, grades, disciplines, and modules?
	Does the content meet the needs of higher education, vocational education, and employment?
The underlying implementation	Are the suggestions about teaching, evaluation, textbook compilation, and resource development reasonable and effective?
	Does the textbook compilation meet the curriculum standards requirements?
	How do teachers and schools use the curriculum standards in their teaching process?
	What are the problems teachers and schools meet in the curriculum standards implementation?
Readability and guidance of curriculum standard text	Is the text structure clear and easy to understand?
	Is the written expression of the text fluent?

interviews of participants in part of Jiangsu, the researchers determined the index and subitem index in the process and provided suggestions for the problems from investigation. The explanation of the data is shown in the histograms and pie charts that follow. In the histograms that follow, the x-axes of histograms represent the participants' different responses (often by region), while y-axes represent the percentage of each type of response (the total percentage is normalized to 100 %). The pie charts give the percentages of different responses of the science teachers in all regions combined.

Table 5.3 Elements of the interviews and research

Category	Subcategory	Description
Course design	Course orientation	The nature and orientation of the high school physics course and attention to high school physics curriculum standards (WTQ1, WTQ20, WSQ2, FJQ1, FJQ2, FJQ3, FJQ4, FTQ1, FTQ2, FTQ3)
	Course objectives	Whether the representations of the three-dimensional objectives in the standards are accurate, reasonable, and practical; whether the objectives of the "horizontal" and "action verbs" are defined clearly and are practical; if it is necessary to adjust the standards (knowledge, skills, and literacy) (WTQ2, FJQ4, FJQ5, FJQ6, FJQ7, FTQ4, FTQP5, FTQ6, FTQ7)
	Curriculum framework	Whether it is reasonable to divide the high school physics course into three series and multiple modules in the curriculum standards, how standards align with curricula and objectives (WTQ3, FJQ8, FJQ9, FJQ10, FTQ8, FTQ9)
Course contents	Content selection	Whether the basic core physics content and physics ways of thinking are covered in the physics course content (WTQ8, WTQ9, WSQ9, FJQ11, FTQ10), whether high school physics course modules and the actual settings and selection are covered in it (WTQ16, WTQ17, WSQ3, FJQ12, FJQ13, FTQ11, FTQ12)
	Course difficulty and breadth	Whether the difficulty level of physical course content is reasonable (WTQ11, WTQ12, FJQ14, FJQ15, FJQ16, FJQ17, FTQ13, FTQ14, FTQ15), which course content is difficult to grasp, which course is important, and why (WSQ4, WSQ5, WSQ13, FTQ25, FTQ26)
	Scope and coordination of course content	The scope and coordination between grades, disciplines, and modules in physics course content across junior and senior high school (WTQ10, WSQ6, WSQ7)
Curriculum implementation	Textbook compilation	Whether the textbook compilation embodies the curriculum standards (WTQ5, WSQ8, WSQ14, FJQ18, FJQ19, FJQ20, FTQ16)
	Implementation suggestions	Whether the physics curriculum implementation embodies the teaching suggestions of the curriculum standards, problems, and difficulties existing in curriculum implementation (WTQ6, WTQ23, WSQ14, FJQ21, FJQ22, FJQ23, FJQ24, FJQ15, FTQ20, FTQ21, FTQ22, FTQ23, FTQ24)
	Curriculum resources	Which course resources are used in curriculum implementation and which textbooks are used, whether the teacher resources are adequate (FTQ19, FTQ17, FJQ26, FJQ27, FTQ31, FTQ32)

(continued)

Table 5.3 (continued)

Category	Subcategory	Description
	Teaching methods	Whether the teaching methods emphasized in curriculum standards are embodied in curriculum implementation, whether scientific inquiry is included in the specific curriculum implementation (WTQ15, WTQ19, WTQ21, WSQ10, WSQ11, WSQ12, FTQ28, FTQ29, FTQ30)
Course evaluation	Evaluation methods	Whether the evaluation recommendations of physics curriculum standards are reasonable and embodied in the implementation, discuss the problems and give suggestions in the evaluation of curriculum implementation (WTQ7, WTQ22, WTQ23, WSQ14, FJQ28, FJQ29, FTQ33, FTQ34, FTQ35, FTQ36)
Curriculum demands	Comprehensibility	The rationality and operability of representation in the curriculum (WTQ4, WTQ13, WTQ14)

W represents the questionnaire, *F* represents interviews, *T* represents teacher, *S* represents the student, *J* represents teaching and research staff, *Q* represents the question. For example, *WTQ12* represents Q12 in the questionnaire for teachers

5.3.3 Results: Achievement of the Physics Curriculum Implementation

5.3.3.1 Rationale for Physics Curriculum

The rationale for the physics curriculum has been accepted by almost all physics teachers and students and most have experienced this curriculum in their teaching and learning. Reflected in the interview data is the importance teachers place on training in order to gain familiarity with the curriculum standards. For example, one teacher said, “It is easy for us to understand the philosophy of physics curriculum standards, while it is hard to put these ideas into teaching practice.”

Based on the interviews, we found that participants basically agreed with the physics curriculum goals, which cover more than just the learning of physics knowledge. Participants believed that the teaching process should also be permeated with physics ideas and methods and that students’ abilities should be promoted. As for the curriculum framework and content, they think “module” is the reflection of selectivity. It is based on the students’ individual differences, the nature of physics, and the physics curriculum content (e.g., students should master physics knowledge, thoughts, and methods).

The implementation of the physics curriculum embodies teachers’ recognition and understanding of diversified teaching methods, though teachers show concerns for team cooperation, experimental inquiry, cooperative learning, and explanation methods. Based on the research into scientific inquiry, content teaching, and their responses to questions about student learning difficulties, we infer that part of the physics curriculum principles have been practiced in a hands-on approach.

5.3.3.2 The Value and Nature of Physics Were Generally Recognized

The first question on the teachers' survey was intended to investigate their understanding of the value of the physics curriculum within the broader discipline of science ("What do you think about the position of the physics curriculum within the science standards?"). The senior high school physics curriculum standards point out that the physics curriculum is "a basic course in science." Figure 5.5 represents the statistical results; different patterns on the bar graph denote teachers from different provinces, which shows the similar tendency. Teachers all agree with the nature and function of the physics curriculum. The percentage of agreement reaches nearly 84 % and ranks the highest, which is also reflected in the interview data. For example, a school principal in Jiangsu Province said, "Physics is a relatively important and prominent curriculum among all basic science curricula. It has a great impact on students' scientific literacy." Combined with his own understanding, a physics researcher in Beijing further elaborates on the physics curriculum's value: "the value of the physics curriculum's goal should be recognized from two aspects: the social perspective and the individual perspective. From the social perspective, physics should serve people, society, and our nation. Life is a process of individual learning, so individuals should build a general knowledge foundation to realize the value of life."

The senior high school physics curriculum should establish foundational knowledge for students, allowing them to recognize individual value and preparing students for a variety of fields in their future careers. Therefore, the physics curriculum pursues not only knowledge foundation but also the development of students' abilities. Comments from the interview data support this, saying "physics curriculum and its value expression are very clear" and "physics value is very (relatively) reasonable."

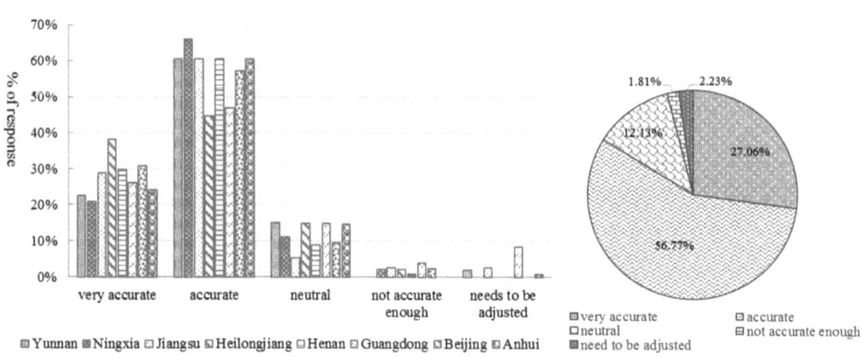

Fig. 5.5 Teachers' understanding of the value of the physics curriculum: the percentage of teachers' responses by response category [Note: the view of value of physics expressed in the physics curriculum standards is very accurate/accurate/neutral/not accurate enough/needs to be adjusted]

5.3.3.3 Three-Dimensional Target of Physics Curriculum

The senior high school physics curriculum standards propose that the physics curriculum aims to enhance students' scientific literacy. While there are multiple definitions of scientific literacy in the educational literature, it is defined in the physics standards as learning basic technical knowledge, understanding scientific methods, developing scientific thinking habits, and promoting problem solving skills in socio-scientific and public affairs (State Council 2006). The term also refers to certain abilities such as the ability to read and understand articles written about science in the media and news, to evaluate arguments based on evidence, and to come to appropriate conclusions based on such arguments. When it comes to specific teaching objectives, the *high school physics curriculum standards* have a three-dimensional target which includes "knowledge and skills," "process and methods," and "emotion, attitude, and value." The physics curriculum determines what teachers expect from their students; therefore, we designed questions about teachers' understanding of the value of the physics curriculum and the objectives of the physics curriculum. When asked "Do you think the three-dimensional target for the physics curriculum is explained accurately and clearly in the *Standards*?", 67 % of teachers said they think it is clear, 20 % said it is accurate, 8 % recommend revising the document, and only 5 % think it is ambiguous or unclear according to the pie chart (see Fig. 5.6). Although the results in different provinces show the same tendency, there are about 22 % of teachers of Jiangsu think the three-dimensional target needs to be adjusted. From the responses to the open-ended questions, suggestions included that the three-dimensional target should be made more operational and specific.

The recognition of the three-dimensional target is also evident in the interview data. For instance, a textbook editor from Beijing believes that "the change of curriculum orientation is one of the changes in curriculum reform. The previous emphasis on 'Double Basics' (basic knowledge and basic skills) is not adequate. Thus, this new round of curriculum reform emerges that advocates a three-dimensional target (knowledge and skills, process and methods, emotion, attitude and value)."

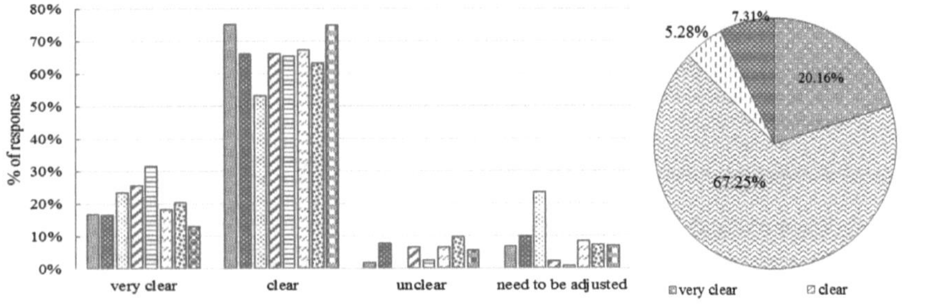

Fig. 5.6 Teachers' understanding the three-dimensional target of physics curriculum

Physics is a course that requires mastering basic knowledge and skills as well as achieving a thorough understanding of the research process and methods. The nature of the physics course within the curriculum orientation makes this three-dimensional target well grounded. Additionally, the three-dimensional target has improved the quality of student learning in the physics courses.

This change is a significant leap in understanding of the nature of physics. The relationship between the three targets (knowledge and skills, process and methods, and attitudes and values) is more specific in the new standards. Even though all three dimensions were included previously, teachers used to emphasize "scientific knowledge and skills." In the new standards, all three components—knowledge and skills, process and methods, and emotion, attitude, and value—are now underscored. As one teacher mentioned, "The dimension of 'knowledge and skills' is the basis. The real meaning of 'process and methods' cannot be grasped before 'knowledge and skills' are mastered and the dimension of 'emotion, attitude and value' is developed on the base of the two steps above. Therefore, the three-dimensional target should be understood in those three steps." Another teacher also stated: "the three-dimensional target based on the new curriculum standards is more specific and detailed, as opposed to the previous knowledge, ability, and moral education goals. In the short term, knowledge and skills are fundamental." This clearly indicates that the teachers have realized the importance of the three-dimensional goals. While this represents one big step forward, it is also clear that teachers still need help in making all three dimensions an integral part of science learning and teaching.

The recognition of the three-dimensional target is also evidenced in the interviews with those who judge the CEE. For instance, one examiner commented that, "Knowledge construction and ability development are combined with the learning process in the new curriculum. The three-dimensional target is significant since it has combined 'emotion, attitude and value' as an independent target, and this is a realization of *education* in a real sense."

5.3.3.4 Physics Curriculum Content and Its Instruction

The classroom is the most active area of reform in relation to the physics curriculum. It is in the classroom that we see the evidence of the depth and breadth of students' knowledge and skills and the relationship between physics and their daily lives. Figure 5.7 shows that 35 % of teachers believe the width and difficulty of knowledge in physics are in line with students' cognitive level, and nearly 61 % think the new curriculum has somewhat of a gap with students' cognitive level, but is somewhat acceptable. It should be noted that there are slight regional differences, for example, not one teacher from Jiangsu Province thinks the width and difficulty of knowledge in physics are unacceptable, while there are about 10 % of teachers from Ningxia who think it is unacceptable.

There is a diversity of teaching methods and modes of information delivery in the physics classroom. There are, respectively, nearly 65 %, 61 %, and 48 % of

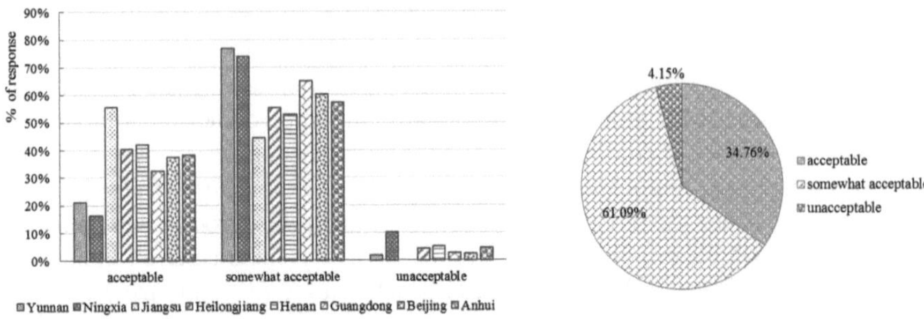

Fig. 5.7 View about rationality of the curriculum width and difficulty. Note: "Acceptable" indicates that the width/difficulty of knowledge is in line with students' cognitive level and acceptable to students; "somewhat acceptable" means that there are some gaps between the knowledge width/difficulty and students' cognitive level and thus is somewhat acceptable to students; unacceptable denotes that there is a large degree of gap between the knowledge width/ difficulty and students' cognitive level and hence is unacceptable to students

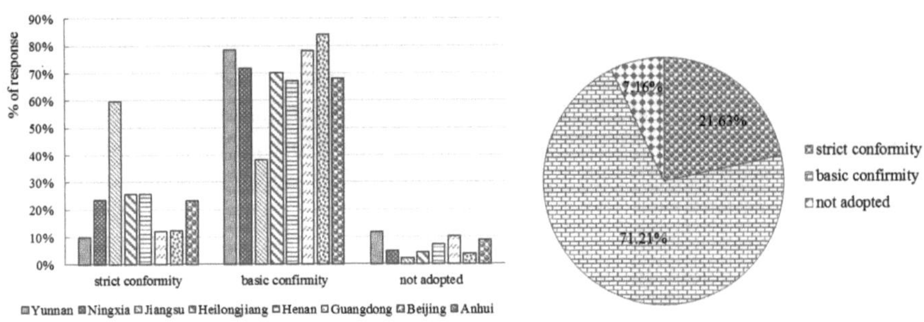

Fig. 5.8 The content of scientific inquiry in physics teaching

teachers who think that teamwork, experimental exploration, and self-regulated learning are effective. The level of fidelity in teachers' implementation of curriculum standards, especially regarding scientific inquiry, was surveyed. We found that 21.63 % of teachers report giving *strict* lessons on scientific inquiry based on the curriculum standards and teaching materials, while 71.21 % have generally followed the curriculum standards and teaching materials (Fig. 5.8).

The changes that have occurred in teaching methods and modes of instruction are revealed in our interviews with teachers and research staff. For instance, a teacher from Beijing stated, "The new curriculum reform emphasizes the students' individual development, which calls upon a significant change in the teaching methods." A teacher from Jiangsu Province said, "Students can get relatively more opportunities of independent inquiry learning, which is now being carried out based on theory." A principal from Henan Province stated that, "Curriculum reform is more likely to be manifested in teaching and learning methods, which explains how the students' initiative and participation could be inspired." Finally, a teacher from Heilongjiang Province believed that "now, teachers pay more attention to scientific inquiry, problem solving, self-regulated learning, and teamwork. The problems for inquiry can be chosen from student experiments, demonstration experiments, and suggested activities from physics textbooks. Students can carry out a research projects by reading related materials, looking for information, doing experiments and discussion." The advocacy of diverse teaching methods such as scientific inquiry is a significant change in the physics curriculum. Only with diverse teaching methods will educators help students develop individuality, innovation, and creativity.

5.3.3.5 Physics Teachers' Professional Development

Whether the physics curriculum standards can be effectively implemented depends on the quality of the teachers in the profession, which has implications for pre- and in-service teachers. Training and professional development should embody the practice of curriculum implementation. In the interviews, teachers suggested that there should be more specific requirements for in-service training, such as hands-on training (in which teachers get practice interacting with the curriculum as they are trained) and case-based training (i.e., training in authentic classroom cases and scenarios).

5.3.4 Results: The Problems of Physics Curriculum Implementation

5.3.4.1 The Operability and Text Expression Accuracy of Physics Curriculum Goals

According to the pie chart, there are about 49 % and 48 %, respectively, of teachers (97 % of teachers) who think the expression of the three-dimensional target is clear and needs only partial adjustment. It is interesting that there are only about 30 % teachers of Ningxia who think the expression of three-dimensional target is clear, while there are about 62 % teachers of Henan who hold the view that the verbs in the expression of goals is clear according to the histogram (see Fig. 5.9). Teachers and research staff also suggest adjusting action verbs: "the three-dimensional target is to a large extent detailed in terms of knowledge, while unspecific and difficult to understand in the other two aspects." Clarifying such verbs would assist teachers and students in identifying what behaviors need to change in order to meet the three-dimensional learning objectives.

Methods and operational goals are sometimes difficult to comprehend according to the curriculum. The goals of the curriculum should match the teachers' understanding and be specific. For instance, phrases such as "to know about the law of universal gravitation" are not specific enough to guide teachers to accurately measure student progress. Some teachers distinguish the curriculum target from

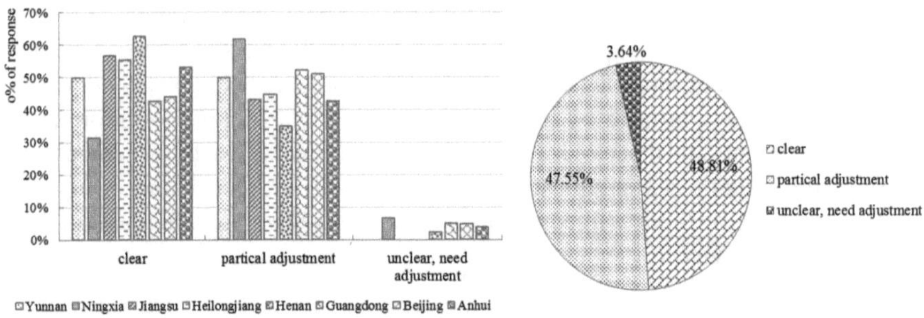

Fig. 5.9 The expression of action verbs in three-dimensional target

teaching goal and put forward that the "the curriculum target, in terms of accuracy of the expression, degree of difficulty, and level of implementation, is relatively precise. But the three-dimensional target in the standards is the target of the curriculum. Teachers need to transform the curriculum target into teaching goals, which should be clearly contained in the curriculum standards."

Judging from the interviews of the teachers, research staff, and teaching material editors, it can be concluded that the expression of the action verbs in the three-dimensional target is so abstract that different teachers have different interpretations. For example, in the "knowledge and skills" dimension, teachers may interpret the action verbs such as "to know" or "to understand" differently based on their own understanding. Therefore, teachers who participated in the survey mentioned that they would like to have more explicit objectives. In addition, the ways of thinking and the ideas we expect students to form should be clearly defined, and the kind of ideas we expect students to develop about physics should be explicitly stated.

5.3.4.2 Missing Both Selectivity and Disciplinary Structure in the Implementation of the Physics Curriculum

Selectivity is stressed in the *senior high school physics curriculum standards*, which would allow students to select the course sequence in which they are most interested (e.g., choosing from series 1 to 3). Students have the freedom to choose from certain modules or learn one module in a series and then begin another series. In the survey, we designed several questions about the understanding of selectivity in the physics curriculum. For example, students were asked to list their choices for learning modules in the 10th and 11th grades. However, almost all of the schools chose Module 3-1 and Module 3-2 for their students. This shows that the school authorities are deciding the modules, so students and teachers in reality do not have the selectivity that was intended by the original curriculum standards. We then asked participants if they think the structure of the physics curriculum (consisting of modules and series) is reasonable. According to the statistics (see Fig. 5.10), nearly

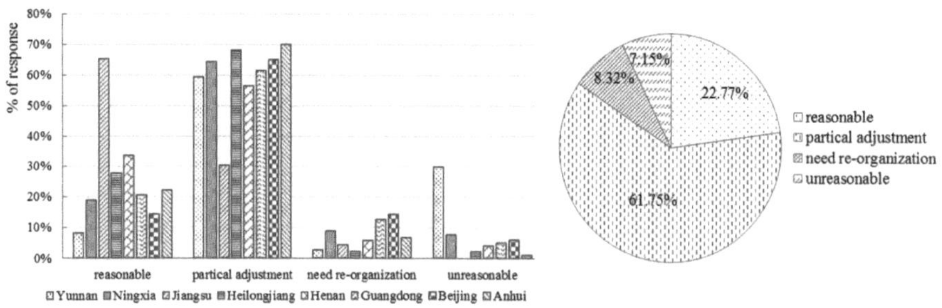

Fig. 5.10 The understanding of the structure of the physics curriculum and module design

62 % of the teachers think the structure of the physics curriculum based on modules is reasonable with only a few that need adjustment, nearly 8 % think modules should be reselected and reorganized, and nearly 7 % believe that the current structure of the curriculum is unreasonable. It is noteworthy that there are about 30 % teachers of Yunnan Province and 0 % teachers of Jiangsu who think the current structure of the curriculum is unreasonable. Further, teachers provide several reasons for their answers, including that the knowledge system is not integral or systematic enough, which violates the basic structure of physics, goes against students' overall understanding of physics, and hinders their performance on the CEE. Comments also indicate that they believe optional modules cannot reflect students' selectivity because the elective modules of strands one are not enough for the requirement of CEE, and the elective modules in strands two are not required. In series 2 (Fig. 5.1), for instance, many teachers raised the same question: "Where does the rationality of Series 2 lie?" The modules are sequential, which means that the knowledge needed for upper modules is gained in the modules that come before. According to the survey, nobody chooses series 2 nationwide, but almost all science students have to study optional Modules 3-1 and 3-2, not necessarily because of interest but because of the requirements of the CEE. When talking about the reasons students may choose Module 3-5, a teacher from Jiangsu Province said that this module is chosen "because of their strong relevance to the compulsory modules." Another comment states that Module 3-3 is not often chosen because "[other modules] benefit students more when they are taking the CEE." The topics covered in Modules 3-4 and 3-5—namely, momentum—are also perceived as easier to master for the CEE.

It was also mentioned in the comments that in reality, it is impossible to keep the framework of physics while allowing for selectivity within the curriculum. Neither teachers nor students are able to choose what they teach or learn, as this is dependent on external evaluation (specifically the CEE). The modular design of the curriculum should consider the logical coherence of the discipline, rather than simply stating that it allows for individual choice. In the survey and interview comments, some experts and teachers mentioned reorganizing the modules.

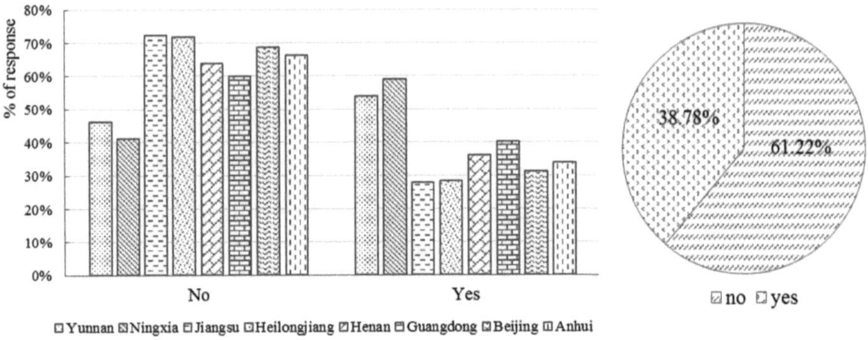

Fig. 5.11 Views on difficult, unfamiliar, or outdated contents in the senior high school physics curriculum standards

5.3.4.3 The Contradictions Among the Difficulty of the Content, Class Time, and the Modules

In the survey we included some questions about the level of difficulty of physics, class time, and curriculum module design. For instance, the 12th question in the teachers' questionnaire focuses on whether the curriculum standards have any difficult, unfamiliar, or outdated content. As shown in Fig. 5.11, nearly 62 % of the teachers do not think that there is any difficult, complicated, unfamiliar, or outdated content in the *senior high school physics curriculum standards*, while 38.78 % believe there is.

Most researchers and teachers think the difficulty of physics is one of the reasons for the declining number of the students who choose physics on the CEE. One teacher said that "Compared to the courses in the past, the new curriculum reform does a good job in handling the difficulty and the academic proficiency test of physics. However, many students still shun physics in the CEE. This is because of the discrepancy of test scores between physics and other subjects [Note: the national average CEE score of physics is generally lower]. Therefore, the contents of physics in optional courses should be adjusted."

Almost all physics teacher mentioned in their comments that the content of the physics course, its difficulty, and the limited amount of class time make it a challenge to teach. They report that it is impossible to finish all of the necessary modules in just two periods a week. "Take electrostatic fields as an example," a physics teacher said, "this is relatively abstract and needs knowledge of mechanics to solve the problems. ... Hence, it takes more than 36 periods to learn Module 3-1."

5.3.4.4 The Consistency and Coordination of the Contents in Each Module, Subject, and Different Grades

We designed the questionnaire to include several questions about the consistency of the physics curriculum (the tenth question for the teachers and the sixth and seventh

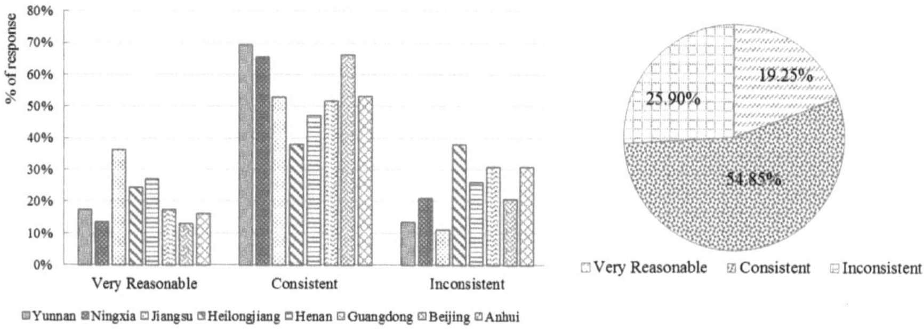

Fig. 5.12 The consistency of the contents, modules, and subjects of the physics curriculum

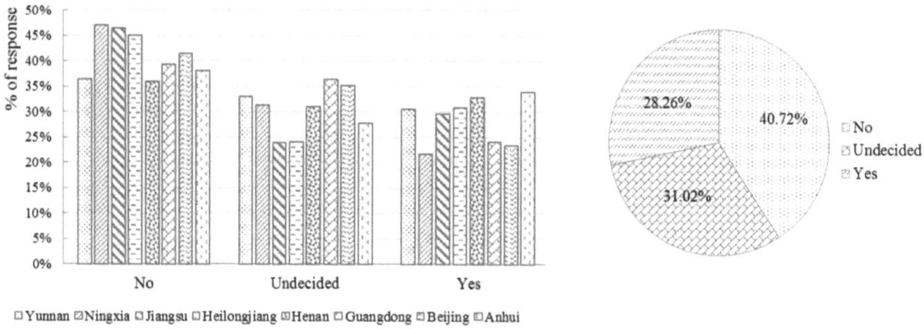

Fig. 5.13 Whether physics and other disciplines are synchronic

questions for the students). The questions are concerned with the consistency of the subjects, grades, and modules. Concerning the consistency of subjects (such as physics and mathematics), the majority of teachers believe that the content is consistent while nearly 26 % of teachers think it is far from reasonable and does not suit the teaching (see Fig. 5.12). It is noteworthy that there are more than 30 % of teachers from Heilongjiang, Guangdong, and Anhui Provinces believe that the content is inconsistent.

Further responses from students on "whether physics and other disciplines are synchronic (e.g., students should understand trigonometry before they learn the topic of force analysis)" show that nearly 41 % of the respondents think that the contents of physics courses and other disciplines are not matched, while nearly 31 % of the students are undecided (Fig. 5.13). According to the histogram, there are about 35 % of students or more in all provinces believe that physics and other disciplines are not synchronic according to the histogram. In the interviews, almost every teacher said that physics and mathematics are less connected or coordinated than they could be. When students begin learning kinematics, especially when applying the parallelogram law to force analysis, it is necessary for them to acquire knowledge of trigonometric functions, in which students of 10th grade are deficient. Also, knowledge of radians is needed when learning angular velocity. These are just

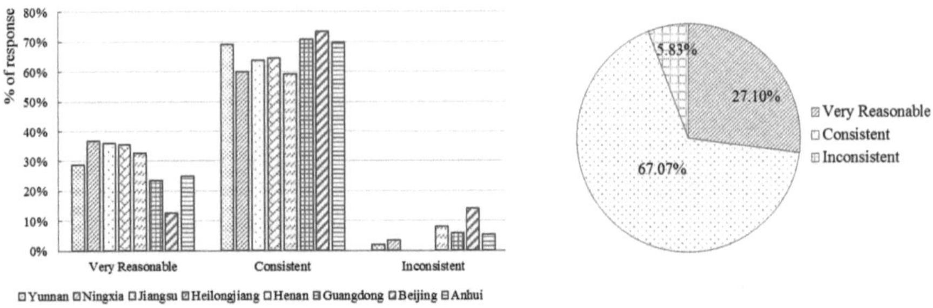

Fig. 5.14 The consistency of physics curriculum in various modules

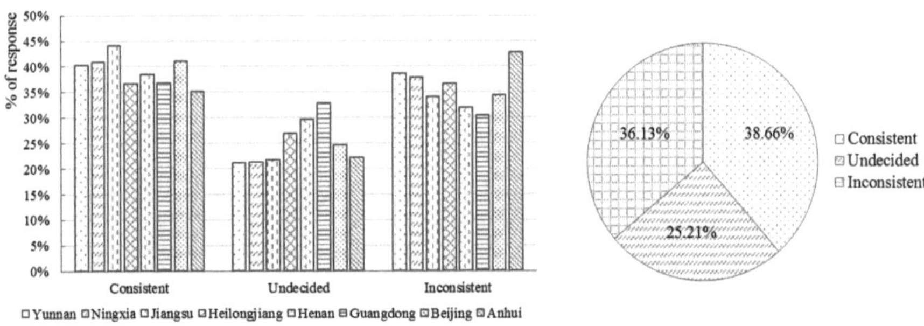

Fig. 5.15 The consistency of the physics knowledge in junior and senior high schools

two examples of the relevance of mathematics to physics, illustrating how connectivity between the subjects plays an important role in students' physics learning.

As for the consistency of thephysics curriculum across various modules and content (Fig. 5.14), nearly 67 % of the teachers think it is very reasonable and consistent, and only about 6 % consider it inconsistent. And the same tendency of different provinces can be seen in the histogram. However, the results from the interviews are actually different from those of the survey, as in the interviews, most teachers stated that they should pay attention to logic (such as the sequence of the presentation of momentum and energy). For example, a teacher from Jiangsu Province said "when it comes to the connection of different modules, there is something unreasonable in the arrangement of Module 3-4 and 3-5. That is, Chap. 1 of Module 3-5 talks about momentum, while the last chapter of Module 3-4 is 'The Theory of Relativity (Light and Electromagnetic Waves).' That is to say, the connection between the two is not so satisfactory, and Module 3-5 then talks about wave-particle dualism, which separates quantum mechanics and the theory of relativity and does not seem logical."

The responses of the students and teachers about consistency in the physics curriculum in junior and senior high school (Fig. 5.15) show that nearly 39 % of the students think there is consistency in the physics curricula across junior and senior high school and nearly 25 % of them are not sure. However, nearly 36 % hold the view that the physics curricula across junior and senior high school are inconsistent, which can be further confirmed in the teachers' interviews, in which several physics teachers pointed out that the consistency of the physics curriculum could be divided according to the coherence of the content and difficulty. The difficulty depends not only on the coherence of the content but also on the methods and ways of thinking within the discipline, which is evidenced in some of the teachers' comments in the interviews. For example, a physics teacher said that "It is a huge transition for first-year senior high school students to learn physics because the qualitative and concrete contents of physics in junior high school have been changed into ones focusing on quantitative, abstract, and logical thinking, yet the key of teaching lies in how teachers guide students to achieve this transition." A teacher from Beijing further mentioned that "The gap between the physics curricula in junior and senior high school is becoming increasingly bigger and the difficulty level should be increased gradually and not abruptly. For example, the kinematic term *displacement* in a senior high school (tenth Grade) physics books is a vector, which is confusing to the students who have learned *distance*, a scalar, in their junior high schools. Besides, the physical symbols in the text books are not consistent. For example, x, l, and s can all stand for displacement. Both elastic force and friction are represented by F, and E not only stands for electric field but also mechanical energy." Another teacher from Henan Province pointed out that "Though the connection and consistency between the physics curricula in junior and senior high school are stressed in physics curriculum standards, teachers often do not pay attention to them. Their teaching content goes beyond students' thinking development and shows the shortage of consideration what students learned in junior high school." The senior high school physics curriculum pays more attention to the cultivation of student capacity, but it is still limited by the evaluation system—the CEE. After the implementation of the new curriculum reform, the two physics curriculum programs yield a much larger gap, and the leap required of learners in both content and cognitive skills requirements accounts for the increasing number of the students who avoid physics in high school.

5.3.4.5 Different Implementation Statuses in Different Cities and Provinces

Scientific inquiry and cooperation are stressed in the physics curriculum and are regarded as effective teaching methods by both students and teachers. However, "scientific inquiry" is commonly just a catchphrase used during implementation and is not yet fully observed in the classroom. For example, though students and teachers agree on the importance of scientific inquiry, teacher-centered methods (e.g., lecture) still account for a large portion of class time. While it is true that students have the opportunity to engage in experimentation, most experiments are traditional, cookbook style labs that students carry out step by step to verify what is in the textbook. Inquiry accounts for a limited portion of their experiences; students report that its application in their physics classroom is less than 15 %. This could be due to the fact that different regions and schools adopt different curricula in addition to what is required for the CEE. Such discrepancies exist for several reasons, including teachers' understanding of inquiry, inquiry-oriented teaching skills, the large class size in some provinces, and the various levels of availability of curriculum resources. For instance, a physics teacher from Heilongjiang Province said that "The experimental instruments for physics in schools, some of which can be date back to the 1980s, have not been updated. The school cannot afford the equipment that plays an important role in the new curriculum reform." A teacher from Henan Province said that "There are more than 60 students in a classroom, and the large number of students and the heavy pedagogical tasks make self-directed learning, cooperative learning, and inquiry learning an empty promise."

According to the interviews, students' basic knowledge and capacity for learning are important foundations for inquiry learning; one teacher states, "It is difficult for students to carry out scientific inquiry because of their lack of awareness of inquiry, which should have been cultivated in junior high school and primary school." Several schools have adopted the model of "Learning Plan Guidance," in which teachers model and guide students in self-directed learning. Nevertheless, teachers possess different understandings of the practice of this model and of teaching via inquiry. The CEE is still the only evaluation criterion that warrants much attention in the implementation of the physics curriculum, which has a great negative impact on the implementation of different teaching methods in practice. A teacher from Ningxia Hui Autonomous Region mentioned in the interviews that "Apparently the implementation of investigative study and scientific inquiry is good, but the form of the CEE is basically not changed, which has caused students and teachers to not be enthusiastic about experiments at all." Therefore, the "all-for-the-exam's-sake" idea becomes one of the main problems with the current implementation of the physics curriculum in high school, which is also shown in the survey results, in which 57.3 % of the teachers agree.

5.3.4.6 Lack of Practicality of the Text of the Physics Curriculum Standards

According to the survey, most researchers and physics teachers think the text of the *senior high school physics curriculum standards* is clear, with an explicit orientation around the three-dimensional target. They also think the content is clear, but the practicality of the text needs modification in several areas. These modifications include improving the clarity of the verbs used in the three-dimensional target (especially in "process and methods" and "emotions and values") and modification of the suggestions for compiling teaching materials. Most suggestions from provincial researchers and teachers (Ningxia, 60 %; Guangdong Province, 59 %; Yunnan Province, 58 %; Beijing, 54 %; and Henan Province, 53 %) about the compilation of teaching materials focus on the lack of feasibility: despite the strong theoretical guidance in the Teaching Guides in the Physics Curriculum Standards, some content can hardly be put into practice. Teachers in all provinces, especially Ningxia Hui Autonomous Region and Yunnan Province, said that the evaluations and the suggestions are acceptable but not necessarily practical.

5.3.5 *Results: Reflections and Suggestions of Survey Respondents*

5.3.5.1 Improve the Three-Dimensional Target Expression of the Physics Curriculum Standards

As previously described, this survey reflects some progress in teachers' understanding of the curriculum objectives surrounding the standards, but there is still a need to further consolidate the meaning of and the relationship between the three-dimensional objectives. In their interview responses, teachers have mentioned that the expressions of "process and methods" and "emotions and values" in the physics curriculum standards should be made more specific and operationally clear, especially the verbs used in those sections. Some examples should also be given to help teachers understand, judge, and apply the curriculum standards. Teacher professional development perhaps is also needed to help teachers better understand the objectives and to provide them with real-life examples that they can use in their own classrooms.

5.3.5.2 Balance the Selectivity and Systematicity of the Contents of the Physics Curriculum

The *physics curriculum standards* tries to present its selectivity by requiring compulsory modules and through offering different series and optional modules. To some extent, the optional physics courses are implemented at the expense of logical coherence of the physics discipline. For instance, both Modules 3-4 and 3-5 include knowledge of lasers, Module 3-4 involves principles of lasers, and Module 3-5 deals with applications of lasers. According to the survey, both students' selectivity and the discipline's methodology are simultaneously neglected in the practical teaching of physics, especially because instead of student interest, the requirements of the CEE determine their choices in selection of the optional modules. Additionally, schools are choosing which elective modules they offer. Therefore, some teachers have put forward several corresponding countermeasures, namely, combining optional Modules 1-1 and 1-2 into Compulsory Module 3. Liberal arts students are required to learn only Compulsory Modules 1, 2, and 3, while science students choose two additional modules from the optional series 3. Series 2 is suggested for deletion because its function could be served by vocational schools.

5.3.5.3 Adjust the Contents of the Modules

While progress has been made, the consistency of the contents in each module and subject and across different grade levels needs to be further improved. The adjustment of each module represents the sequential adjustment of content within the modules. For instance, content requiring students to "understand the experimental gas laws" or to "know the ideal gas model by experiments" relies on information about molecular dynamic theories and knowledge of statistics.

5.3.5.4 Increase the Class Periods

Physics is considered one of the most important and basic scientific disciplines; however, the survey results show that teachers and students alike do not think enough quality instructional time is attributed to the subject. It may be that the number of class periods dedicated to the subject should be increased. According to the survey, class time dedicated to physics varies across the provinces and cities. The teachers themselves suggested that three to four periods per week should be planned in the standards, as opposed to two periods per week as currently advised. More time would allow teachers to effectively implement the three-dimensional learning objectives and allow for more comprehensive instruction.

5.3.5.5 Improve the Resources of Physics Courses and Encourage Low-Cost Resource Development

Currently, there is little emphasis on experimentation in physics teaching. This is due to poor conditions in classroom laboratories as well as the lack of equipment needed to complete physics experiments. Both large class size and nonuniformity of student abilities are also major challenges. Solving these problems depends on systematic changes such as increasing education budgets and improving the physics curriculum to encourage development of low-cost resources. In addition, introducing and emphasizing cooperative learning and exploration in primary and middle school will increase students' abilities prior to their arrival in high school.

5.3.5.6 Improve the Guidance of the Physics Curriculum Standards Toward the CEE

Process-oriented evaluation is emphasized in the *physics curriculum standards*; however, the CEE is the only evaluation index used in teaching.[1] As a result, the syllabus of the CEE becomes the sole focus for most teachers. The relationship between the curriculum standards, the exam description, and the CEE should be made explicit in the revision of the curriculum standards. Just as some teachers have suggested, combining the assessment of curriculum implementation with the CEE should be the goal of the curriculum. The reformed evaluation and assessment methods will be embraced by more teachers if they are consistent with the CEE. While reforming the mechanisms of the CEE, we should also improve the evaluation guidelines contained in the *physics curriculum standards*.

5.4 The Achievements, Challenges, and Reflections of the Science Curriculum Reform in Senior High School

During the decade of curriculum reform in elementary education, some achievements have been made in the science curriculum, such as emphasizing curriculum research, the curricular system, and curricular implementation. Looking back at the decade of science curriculum reform in senior high school, we have also identified several achievements:

1. We now know that most of the teachers surveyed support the idea of science curriculum reform and some curricular concepts, such as "promoting inquiry learning" and "emphasizing the connection with the real world," which are currently stressed in the curriculum standards for biology, chemistry, and physics. These concepts contribute to the effort to improve scientific literacy outlined in the curriculum standards by emphasizing the explorations of the scientific problems related to life, society, and the integration of science and technology.

[1] In theory, given the selective nature of CEE, they do not need to be aligned with the high school curriculum standards. The exit exams (for graduation purposes) should be aligned with the curriculum standards. But in reality, Chinese students, parents, and teachers all believe that the failure or success of students' high school education is solely determined by the performance on CEE. A mismatch between the curriculum standards and CEE would certainly undermine the implementation of a standards-based curriculum. CEE reform is now taking place, first in Shanghai and Zhejiang as the pilot sites. Take the CEE reform of Zhejiang Province as an example; students can choose three subjects from seven subjects (politics, history, geography, physics, chemistry, biology and technology) besides three common required subjects (Chinese, mathematics, and the second language). Students will be enrolled according to the major they choose. The high school examination and comprehensive quality evaluations, as well as the achievement of CEE, will be considered for enrollment.

2. The standards have established the scientific learning field, consisting of modules forming the "compulsory + optional" curriculum structure, which allows for selectivity. For example, the chemistry course is made up of two compulsory modules and six optional modules, biology is composed of three compulsory modules and three optional modules, and physics includes two compulsory modules and three series of optional modules.
3. The content in the science courses is based on the various needs of students' development. In other words, the compulsory courses are consistent with the science curriculum content in junior high school and still allow for student choice in the optional modules, though more autonomy could be given in this matter. Textbook considerations also make it possible to meet a diversity of levels based on students' differential development.
4. Diverse learning styles are recognized in the science standards, and a column titled "Suggestions on Activities and Research" has been set forth in the content standards, providing students with an abundance of projects for inquiry learning, self-regulated learning, and cooperative learning. Diverse columns such as "Thinking and Discussion," "STS," and "Science Walk," which were found in different sets of textbooks, are helpful for students to acquire knowledge and skills using various activities.
5. Although the CEE still plays a leading role in evaluation and assessment, the selection of students' examination forms and methods is diverse, such as the current "trinity examination" (it combines candidates' performance of high school evaluation test, CEE, and recruitment interview by the university, in which different universities may have their own interview topics) and self-enrollment examination. Also, the CEE reform program which was proposed by the Ministry of Education recently signaled a reform of College Enrollment System.
6. The curriculum reform in senior high school science has attracted the attention of researchers to the problems in the process of the reform. The study of science curriculum and teaching will contribute to the effective implementation of science curriculum reform. Both experts in discipline-based science education and teachers have actively participated in studies related to curriculum reform, implementation, and teaching. Though there is still inconsistency in terms of which subjects attract the most attention, more efforts have been made to investigate science teachers' adaptability to the process of curriculum reform (including teachers' understanding and participation in the curriculum reform and their resistance to reform efforts) and to explore changes in students' learning styles within the framework of curriculum reform.

Science curriculum reform in senior high school has brought about significant changes and promising perspectives. However, there are still some problems and difficulties, which include the following:

1. There are some deviations between science curriculum concepts and implementation of the curriculum. Science teachers have varied views on the science curriculum and learning methods. Although creative teaching methods have

been put forward in the curriculum reform, some deviations have emerged, such as illusory student autonomy, blind inquiry (questions about inquiry deriving from teachers, formalized inquiry procedures, and a lack of analysis of inquiry results), low levels of student cooperation (a deficit in the ability of students to engage in cooperation or cooperative learning), and teacher-centered conversations (unequal conversations between students and teachers, i.e., teachers' opinions are respected over those of students) (Guan 2011). The structure of the physics curriculum includes student selection of elective courses, though some courses that were formerly elective have become compulsory, removing some student choice (e.g., all first-year students must study Module 3-1 whether they choose science or liberal arts). Additionally, the reformed, multitrack curriculum structure has not yet been fully implemented, which also reflects the contradiction between the new model of senior high school curriculum reform and the traditional curriculum organization model in high schools (Zhang 2006).

2. The uneven distribution of curriculum resources further widens the educational gaps between urban and rural areas and different regions. The lack of laboratory instruments has an impact on the implementation of experimental teaching, and a lack of funding makes it impossible for schools in the most economically underdeveloped areas to afford necessary equipment. Without the proper materials and resources, there will continue to be a gap in education between cities, provinces, and even schools.
3. Although science teachers approve of the idea of diverse evaluation methods, their practical teaching cannot satisfy students' disparate abilities because of two factors, both of which are related to the status afforded to the CEE. First, there has been no change in the function of the CEE as a paper-and-pencil test, and second, the society, schools, and parents all pay the highest attention to the CEE. The selection of modules in the CEE also determines to some degree what will be taught and at what level the concepts will be taught. Until this situation changes, teaching cannot meet the needs of diverse student populations.

To further promote the science curriculum reform at the senior high school level, we should reflect on and discuss the achievements and problems of the reform from different angles. Educational administrative departments can carry out the curriculum reform to help make scientific and democratic decisions, promote implementation of the curriculum reform, and construct curriculum resources. For instance, Zhejiang Province has been carrying out comprehensive science curriculum in junior high school for 20 years, and now it serves to enhance the curriculum reform in senior high school. Therefore, the educational administrative departments at different levels should play leading roles in emphasizing the curriculum reform at the senior high school level. Lu and Zhang (2000) sums this up by saying, "educational administrative departments should be the best link between theoretical studies and the practice of the reform."

Because science teachers are the ones who carry out curriculum reform, many efforts require teachers to improve their skills and abilities in science curriculum implementation, curriculum design and development, and curriculum and teaching evaluation. Each of these areas depends on the active participation of teachers in the curriculum reform. The research into science education and curriculum teaching instructions is of great importance for carrying out curriculum reform. Although current science education research looks at many topics, science education researchers still need to increase the weight of importance of their studies into the practical issues involved in curriculum reform in order to improve science curriculum reform efforts like those in China.

Evaluation, as the core and most direct motivation in the reform, should be based on the current reality of the CEE. Gradually, an examination system can be established with classified examinations, comprehensive evaluation, and diverse admission. By learning and developing advanced evaluation technology, we will gradually break away from a "test scores only" evaluation and ultimately adopt diversified evaluation methods.

In this chapter we have provided an overview of science curriculum reform and implementation at senior secondary schools in China, using the physics curriculum standards implementation as an example. As the research shows, there are both achievements and problems in science curriculum reform efforts and in the implementation of these reforms. For further improvement, more research and an increase in teacher professional development are needed to help teachers better understand scientific literacy, inquiry, and the meaning and relationship of the three-dimensional objectives. A better transition from the junior high school science curriculum (grades 7–9) to the senior high school science curriculum (grades 10–12) needs to be achieved. At the policy level, leaders in national curriculum and standardized assessments need to be informed about the necessary consistency of curriculum standards, textbooks, teaching practice, and college entrance evaluation. Given the high attention to the CEE across the country, the reform of this high-stakes testing is the key to any significant reform in curriculum and instruction. Assessment developers and researchers should also explore alternative ways of assessing scientific inquiry. At the pre-classroom level, teachers should be encouraged to try diverse teaching methods in their practice and improve their teaching ability through reflection and other professional development forms. One way is to provide teachers with more real-life or exemplary teaching examples so they can develop a fuller understanding of scientific inquiry and know how they can implement it effectively in classrooms. We are hopeful to further improve Chinese science education practice and research through both internal and cross-cultural collaboration and exchange.

Acknowledgments We wish to thank the Ministry of Education (DHA140278), CNASE (B011-140512L12), and CSC for their support of the research. We also would like to thank the teachers and students who participated in the investigation and interviews.

References

Bao, Z. T. (2004). *Senior secondary school physics textbook* (普通高中课程标准实验教科书). Guangzhou: Guang Dong Education Press. (In Chinese).

Chen, X. M., & Wu, Z. R. (2005). *Senior secondary school physics textbook* (普通高中课程标准实验教科书). Beijing: Educational Science Press. (In Chinese).

Gao, S. (2009). *The survey and comparative study of senior high school physics module implementation* (普通高中物理模&#x5757课程教学实施调查及比较研究). Master dissertation. (In Chinese).

Gao, L. B., & Huang, S. B. (2005).On module curriculum teaching in high school (谈高中模&#x5757课程的教学). *Educational Science Research*, *1*, 54–57. (In Chinese).

Guan, S. L. (2011). On our senior high school curriculum reform (刍议我国普通高中课程改革). *Curriculum and Teaching*, *12*, 31–35. (In Chinese).

Huang, X. (2009). Setting of NOS teaching objective (科学本质理解的教学目标的设定). *Jiangsu Educational Research*, *6A*, 37–41.

Kong, L. P., Liu C. K., & Zhang F. C. (2007). Explore, improve and perfect: A review of ShanDong Province high school chemistry curriculum implementation (探索 提高 完善——山东省高中化学新课程实施情况回&#x7738). *Chemical Education*, *3*, 20–24. (In Chinese).

Li, Y. (2006). *The development and implementation of school-based biology curriculum* (高中生物校本课程开发与实施). Master thesis, East China Normal University. (In Chinese).

Li, L. P. (2008). *Analysis and comparison of the physics curriculum content of senior high school and the science curriculum contents of junior high school* (新课标下初中科学课程物理内容与高中物理课程内容的分析比较研究). Master dissertation, Center China Normal University. (In Chinese).

Li, G. F. (2010a). The deviation and its correction of curriculum difficulty model: A discussion with Professor Shi Ningzhong (课程难度模型运用中的偏差及其修正——与史宁中教授等商&#x69B7). *Shanghai Education and Research*, *3*, 46–49. (In Chinese).

Li, G. F. (2010b). Question on the verb of curriculum standards (对课程标准中"行为动词"的质疑). *Shanghai Education and Research*, *6*, 54–56. (In Chinese).

Li, G. F., & Hu, W. P. (2011). Report on high school biology curriculum implementation in Northwestern China (西北地区普通高中生物课程实施情况调查报告). *Curriculum, Teaching Material and Teaching Methods*, *10*, 80–85. (In Chinese).

Liu, Y. B., Shen, L., Li C. H., & Yan, L. Z. (2009). School-based implementation of the new high school chemistry curriculum (高中化学新课程的校本化实施). *Chemical Education*, *6*, 6–9. (In Chinese).

Lu, D., & Zhang, T. K. (2000). On our basic education curriculum reform trends (试论我国基础教育课程改革的趋势). *Curriculum, Textbook, Teaching Method*, *2*, 1–5.

Ma, Y. P. (2001). The curriculum implementation and its role in curriculum reform (课程实施及其在课程改革中的作用). *Curriculum Textbook, Teaching Method*, *9*, 18–23. (In Chinese).

Meng, D. L. (2008). *Research on development and utilization of resources in new high school biology curriculum implementation* (高中生物新课程实施中资源开发与利用的调查研究). Master's degree thesis, Northeast Normal University. (In Chinese).

Ministry of Education. (2001a). *National biology curriculum standards for senior secondary school (trial version)* (高中生物课程标准 (实验稿)). Beijing: People's Education Press. (In Chinese)

Ministry of Education. (2001b). *National chemistry curriculum standards for senior secondary school (trial version)* (高中化学课程标准 (实验稿)). Beiiing: People's Education Press. (In Chinese)

Ministry of Education. (2003). *National physics curriculum standards for senior secondary school (trial version)* (高中物理课程标准 (实验稿)). Beijing: People's Education Press. (In Chinese).

Shu, B. R., & He, R. W. (2004). *New senior physics textbook* (普通高中课程标准实验教科书). Shanghai: Shanghai Technological Education Press. (In Chinese)

Sun, G. S. (2008). Problems and countermeasures in high school chemistry curriculum implementation (高中化学新课程实施过程中的问题及对策). *Teaching and Management*, *11*, 62–64. (In Chinese).

The Central People's Government of the People's Republic of China. (2006). *Action program for scientific literacy of the entire people* (全民科学素质行动计划纲要). http://www.gov.cn/jrzg/2006-03/20/content_231610.htm. (In Chinese).

The People's Government of Zhejiang Province. (2014). *The reform program of Zhejiang college entrance examination* (浙江省高考改革方案). (official document). (In Chinese).

Wang, H. X. (2012a). *Chemistry curriculum standards and textbook analysis in high school* (高中化学课程标准与教材分析). Beijing: Science Press. (In Chinese).

Wang, H. X. (2012b). *The consistency of high school physics content standards and academic proficiency examination* (高中物理内容标准和学业水平考试的一致性研究). Master Dissertation, Southwest University. (In Chinese).

Xiao, X., & Xing, H. J. (2011). Externalization research of scientific method in senior high school physics curriculum standards (高中《物理课程标准》中的科学方法显化研究). *Journal of Capital Normal University (Natural Science)*, *4*, 29–33. (In Chinese).

Xu, J. (2007). *Consistency study of senior high school physics curriculum content and university physics curriculum content* (普通高中物理课程内容与大学物理课程内容的适切性研究). Master dissertation, Southwest University. (In Chinese).

Zhang, S. S. (2006). Practice and thoughts of senior high school curriculum reform (普通高中课程改革的实践与思考). *Curriculum, Textbook, & Teaching Method, 7*, 17–21.

Zhang, L. J. (2008). *Difficulties and countermeasures in high school biology curriculum implementation* (高中生物新课程实施中的困难及对策). Master dissertation, Hunan Normal University. (In Chinese).

Zhang, W. S., Zhang, D. C. & Peng, Q. C. (2006). *New senior physics textbook* (普通高中课程标准实验教科书). Beijing: People's Education Press. (In Chinese).

Chapter 6
Examining the Senior Secondary School Chemistry Curriculum in China in View of Scientific Literacy

Bing Wei and Bo Chen

6.1 Introduction

The latest round of curriculum reform in China was initiated in 1999 after a long period of ferment (Huang 2004). The ambition of this reform was to raise the quality of education in order for students to meet the challenges of new times. Specially, cultivating students' creativity and practical abilities has been a major concern in this round of curriculum reform (CCCPC and SC 1999). As part of the national curriculum, the Senior Secondary School Chemistry Curriculum (SSSCC, for Grades 10–12) came into a new era when its national standards were released by the Ministry of Education (MoE) in 2003. The official document argues that the SSSCC is an indispensable component of school science education and plays an important role in further raising the level of scientific literacy of senior secondary school students (MoE 2003a). Furthermore, it asserts that the "SSSCC is linked up with the chemistry or integrated science curricula at the compulsory education stage and therefore it is, in essence, a type of general education-oriented curriculum"[1] (MoE 2003b, p. 1). More importantly, according to the curriculum designers (Wang and Wang 2004), the curriculum goals are defined by the three dimensions in view of scientific literacy: (1) knowledge and skills, (2) processes and methods, and

[1] According to the Outline of Curriculum Reform of Basic Education (MoE 2001), at the junior secondary school level (Grades 7–9), the science curriculum can take either of two forms: separate curriculum (biology, chemistry, or physics) or integrated science curriculum.

B. Wei (✉)
Faculty of Education, University of Macau, E33, Av. da Universidade, Taipa, Macau, China
e-mail: bingwei@umac.mo

B. Chen
School of Chemistry and Chemical Engineering, Guangzhou Higher Education Mega Center, Guangzhou University, 230 Wai Huan Xi Road, Guangzhou, China

L.L. Liang et al. (eds.), *Chinese Science Education in the 21st Century: Policy, Practice, and Research*, Contemporary Trends and Issues in Science Education 45, DOI 10.1007/978-94-017-9864-8_6

(3) emotions, attitudes, and values. In addition, according to the national standards, the SSSCC is made up of required and selective course modules. The required course modules are *Chemistry 1* and *Chemistry 2*, which are required for all senior secondary school students, and the six selective course modules are *Chemistry and Daily Lives*, *Chemistry and Technology*, *Particulate Structure and Properties of Substance*, *Chemical Reaction Mechanism*, *Basic Organic Chemistry*, and *Experimental Chemistry*, which are provided for students based on their needs and interests (MoE 2003a).

In China, textbooks are usually thought of as a constituent element of the national curriculum development (Wu et al. 1992). It is especially recognized in the circle of chemistry education in China that textbooks can be seen as the substantiation of the curriculum, and the ideas of new curriculum should be delivered to practicing teachers through textbooks (Wang 2010). Therefore, in the whole process of curriculum reform, no effort has been spared to compile and publish new textbooks (Wei 2012). Up to now, there have been three series of senior secondary school chemistry textbooks that were written according to the national standards of the SSSCC, have passed the official review, and are currently used in schools (Wang 2010). These three series of chemistry textbooks were edited by Song (2006) and Wang (2006, 2007), respectively. There are eight books for each series with two books for the required modules (*Chemistry 1* and *Chemistry 2*) and six books for the selective modules. As mentioned above, the notion of scientific literacy is the core theme in the new SSSCC. In this chapter, we are interested in examining to what extent the themes of scientific literacy have been reflected in both the national standards of the SSSCC and the three series of senior high school chemistry textbooks. Specifically, the research questions were posed as follows: (1) To what extent are the themes of scientific literacy distributed in the national standards of the SSSCC? (2) To what extent are the themes of scientific literacy distributed in the three series of senior secondary school chemistry textbooks? (3) What are the differences in the distributions of the themes of scientific literacy among the three series of senior secondary school chemistry textbooks? (4) What are the differences in the distributions of the themes of scientific literacy between the national standards and the textbooks of the SSSCC? (5) What are the differences in the distributions of the themes of scientific literacy between the current and traditional senior secondary school textbooks? and (6) What are the differences in the distributions of the themes of scientific literacy between theoretical and descriptive contents in the senior secondary school textbooks?

6.2 Scientific Literacy and Curriculum Balance

The term scientific literacy was coined in the 1950s, and from its very beginning, it was rooted in the view of general education rather than specialized education (Bybee 1997a). That is to say, scientific literacy is oriented to "all" students or the future citizens in the society. As such, it has been used as a slogan to provide a symbol for contemporary reform and serves to unify people interested in science education reforms (Bybee 1997b). Since there is no correct interpretation of a slogan, it is little wonder that different individuals interpret scientific literacy in different ways (Laugksch 2000). As for school science curriculum in particular, scientific literacy is often defined in relation with varied curricular issues, such as purposes, content, and outcomes of science education. Champagne and Lovitts (1989), for example, observed that responses to the question "What does it mean to be scientifically literate?" usually include three kinds of descriptions: the behaviors of a scientifically literate person in a variety of contexts; the mental state of a scientifically literate person in terms of knowledge, skills, and disposition; and references to educational experiences that are assumed to produce a scientifically literate person. When examining the various interpretations of scientific literacy in the 1970s, Bybee (1997a) identified four perspectives of scientific literacy: "content" (Agin 1974), "context" (Shen 1975), "experiences" (NSTA 1971), and "outcomes" (Showalter 1974). Except for "context," the other three perspectives, "outcomes," "content," and "experiences," correspond to the three descriptions of Champagne and Lovitts (1989), respectively. It seems that the "content" perspective permeates the other three ones, as it aims to provide answers to the key question of "what constitutes science curriculum" in terms of scientific literacy. A typical example of the definition of scientific literacy in this perspective is the science content suggested by National Research Council (NRC 1996): (1) unifying concepts and processes, (2) science as inquiry, (3) physical science, (4) life science, (5) earth and space science, (6) science and technology, (7) science in personal and social perspectives, and (8) history and nature of science.

Obviously, the notion of scientific literacy expresses a set of themes that include traditional subjects and also the content cutting across these subjects, such as the nature and history of science, the relation between science and technology, and science-related social issues. According to Roberts (2007), these varied themes can be grouped into two types: science subject matter and situations in which science can play a role in other human affairs. In order to develop a science curriculum, various parts of science should come together to form some sort of a coherent whole so as to achieve scientific literacy. Specifically, for instance, Wellington (2000) suggests that a balanced curriculum in science should contain these elements: (1) a study of the content and concepts of the sciences; (2) consideration of the practices and processes of science; (3) study of the links between science, technology and society; and (4) consideration of the history and nature of science. Chiappetta et al. (1991a) argue that science textbooks should assist in the development of a scientifically literate society. They assert that in order to accomplish this task, the content of textbooks should provide a curriculum balance which stresses fairly equal proportions of knowledge, investigation, thinking, and the interaction between science, technology, and society. Furthermore, they developed a quantitative method to examine the content of different types of science textbooks for their relative emphases on four themes of scientific literacy: (1) science as a body of knowledge, (2) science as a way of investigating, (3) science as a way of thinking, and (4) the interaction among science, technology, and society. Applying this method to analyze chemistry textbooks commonly used in the United States,

Chiappetta et al. (1991b) found that the majority of texts analyzed placed most emphases on "science as a body of knowledge," followed by "science as a way of investigating," while very little text was devoted to "the interaction among science, technology, and society" and even less to "science as a way of thinking."

By using the same method in analyzing physics textbooks in Australia, Wilkinson (1999) has found that the majority of the textbooks stress "science as a body of knowledge," place some emphases on "science as a way of investigating," and have little emphasis on "science as a way of thinking." Texts produced for a new physics course after 1990 are found to place more emphasis on the theme of "the interaction between science, technology, and society" than texts produced prior to 1990. The possible influence of science, technology, and society (STS) movement on this change was noted by Wilkinson, but no further analysis was made. Based on more recent conceptions of scientific literacy and the philosophy of science, BouJaoude (2002) modified the analytical framework proposed by Chiappetta et al. (1991a, b) in two ways: including the personal use of science in the theme of "interaction of science, technology, and society" and replacing the theme of "science as a way of thinking" with "science as a way of knowing." More significantly, BouJaoude (2002) extended his research to curriculum documents and distinguished them at different levels: general objectives (goals), introductions, objectives, instructional objectives, and learning activities. He found that the Lebanese curriculum emphasizes "the knowledge of science," "the investigative nature of science," and "interaction of science, technology, and society" but neglects "science as a way of knowing." BouJaoude's research demonstrated that the analytical framework can be used to analyze not only science textbooks but science curriculum documents also.

Even though these aforementioned studies were based on the concept of curriculum balance, neither of them, however, gave an exact definition of the desired state of curriculum balance. In order to effectively evaluate the extents of the SSSCC and the three series of senior secondary school chemistry textbooks in terms of curriculum balance, we argue that the themes of scientific literacy should be weighed equally at varied levels of this curriculum. The basis of this argument is that there should not be a difference among the values of varied curriculum emphases, and every curriculum emphasis is equally important in terms of scientific literacy (Roberts 1988, 2007). This ideal state will be used as a criterion to analyze the real state of curriculum balance in this chapter. In the following section, we will describe three issues involved in the methodology of this chapter.

6.3 Method

6.3.1 Analysis Targets

As mentioned earlier, this chapter aims to examine the distribution of the themes of scientific literacy in the senior secondary school chemistry curriculum, which includes the official curriculum document released by the MoE (2003a), that is, the *Senior Secondary School Chemistry Curriculum Standards* (SSSCCS), and the accompanying chemistry textbooks. Therefore, the SSSCCS, and the three series of the new chemistry textbooks compiled under the guide of the SSSCCS, constitute the main analysis targets in this chapter. In the SSSCCS, curriculum objectives are presented in three dimensions: they are "knowledge and skills," "processes and methods," and "emotions, attitudes, and values," with each one containing several items. Content standards comprise two parts: "teaching objectives" and "recommended activities and investigation" (for short "activities"). The contents of these two parts are presented item by item under secondary themes. In brief, the SSSCCS were analyzed at three levels: "curriculum objectives," "teaching objectives," and "activities." For textbooks, it should be noted that only those for the two required course modules (i.e., *Chemistry 1* and *Chemistry 2*) in the SSSCC were analyzed in this chapter. This was based on the consideration that required course modules are taken by all senior secondary school students, which fits correctly with the meanings of scientific literacy. Furthermore, in order to explore the changing tendency of the embedding of the idea of scientific literacy in textbooks, Volume 1 in the series of Wu and Hu (2003) was also included in the analysis targets. This textbook volume was compiled by the People Education Press (PEP)[2] under the guide of the teaching syllabus of senior secondary school chemistry (State Education Commission 1996) rather than the SSSCCS (MoE 2003a) and thus was taken as a representative of traditional chemistry curriculum in this chapter. It should be noted that as the content distributions and layouts in traditional chemistry textbooks were very different from those in the new chemistry books, we only selected the representative topics to make comparisons between the current chemistry textbooks and traditional ones. Textbooks analyzed in this study are listed in Table 6.1.

In addition, as we know, chemistry subject matter can be generally categorized into theoretical and descriptive parts. In order to examine the differencesof the distributions of the themes of scientific literacy between these two types of subject knowledge, we took the topics of "atomic structure" and "sulfur and its compounds" as typical examples of theoretical and descriptive chemistry, respectively, and made comparisons between them. The selection of these two topics was based

[2] Since the 1950s, the PEP has been the designated national education press that produces the national unified syllabi and textbooks directly under the leadership of the Ministry of Education (Wu et al. 1992). Although its status as a monopoly has decreased in recent years, textbooks produced by the PEP are still popular in primary and secondary schools in China.

Table 6.1 The textbooks of the senior secondary school chemistry analyzed in this study

Chief editor	Title	Publisher	Publication date	Edition
Wu and Hu	Chemistry volume 1	PEP	June 2003	1st
Song	Chemistry 1	PEP	June 2006	2nd
	Chemistry 2	PEP	May 2006	2nd
Wang	Chemistry 1	JEP	June 2006	3rd
	Chemistry 2	JEP	June 2006	3rd
Wang	Chemistry 1	SSTP	July 2007	3rd
	Chemistry 2	SSTP	July 2007	3rd

PEP People's Education Press, *JEP* Jiangsu Education Press, *SSTP* Shandong Science and Technology Press

on the consideration that they are important elements both in traditional and current curricula.

6.3.2 Instrument

In order to answer the research questions, a framework of scientific literacy adapted from Chiappetta et al. (1991a) and BouJaoude (2002) was employed to analyze the SSSCCS and the senior secondary school chemistry textbooks. The validity of the framework has been ensured through its conceptual and empirical bases. The framework of scientific literacy is shown in Table 6.2.

6.3.3 Data Analysis

The unit of analysis of the SSSCCS was the items presented in "curriculum objectives," "teaching objectives," and "activities." Each item was analyzed and categorized using the framework described above. When an item contained more than one theme of scientific literacy, each theme involved was counted once. The item was not categorized if it was not involved in any of the four themes of scientific literacy.

The elements of a textbook (units of analysis) that were used in the analysis included complete paragraphs, figures, tables, marginal comments, complete steps in laboratory or hands-on activity, and specific sections. Prefaces, tables of contents, appendixes, and exercises in a textbook, however, were not analyzed in this chapter. Each unit of analysis was analyzed and categorized using the framework described above. When a unit of analysis contained more than one theme of scientific literacy, each theme involved was counted once. The unit of analysis was not categorized if it was not involved in any of the four themes of scientific

Table 6.2 The framework of scientific literacy

Category	Subcategory	Illustration
I. The knowledge of science	Facts, concepts, principles, and laws	Check this category if the intent of the text is to present, discuss, or ask the student to recall information, facts, concepts, principles, laws, theories, etc. It reflects the transmission of scientific knowledge where the student receives information
	Hypotheses, theories, and models of science	
II. The investigative nature of science	Requires students to answer a question through the use of materials	Check this category if the intent of the text is to stimulate thinking and doing by asking the student to "find out." It reflects the active aspect of inquiry and learning, which involves the student in the methods and processes of science such as observing, measuring, classifying, inferring, recording data, making calculations, experimenting, etc.
	Requires students to answer a question through the use of charts, tables, etc.	
	Requires students to reason out an answer	
	Requires students to make a calculation	
	Engages students in a thought experiment or activity	
	Requires students to communicate through the use of a variety of means, such as writing, speaking, and using graphs, tables, and charts	
III. Science as a way of knowing	Describes how scientists experiment	Check this category if the intent of the text is to illustrate how science in general or a certain scientist in particular went about "finding out." This aspect of the nature of science represents thinking, reasoning, and reflection, where the student is told about how the scientific enterprise operates
	Shows the historical development of an idea	
	Emphasizes the empirical nature and objectivity of science	
	Illustrates the use of assumptions	
	Shows how science proceeds by inductive and deductive reasoning	
	Gives cause and effect relationships	
	Discusses evidence and proof	
	Presents the scientific methods and problem-solving steps	
	Role of self-examination in science	
IV. Interaction of science, technology, and society	The usefulness of science and technology to society	Check this category if the intent of the text is to illustrate the effects or impacts of science on society. This theme of scientific literacy pertains to the application of science and how technology helps or hinders humankind. In addition, it involves social issues and careers
	The negative effects of science and technology on society	
	Social issues related to science or technology	
	Careers and jobs in scientific and technological fields	
	Personal use of science to make everyday decisions, solve everyday problems, and improve one's life	
	Science-related moral and ethical issues	

Adapted from Chiappetta et al. (1991a) and BouJaoude (2002)

literacy or it was only involved in review questions, vocabulary words, or goal and objective statements.

After the categorization of units of analysis was completed, then the descriptive statistics were adopted, that is, the frequency and the percentage of occurrence of each theme were calculated. As argued above, the ideal state of curriculum balance is that the themes of scientific literacy should be weighed equally. Therefore, the less the difference among the percentages of the four themes of scientific literacy is, the better the issue of curriculum balance is treated; the greater the difference among the percentages of the four themes is, the worse the issue of curriculum balance is treated.

To ensure the reliability of the results, the two researchers met several times to discuss the framework in order to reach a common understanding of its components and subcomponents. Following these meetings, they conducted the analysis and categorization independently with the result of an inter-rater agreement of 90 %, then discussed the discrepancies, and reached consensus on categorization. It is worth noting that due to the analysis method in this study, the sums of frequencies are not always equal to the number of units of analysis, and the total percentage is not always equal to 100 %.

6.4 Results

6.4.1 The Distribution of the Themes of Scientific Literacy in the SSSCCS

The distribution of the themes of scientific literacy in the SSSCCS at different levels is presented in Table 6.3.

As seen in Table 6.3, the emphasis in curriculum objectives is the theme of "the investigative nature of science." The percentage of this theme reaches as high as 41.67 %, while the percentages of the other three ones are all only 16.67 %. In teaching objectives, "the knowledge of science" is treated with the highest weight

Table 6.3 The distribution of the themes of scientific literacy in the SSSCCS

		Themes of scientific literacy				
Level		I	II	III	IV	Total items
Curriculum objectives	Frequency	2	5	2	2	12
	Percentage (%)	16.67	41.67	16.67	16.67	
Teaching objectives	Frequency	20	8	6	14	34
	Percentage (%)	58.82	23.53	17.65	41.18	
Activities	Frequency	7	20	2	12	40
	Percentage (%)	17.50	50.00	5.00	30.00	

Themes of scientific literacy: I = the knowledge of science; II = the investigative nature of science; III = science as a way of knowing; IV = interaction of science, technology, and society

(58.82 %), followed by "interaction of science, technology, and society" (41.18 %) and "the investigative nature of science" (23.53 %). "Science as a way of knowing" is given the lowest weight (17.65 %). For activities, "the investigative nature of science" has the highest proportion (50.00 %). The frequency of "interaction of science, technology, and society" (30.00 %) is slightly higher than that of "the knowledge of science" (17.50 %). "Science as a way of knowing" is still in the lowest proportion (5.00 %).

From the results above, it can be concluded that the percentage distributions of the themes of scientific literacy are different among curriculum objectives, teaching objectives, and activities in the SSSCCS. The issue of curriculum balance is treated best in "teaching objectives" (58.82 %, 23.53 %, 17.65 %, and 41.18 %, respectively, to each theme) and worst in "activities," with "curriculum objectives" being between the two. Overall, of the four themes of scientific literacy, "the knowledge of science," "the investigative nature of science," and "interaction of science, technology, and society" are emphasized in the SSSCCS, while "science as a way of knowing" is neglected.

6.4.2 The Distributions of the Themes of Scientific Literacy in New Senior Secondary School Chemistry Textbooks

The distributions of the themes of scientific literacy in the three series of new senior secondary school chemistry textbooks are presented in Table 6.4.

Table 6.4 The distributions of the themes of scientific literacy in the three series of new senior secondary school chemistry textbooks

Textbooks		Themes of scientific literacy				Total units of analysis
		I	II	III	IV	
Chemistry 1 (Song 2006)	Frequency	233	117	7	50	435
	Percentage (%)	53.56	26.90	1.61	11.49	
Chemistry 2 (Song 2006)	Frequency	212	94	8	70	416
	Percentage (%)	50.96	22.60	1.92	16.83	
Chemistry 1 (Wang 2006)	Frequency	198	116	15	58	408
	Percentage (%)	48.53	28.43	3.68	14.22	
Chemistry 2 (Wang 2006)	Frequency	225	107	7	55	426
	Percentage (%)	52.82	25.12	1.64	12.91	
Chemistry 1 (Wang 2007)	Frequency	279	139	15	132	604
	Percentage (%)	46.19	23.01	2.48	21.85	
Chemistry 2 (Wang 2007)	Frequency	217	117	17	87	452
	Percentage (%)	48.01	25.88	3.76	19.25	

Themes of scientific literacy: I = the knowledge of science; II = the investigative nature of science; III = science as a way of knowing; IV = interaction of science, technology, and society

Table 6.4 shows that the percentage distributions of the four themes of scientific literacy in the six textbooks have similar trends, which place most emphases on "the knowledge of science," followed by "the investigative nature of science," while very little is devoted to "interaction of science, technology, and society" and even less to "science as a way of knowing." The percentages of "the knowledge of science" for the analyzed textbooks range from 46.19 % (Chemistry 1, Wang 2007) to 53.56 % (Chemistry 1, Song 2006). This implies that the contents of new senior secondary school chemistry textbooks are more concerned about scientific facts, concepts, principles, theories, laws, etc. Conversely, the proportions of "science as a way of knowing" in the three series are very low, ranging from 1.61 % (Chemistry 1, Song 2006) to 3.76 % (Chemistry 2, Wang 2007). It suggests from these results that none of the three series deals well with the issue of curriculum balance. By comparison, however, the issue of curriculum balance is treated best in Wang's (2007) and worst in Song's (2006), with Wang's (2006) being between the two. On the whole, the new textbooks are worse than the SSSCCS in dealing with the issue of curriculum balance.

6.4.3 The Distributions of the Themes of Scientific Literacy on the Topic of "atomic structure" in the Different Series of Chemistry Textbooks

The distributions of the themes of scientific literacy on the topic of "atomic structure" in the different series of chemistry textbooks are presented in Table 6.5.

As shown in Table 6.5, the traditional series (Wu and Hu 2003) stresses the first theme, "the knowledge of science." The percentage of this theme reaches as high as 78.13 %, while the percentages of the second through fourth themes are only

Table 6.5 The distributions of the themes of scientific literacy on the topic of "atomic structure" in the different series of chemistry textbooks

Textbooks		Themes of scientific literacy				Total units of analysis
		I	II	III	IV	
Wu and Hu (2003)	Frequency	25	2	3	1	32
	Percentage (%)	78.13	6.25	9.38	3.13	
Song (2006)	Frequency	16	3	0	2	25
	Percentage (%)	64.00	12.00	0	8.00	
Wang (2006)	Frequency	21	8	8	2	42
	Percentage (%)	50.00	19.05	19.05	4.76	
Wang (2007)	Frequency	15	12	5	6	39
	Percentage (%)	38.46	30.77	12.82	15.38	

Themes of scientific literacy: I = the knowledge of science; II = the investigative nature of science; III = science as a way of knowing; IV = interaction of science, technology, and society

6.25 %, 9.38 %, and 3.13 %. This obviously shows that the distribution of the four themes of scientific literacy is fairly unequal in this series. Compared with the traditional one, the proportions of "the investigative nature of science" and "interaction of science, technology, and society" have been added in the three new series. For "science as a way of knowing," although there was no content of this theme in Song's (2006), the proportions of this theme in Wang's (2006) (19.05 %) and Wang's (2007) (12.82 %) are considerable. Overall, the current series are better than the traditional one in dealing with the issue of curriculum balance. As far as three series of new textbooks are concerned, the issue of curriculum balance is treated best in Wang's (2007) (38.46 %, 30.77 %, 12.82 %, and 15.38 %, respectively) and worst in Song's (2006), with Wang's (2006) being between the two.

6.4.4 The Distributions of the Themes of Scientific Literacy on the Topic of "sulfur and its compounds" in the Different Series of Chemistry Textbooks

The distributions of the themes of scientific literacy on the topic of "sulfur and its compounds" in the different series of chemistry textbooks are presented in Table 6.6.

Table 6.6 shows that no treatment is given to the theme of "science as a way of knowing" in any of the four series. This means that neither the traditional series nor the new ones deal with the issue of curriculum balance well. In other words, the problem of the imbalanced distribution of the themes of scientific literacy in the traditional series has not been solved in the new ones. Comparing the new series with the traditional one, "the investigative nature of science" seems to have

Table 6.6 The distributions of the themes of scientific literacy on the topic of "sulfur and its compounds" in the different series of chemistry textbooks

Textbooks		Themes of scientific literacy				Total units of analysis
		I	II	III	IV	
Wu and Hu (2003)	Frequency	28	16	0	8	54
	Percentage (%)	51.85	29.63	0	14.81	
Song (2006)	Frequency	22	13	0	12	48
	Percentage (%)	45.83	27.08	0	25.00	
Wang (2006)	Frequency	16	14	0	8	40
	Percentage (%)	40.00	35.00	0	20.00	
Wang (2007)	Frequency	20	12	0	11	43
	Percentage (%)	46.51	27.91	0	25.58	

Note. Themes of scientific literacy: I = the knowledge of science; II = the investigative nature of science; III = science as a way of knowing; IV = interaction of science, technology, and society

changed little, while "interaction of science, technology, and society" in the new series has gained at the expense of "the knowledge of science."

6.4.5 The Comparison of Theoretical Chemistry and Descriptive Chemistry With Respect To the Distribution of the Themes of Scientific Literacy

As mentioned earlier, we take the topics of "atomic structure" and "sulfur and its compounds" as examples of theoretical chemistry and descriptive chemistry, respectively, in this chapter. The comparison of these two topics with respect to the distribution of the themes of scientific literacy is presented in Table 6.7.

As seen in Table 6.7, for the theme of "the knowledge of science," the proportion in "atomic structure" is higher than that in "sulfur and its compounds" in all series except Wang's (2007). For "the investigative nature of science," the proportion in "sulfur and its compounds" is higher than that in "atomic structure" in all series except Wang's (2007). For "science as a way of knowing," there is no content of this theme in the topic of "sulfur and its compounds" in any of the four series. Conversely, there are some contents on this theme in the topic of "atomic structure" in all series except Song's (2006). Like the situation of "the investigative nature of science," the proportion in "sulfur and its compounds" is higher than that in "atomic structure" in all series in "interaction of science, technology, and society." It can be inferred from this result that theoretical chemistry is better than descriptive chemistry in dealing with the issue of curriculum balance in both the traditional and new series.

Table 6.7 The comparison of the topics of "atomic structure" and "sulfur and its compounds" with respect to the percentage distribution of the themes of scientific literacy in different series of chemistry textbooks

		Themes of scientific literacy			
Textbooks	Topics	I (%)	II (%)	III (%)	IV (%)
Wu and Hu (2003)	Atomic structure	78.13	6.25	9.38	3.13
	Sulfur and its compounds	51.85	29.63	0	14.81
Song (2006)	Atomic structure	64.00	12.00	0	8.00
	Sulfur and its compounds	45.83	27.08	0	25.00
Wang (2006)	Atomic structure	50.00	19.05	19.05	4.76
	Sulfur and its compounds	40.00	35.00	0	20.00
Wang (2007)	Atomic structure	38.46	30.77	12.82	15.38
	Sulfur and its compounds	46.51	27.91	0	25.58

Themes of scientific literacy: I = the knowledge of science; II = the investigative nature of science; III = science as a way of knowing; IV = interaction of science, technology, and society

6.5 Conclusion and Discussion

In this chapter, we have analyzed the distribution of the themes of scientific literacy in the new senior secondary school chemistry curriculum in China. A general conclusion drawn from the findings of this chapter is that the issue of curriculum balance has not been well treated in this curriculum. Specifically, five points should be noted. Firstly, the theme of "science as a way of knowing" is greatly lacking in the SSSCCS, and the percentage distributions of the themes of scientific literacy are different among "curriculum objectives," "teaching objectives," and "activities" in this official document. Secondly, the percentage distributions of the four themes of scientific literacy in the three series of the new senior secondary school chemistry textbooks have similar trends, but none of the three series deals well with the issue of curriculum balance. Thirdly, by comparison, the issue of curriculum balance is treated best in Wang's (2007) and worst in Song's (2006), with Wang's (2006) being between the two, and the new textbooks are worse than the SSSCCS in dealing with the issue of curriculum balance. Fourthly, compared with the traditional one, the themes of "the investigative nature of science" and "the interaction of science, technology, and society" have been increased in the three new series. Fifthly, when taking the topics of "atomic structure" and "sulfur and its compounds" as examples of theoretical chemistry and descriptive chemistry, respectively, we found that the former is better than the latter in dealing with the issues of curriculum balance in both traditional and new textbooks.

From the results of this study, we can infer that the issue of curriculum balance has associations with science curriculum policy, the authorship effect of curriculum documents and science textbooks, and the nature of subject content knowledge (such as theoretical or descriptive chemistry content in this study). It is beyond the scope of this chapter to discuss each of these factors. In short, similar to Chiappetta et al.'s (1991a, b) and Wilkinson's (1999) studies, this chapter has shown that the issue of curriculum balance has not been dealt with well in science textbooks. The most remarkable defect is that "science as a way of knowing" is greatly lacking. Although the themes of "the investigative nature of science" and "interaction of science, technology, and society" have gradually grown in new science textbooks under the influence of the movements of "Scientific Inquiry" (Abd-El-Khalick et al. 2004) and "STS" (Solomon and Aikenhead 1994), respectively, the theme of "science as a way of knowing" is still lacking in recent years. At this point, our research is consistent with those studies which focused on examining representations of the nature of science in science textbooks (e.g., Abd-El-Khalick et al. 2008; Wei et al. 2013). Furthermore, we found in this study that the theme of "science as a way of knowing" is lacking both in textbooks and in science curriculum documents. Linking with BouJaoude's (2002) study, which was conducted in Lebanon, a current situation in developing countries is revealed: this theme is heavily marginalized in science curriculum documents. The reasons for this situation need further exploration.

In China, as mentioned earlier, the goals of the senior secondary school chemistry curriculum are defined by three dimensions in view of scientific literacy: (1) knowledge and skills, (2) processes and methods, and (3) emotions, attitudes, and values. Comparing these three dimensions with the four themes of scientific literacy, it can be seen that "knowledge and skills" and "processes and methods" are roughly equal to the themes of "the knowledge of science" and "the investigative nature of science," respectively, and some viewpoints of "interaction of science, technology, and society" are encompassed in the dimension of "emotions, attitudes, and values." However, the theme of "science as a way of knowing" is missing. The reasons must be multiple. One definite reason is that scientific epistemology and the contemporary scholarship of the nature of science are neglected by chemistry curriculum designers. As Driver et al. (1996) argued, the nature of science has great potential contribution to one's scientific literacy. Thus, we cannot imagine what a scientifically literate person would be like if he or she is ignorant of the nature of science. This is the reason that the theme of "science as a way of knowing" is included in the framework when discussing the issue of curriculum balance in view of scientific literacy. In the future revision of senior secondary school chemistry curricula, curriculum designers and textbook writers should treat the issue of curriculum balance better so as to achieve the balanced development of scientific literacy of students. When dealing with the issue of curriculum balance, our suggestion is that, as a first step, chemistry curriculum designers should fully and comprehensively grasp the meanings of scientific literacy imported from the West, especially those informed by the contemporary development of the philosophy of science, and then transmit these meanings into chemistry curriculum documents and chemistry textbooks. More specifically, the theme of "science as a way of knowing" should be included as a dimension of the curriculum goals presented in national standards, and increased history of science and nature of science content should be included in textbooks. Finally, we must say that this chapter was based on frameworks in the Western literature, but we ignored the interpretations and perceptions of chemistry curriculum designers and textbook writers on the notions of "scientific literacy" and "science as a way of knowing." This is a limitation of this chapter. Furthermore, we must acknowledge that this study focused on only the national standard document and textbooks; further investigation of classroom instruction and assessments may reveal different pictures of curriculum balance.

References

Abd-El-Khalick, F., BouJaoude, S., Duschl, R., Lederman, N., Hofstein, R., Niaz, M., et al. (2004). Inquiry in science education: International perspectives. *Science Education, 88*, 397–419.

Abd-El-Khalick, F., Waters, M., & Le, A.-P. (2008). Representations of nature of science in high school chemistry textbooks over the past four decades. *Journal of Research in Science Teaching, 45*, 835–855.

Agin, M. (1974). Education for scientific literacy: A conceptual frame of reference and some application. *Science Education, 58*, 403–415.

BouJaoude, S. (2002). Balance of scientific literacy themes in science curricula: The case of Lebanon. *International Journal of Science Education, 24*(2), 139–156.
Bybee, R. W. (1997a). *Achieving scientific literacy: From purposes to practices*. Portsmouth: Heinemann.
Bybee, R. W. (1997b). Toward an understanding of scientific literacy. In W. Graber & C. Bolte (Eds.), *Scientific literacy* (pp. 37–68). Kiel: IPN.
Central Committee of the Communist Party of China (CCCPC), & The State Council (SC). (1999). *Decisions on deepening education reform and fully promoting the quality education* (关于深化教育改革全面推进素质教育的决定). Beijing: People's Education Press. (In Chinese).
Champagne, A. B., & Lovitts, B. E. (1989). Scientific literacy: A concept in search of a definition. In A. B. Champagne & B. E. Lovitts (Eds.), *This year in school science 1989: Scientific literacy* (pp. 1–14). Washington, DC: AAAS.
Chiappetta, E. L., Fillman, D. A., & Sethna, G. H. (1991a). A method to quantify major themes of scientific literacy in science textbooks. *Journal of Research in Science Teaching, 28*, 713–725.
Chiappetta, E. L., Sethna, G. H., & Fillman, D. A. (1991b). A quantitative analysis of high school chemistry textbooks for scientific literacy themes and expository learning aids. *Journal of Research in Science Teaching, 28*, 939–951.
Driver, R., Leach, J., Millar, R., & Scott, P. (1996). *Young peoples' images of science*. Bristol: Open University Press.
Huang, F. (2004). Curriculum reform in contemporary China: Seven goals and six strategies. *Journal of Curriculum Studies, 36*(1), 101–115.
Laugksch, R. C. (2000). Scientific literacy: A conceptual overview. *Science Education, 84*, 71–94.
Ministry of Education (MOE). (2001). *An outline for the basic education curriculum reform* (基础教育课程改革纲要). (official document). (In Chinese).
Ministry of Education (MOE). (2003a). *National chemistry curriculum standards for senior secondary school (trial version)* (普通高中化学课程标准(实验)). Beijing: People's Education Press. (In Chinese).
Ministry of Education (MOE). (2003b). *Program of general senior secondary school curriculum reform (trial version)* (普通高中课程改革方案(实验)). (official document). (In Chinese).
National Research Council (NRC). (1996). *National science education standards*. Washington, DC: Author.
National Science Teachers Association (NSTA). (1971). *School science education for the 1970s: A NSTA position statement*. Washington, DC: Author.
Roberts, D. A. (1988). What counts as science education. In P. Fensham (Ed.), *Development and dilemma in science education* (pp. 27–55). Philadephia: The Falmer Press.
Roberts, D. A. (2007). Scientific literacy/science literacy. In S. K. Abell & N. G. Lederman (Eds.), *Handbook of research on science education* (pp. 729–780). London: Lawrence Erlbaum Associates, Publishers.
Shen, B. P. (1975). Science literacy: The public need. *The sciences,* Jan–Feb., 27–29.
Showalter, V. (1974). What is unified science education? Program objectives and scientific literacy. *Prism II, 2*, 1–6.
Solomom, J., & Aikenhead, G. (Eds.). (1994). *STS education: International perspectives on reform*. New York: Teachers College Press.
Song, X. (Ed.). (2006). *Chemistry, book 1, book 2* (化学1, 化学2) (2nd ed.). Beijing: People's Education Press. (In Chinese).
State Education Commission (SEdC). (1996). *The chemistry syllabus of senior secondary schools* (普通高中化学教学大纲). Beijing: People Education Press. (In Chinese).
Wang, Z. (Ed.). (2006). *Chemistry, book 1, book 2* (化学1, 化学2) (3rd ed.). Nanjing: Jiangsu Education Press. (In Chinese).
Wang, L. (Ed.). (2007). *Chemistry, book 1, book 2* (化学1, 化学2) (3rd ed.). Jinan: Shandong Science & Technology Press. (In Chinese).

Wang, L. (2010). Progress and reflection of secondary chemistry curriculum reform in the past ten years (part A) (基础化学教育课程改革10年进展与反思(上)). *Chinese Journal of Chemical Education, 31*(4), 15–21. (In Chinese).

Wang, Z., & Wang, L. (Eds.). (2004). *Illustrations of the national standards of general senior secondary schools* (普通高中化学课程标准解读). Wuhan: Hubei Education Press. (In Chinese).

Wei, B. (2012). Chemistry curriculum reform in China: Policy and practice. In H. Yin & C. J. Lee (Eds.), *Curriculum reform in China: Changes and challenges* (pp. 95–109). New York: Nova Science Publishers.

Wei, B., Li, Y., & Chen, B. (2013). Representations of nature of science in selected histories of science in the integrated science textbooks in China. *School Science and Mathematics, 113*(4), 1–11.

Wellington, J. (2000). *Teaching and learning secondary science: Contemporary issues and practical approaches*. London: Routledge.

Wilkinson, J. (1999). A quantitative analysis of physics textbooks for scientific literacy. *Research in Science Education, 29*(3), 385–399.

Wu, Y., & Hu, M. (Eds.). (2003). *Chemistry volume 1* (化学第一册). Beijing: People's Education Press. (In Chinese).

Wu, L., Chen, H., & Lv, D. (1992). Theories and practice in making teaching materials of primary and secondary schools in new China (新中国中小学教材编制的理论与实践). In The Chinese Association of International Education Exchange (Ed.), *Curriculum development and social progress: Selected papers in the international symposium on curriculum development and social progress* (pp. 15–22) (课程发展与社会进步:课程发展与社会进步国际研讨会论文选编). Beijing: People's Education Press. (In Chinese).

Part III
Environmental and Socioscientific Issues in Chinese Science Education

Gavin W. Fulmer

Editor's Introduction: Part III

The two chapters in this section comprise a concise and information introduction to current work in China on environmental and sustainability education and in research on socioscientific issues (SSIs). Readers will note that the authors for both chapters are from Hong Kong institutions, although the chapters do include discussion of related topics outside of Hong Kong. While there is an ongoing body of curriculum on environmental and sustainability education in mainland China (as discussed in Chap. 7), there is also potential for environmental science education to be ignored because its content is not included in the National College Entrance Examinations, which prioritize physics, chemistry, and life sciences. Furthermore, while this section demonstrates that there is a national curriculum document on environmental science, there seems to be little explicit attention to the concept of SSI outside of Hong Kong (see Chap. 8). This contrasts not only with work in the West but also in other East Asian contexts, including Hong Kong, Taiwan (e.g., Wu and Tsai 2011), or Korea (e.g., Lee et al. 2006).

According to Zeidler and colleagues (2005), "SSI education aims to stimulate and promote individual intellectual development in morality and ethics as well as awareness of the interdependence between science and society" (p. 360). Thus, it is fundamentally different from science-technology-society (STS) education, which emphasizes the connections between science and society but does not emphasize the students' developing capacities to use scientific ideas when making social, moral, and ethical decisions as a citizen. Indeed, SSI addresses some of the concerns that were raised in the STS movement, but focuses on addressing them from a cognitive and developmental perspective rather than a policy advocacy perspective (Sadler and Dawson 2012). In this way, work on SSI is arguably more closely connected to applications in curriculum, pedagogy, and assessment than STS alone. Likewise, researchers who focus on SSI point to its promise for giving authentic and engaging contexts for scientific problems, leading to greater motivation among students (e.g., Dori et al. 2003) and a stronger focus on argumentation and evidence (e.g., Zohar and Nemet 2002).

In keeping with the distinguishing features of SSI, a central element in SSI education and research is its examination of the intersections of scientific knowledge, school science education, and pressing social, political, and economic decisions that students may face as citizens. In the USA, for example, this reflects the political and social situation around hot-button issues such as climate change—thus, the connection to environmental science and sustainability education—which is one of the significant topics within SSI. Yet other social and scientific topics are also fruitful areas for SSI, such as societies' energy needs and production and public health policies. Indeed, in Chap. 8, YC Lee explores a variety of different SSI topics that are similar to or that could be adapted to the US context or other international contexts.

From an international perspective, the relatively slow uptake of research on SSI in mainland China may reflect cultural and political hurdles in the way of adoption of SSI research. SSI education is broadly about empowering students to become citizens who can engage in discussions on science-informed social and public policy. Yet there are different political, social, and cultural contexts for mainland China than for Hong Kong. Though Hong Kong shares with mainland China a cultural tradition influenced by Confucian values (Hofstede 2001), the political and social environment in Hong Kong is much more strongly influenced by Western principles. As Lee notes in Chap. 8, this difference "has implications for the flow of information and negotiation by the public on SSIs" (p. 9).

There is no single solution to broadening interest in SSI among science education scholars in China. Barriers of distance and language may have delayed dissemination of major research on SSI to science educators and scholars in China. For example, scholars from mainland Chinese institutions make up a relatively small number of presenters and attendees at NARST annual meetings, particularly compared to the robust involvement of scholars from other nations such as Turkey, South Korea, and Taiwan. This is especially surprising when considering the sheer number of universities and science education researchers within mainland China. However, recent environmental issues (e.g., air pollution and smog in Beijing and other major cities) appear to be raising popular awareness of the value of protecting the environment. It is anticipated that environmental science education could play a more visible role in mainland China. This could even provide opportunities for increased attention to SSI, despite the lack of any teaching standards on SSI itself. Thus, there is a great deal of potential for communication and collaboration with mainland Chinese scholars about SSI from counterparts around the world.

Chapter Introductions

Chapter 7 addresses the transitions in definitions and educational policy around environmental education in Hong Kong and mainland China, with comparison to Taiwan. The chapter charts the transition in these education policies toward the terminology of education for sustainable development (ESD). The authors provide a comparison of the historic and current policies on education on the environment for Chinese communities of mainland China, Hong Kong, and Taiwan. The findings show a shift from a focus on environmental protection, then to public engagement, and now to environmental sustainability in social and economic development (a core idea of ESD). This shift reflects key publications by UNESCO on education and the environment (e.g., UNESCO 1978, 1992, 2006), with varying time and structure for each respective education system. This underscores how curriculum and policy in all three systems pursue their own paths but are simultaneously informed and influenced by global initiatives and trends. Furthermore, the authors describe the different ways that each system seeks to infuse environmental education into its curriculum and the role of central governments or nongovernmental organizations in achieving such a vision. For example, in mainland China, the central government plays the largest role in supporting ESD, whereas Taiwan has a combination of soft policy supports from the Ministry of Education as well as activity by nongovernmental organizations.

Chapter 8 addresses work on SSI in Hong Kong, including findings from a parallel study in southeastern mainland China. The author presents findings from four studies about students' reasoning in SSIs using a variety of contexts: a smoking ban in restaurants, conservation of wild bats encroaching on human dwellings, control of whaling, and procedures for slaughter of chickens to manage risk of bird flu. The whaling context was unfamiliar to the Hong Kong students but allows comparison with previous studies in the UK and Sweden, whereas the three other contexts are relatively familiar for the students. Across contexts, Lee concludes that students did not necessarily develop strong arguments, with inadequate attention to scientific data as evidence for their SSI arguments. These findings are consistent with previous work in Western settings that show that students vary in the quality of their arguments (see Sadler 2004, for a review) and that there are complex social factors within classrooms that can affect students' developing arguments (e.g., Yoon 2008). However, Lee finds that Hong Kong students' views are widened by sharing, unlike in Western settings where students may become more obstinate upon sharing and reflection (e.g., Svenson et al. 2009).

References

Dori, Y. J., Tal, R., & Tsaushu, M. (2003). Teaching biotechnology through case studies: Can we improve higher-order thinking skills of non-science majors? *Science Education, 87*, 767–793.

Hofstede, G. H. (2001). *Culture's consequences: Comparing values, behaviors, institutions, and organizations across nations* (2nd ed.). Thousand Oaks: Sage.

Lee, H., Abd-El-Khalick, F., & Choi, K. (2006). Korean science teachers' perceptions of the introduction of socio-scientific issues into the science curriculum. *Canadian Journal of Math, Science & Technology Education, 6*(2), 97–117.

Sadler, T. D. (2004). Informal reasoning regarding socioscientific issues: A critical review of research. *Journal of Research in Science Teaching, 41*(5), 513–536. doi:10.1002/tea.20009.

Sadler, T. D., & Dawson, V. (2012). Socio-scientific issues in science education: Contexts for the promotion of key learning outcomes. In B. J. Fraser, K. G. Tobin, & C. J. McRobbie (Eds.), *Second international handbook of science education* (pp. 799–810). Dordrecht: Springer.

Svenson, O., Salo, I., & Lindholm, T. (2009). Post-decision consolidation and distortion of facts. *Judgment and Decision Making, 4*(5), 397–407.

UNESCO. (1978). *Intergovernmental conference on environmental education—Tbilisi (USSR)—Final report*. Paris: UNESCO. Retrieved from http://unesdoc.unesco.org/images/0003/000327/032763eo.pdf

UNESCO. (1992). *UN conference on environment and development: Agenda 21*. Geneva: UNESCO.

UNESCO. (2006). *Framework for the UNDESD international implementation scheme*. Paris: UNESCO.

Wu, Y.-T., & Tsai, C.-C. (2011). High school students' informal reasoning regarding a socio-scientific issue, with relation to scientific epistemological beliefs and cognitive structures. *International Journal of Science Education, 33*(3), 371–400. doi:10.1080/09500690903505661.

Yoon, S. (2008). Using memes and memetic processes to explain social and conceptual influences on student understanding about complex socio-scientific issues. *Journal of Research in Science Teaching, 45*, 900–921.

Zeidler, D. L., Sadler, T. D., Simmons, M. L., & Howes, E. V. (2005). Beyond STS: A research-based framework for socioscientific issues education. *Science Education, 89*(3), 357–377.

Zohar, A., & Nemet, F. (2002). Fostering students' knowledge and argumentation skills through dilemmas in human genetics. *Journal of Research in Science Teaching, 39*, 35–62.

Chapter 7
Challenges and Opportunities for Environmental Education Toward Education for Sustainable Development in Chinese Communities

Irene Nga-yee Cheng and Winnie Wing-mui So

7.1 Introduction

The previous discussion of the relationship between science education and environmental education (Gough 2002) has asserted that, historically, science education functions as a host for environmental education, and it is not uncommon to find that environment-based curriculum for basic education is in science subject matter areas. The social aspects of the science curriculum should not be neglected in conserving a role for the science curriculum within environmental education (Dillon 2014). Science, as with science education, represents a variety of perspectives such as cultural, political, economic, and theological (Cobern 1998). This multi-perspective nature of science and science education supports the underlying philosophy of environmental education to cultivate environmental ethics and responsibility for shaping and managing the environment. Science education, in this regard, assumes the role to help students understand the complexities and interconnectedness of the structures within which they live so as to develop them a more holistic view to make sense of the world and align with the social constructivist inclinations among the goals of environmental education (Duschl and Osborne 2002; Hart 2007). This chapter focuses on environmental education in order to supplement other discussions related to science education.

Given the numerous environmental problems facing humankind, environmental education (EE) has become seen as an important strategy for the development of an environmentally sustainable society. Teachers' knowledge, attitudes, and behavior are believed to contribute significantly to the education and quality of students

I.N.-y. Cheng (✉) • W.W.-m. So
Room 32, 1/F, Block D3, Department of Science and Environmental Studies, The Hong Kong Institute of Education, 10 Lo Ping Road, Tai Po, Hong Kong
e-mail: nycheng@ied.edu.hk

L.L. Liang et al. (eds.), *Chinese Science Education in the 21st Century: Policy, Practice, and Research*, Contemporary Trends and Issues in Science Education 45, DOI 10.1007/978-94-017-9864-8_7

(Hestness et al. 2011; Tuncer et al. 2009). To ensure a firm groundwork for EE, teachers should equip themselves with adequate environmental know-how, proper attitudes, and appropriate lifestyle regarding the environment (Moseley et al. 2014; Yavetz et al. 2014). However, EE has undergone continuous changes in its conceptualization. Are the current practices of EE in the Chinese communities of mainland China, Hong Kong, and Taiwan concurrent with such changes? In this changing context, what problems would teachers face and how would teacher education programs be developed to help teachers overcome these problems? This chapter attempts to examine each of the above issues.

7.2 From EE to Education for Sustainable Development (ESD)

A review of the development of EE provides a good starting point to better understand the context wherein teachers work and handle problems in environmental teaching. EE was called upon as a means to address environmental issues worldwide in the 1972 United Nations Conference on the Human Environment in Stockholm. This was followed in 1975 by the Belgrade Charter, which was designed as a global framework for EE (Belgrade Charter 1975). The Tbilisi Declaration of the Conference on EE in 1977 stated further that the ultimate aim of EE is to "enable people to understand the complexities of the environment" and suggested an interdisciplinary approach to EE (UNESCO 1978, p. 12).

The *Agenda 21* of the Rio Summit in 1992 set out broad proposals for reorienting education, increasing public awareness, and promoting training toward sustainable development, suggesting a shift from EE to education for sustainability (EfS) (Tilbury and Wortman 2006). McKeown and Hopkins (2003), on the other hand, identified that although the words "social," "political," and "economic" appear in the Tbilisi document, they do not appear as frequently as the word "environment." They therefore argue that the preponderance of references in the Tbilisi documents is to environmental problems and the impact of humans on the natural environment and much less to the quality of life in terms of society and economics when compared with *Agenda 21*.

It was in the 1980s that the term *sustainable development* was used to link economic growth and environmental preservation (Tilbury 1995). In the United Nations Decade of Education for Sustainable Development (UNDESD; 2005–2014) International Implementation Scheme, it was clearly mentioned that sustainable development should not be equated to EE, and that the former encompassed the latter, setting it in the broader context of sociocultural factors (UNESCO 2006). On the other hand, ESD and EfS (Parliamentary Commissioner for the Environment 2004) are different in the sense that the latter gives greater prominence to the root social, political, and economic causes of the environmental situation (Tilbury 1995).

Nevertheless, the Declaration of Thessaloniki (UNESCO-EPD 1997) stated that "the concept of sustainability encompasses not only environment but also poverty, population, health, food security, democracy, human rights, and peace" (p. 2). Leal Filho and Behrens (2003) also contended that EfS complements a number of other fields such as EE, global education, economics education, development education, multicultural education, conservation education, outdoor education, and global change education. While the intent of EE is to learn to see the whole picture surrounding a separate problem (Meadows 1990), the intent of EfS is to learn to understand the parts played by aesthetic, social, economic, political, historical, and cultural elements in environmental and developmental problems (UNESCO 1992). The curriculum of EfS is concerned with how people interact with their total environment and with addressing environmental problems holistically (Tilbury 1995). That is why "a holistic and interdisciplinary approach which brings together the different disciplines and institutions while retaining their distinct identities" was suggested to be used in addressing sustainability (UNESCO-EPD 1997, p. 2).

As we have shown, EE, EfS, and ESD have different scopes and approaches to humans' views and interactions with the environment. In this chapter, we would like to clarify these differences: (1) EE focuses more on the transmission of environmental knowledge related to the causes and effects of environmental problems, while ESD and EfS pay more attention to the relationships between humans and the environment and the cultivation of environmental citizenship for the betterment of our future environment, and (2) when compared with EE, ESD and EfS require interdisciplinary and holistic approaches in teaching as they are set in a broader context of sociocultural factors. Teaching strategies such as outdoor experiential learning and issue-based inquiry learning are therefore seen as more appropriate means than didactic approaches in the promotion of sustainability. The transition from EE to ESD or EfS revealed not only a broader sense of, as McKeown and Hopkins (2003) asserted, understanding the environment from environmental protection to addressing the needs of both the environment and society (pp.119–120) but also a paradigm shift in the curriculum design, teaching emphasis, and teaching approaches adopted in classrooms.

Different communities are at diverse stages of development in this continuum from EE at one end to ESD at the other. In the following paragraphs, we examine the development of the concepts in the Chinese communities and discuss whether such conceptual change would pose any challenges to their curriculum implementation. Since "ESD" is the terminology used most often at the international level and in United Nations documents (McKeown and Hopkins 2003), to avoid lexical confusion, "ESD" will be adopted in this chapter as an all-embracing concept of education which relates to the promotion of sustainability.

7.3 The Development of ESD in Chinese Communities

This section reviews the development of the concept of ESD in mainland China, Hong Kong, and Taiwan in the last few decades. The three communities, however, have diverse history of development and experienced major changes in their EE practices at different critical points in time that can make the comparison difficult. To facilitate a parallel and structured comparison between the three communities, the timeline of development would be split into two major phases, pre- and post-1990s, given that the reorienting of education from EE to EfS proposed by the Rio Summit in 1992 seemed to be the turning point of many EE policies practiced in Chinese communities.

7.3.1 Mainland China

Scholars used to divide the development of EE in mainland China into three phases, namely, the "expert" phase, the "red" phase, and the sustainability phase (Lee and Tilbury 1998). Chronologically, the "expert" and the "red" phases (from 1973 to 1992) fell within the period prior to the 1990s, whereas the sustainability phase started from 1993 until now. Their developments are traced in the period before and after the 1990s, respectively.

7.3.1.1 Pre-1990s Development

The Chinese government started its work on EE with the first National Meeting in Environmental Protection held in Beijing in 1973, being in fact a response to the initiative of EE in the 1972 Stockholm Conference. Its emphasis was on educating the public about the tremendous environmental problems in mainland China. Later, in 1981, a paper called *The Decisions on Enhancing Environmental Protection Work at the Present Time* was issued by the State Council as the basis for EE. A training school for officials and in-service staff working on the matter of the environment was also set up in Qinhuangdao in the same year. What is more, environmental protection was first included in the primary and secondary curricula, and two pilot schools in Beijing and Guangzhou started to implement the proposition in the beginning of the 1980s (Wasmer 2005). Nevertheless, Lee et al. (2006) criticized the fact that EE by that time focused mainly on environmental protection propaganda. It was assumed in this phase that knowledge was the key to environmental protection. It was also believed that by raising the level of public knowledge and increasing the number of trained experts in the field, environmental deterioration would be arrested (Lee and Tilbury 1998). This period from 1973 to 1983 was termed the "expert" phase.

Since the late 1980s, there has been more focus on the improvement of environmental awareness and public participation in environmental protection (Lee and Tilbury 1998). Congresses on EE in primary and secondary schools took place, and the State Environmental Protection Agency (SEPA) was established in 1988. Environmental education was officially included in the 9-year compulsory education plan in 1987 (State Education Commission 1994, p. 21). Later, in 1990, EE was introduced as an optional subject in higher secondary schools (Wasmer 2005). The period 1983–1992 was seen as a transitional period from the focus on educating the public about scientific knowledge to solve environmental problems which was reoriented toward a tendency to improve environmental awareness and participation of the public (Lee and Tilbury 1998). In the pre-1990s, EE was included in the formal school curriculum as one single subject among others in the school curriculum, and environmental problems were seen as detached from economic and social dimensions.

7.3.1.2 Post-1990s Development

As noted by Wasmer (2005), EE was officially integrated in economic and social contexts since 1993 when it was integrated in the new curriculum guidelines. The year 1993 marked the beginning of the sustainability phase in mainland China. *China's Agenda 21 – White Paper on China's Population, Environment, and Development in the Twenty-First Century* was also approved in 1994 in which EE was considered as a way to improve sustainability. Echoing the process to formalize EE and to acknowledge the importance of the mastery of basic knowledge of environmental protection by students, the green school program has been launched since 2000 in response to the *Outline of Environmental Promotion and Education Activities* (1996–2010), which was jointly released by SEPA, the Ministry of Communications, and the State Education Commission, now the Ministry of Education (Lee and Huang 2009; Zhang 2004; Xiao 2002). In 2003, the Ministry of Education issued *Syllabuses for Special Topics Education on Environmental Education for Primary and Middle School Students*. The document stated that EE should penetrate every subject. Each subject should, by adopting the approach of special topics education, facilitate students' appreciation of and concern about nature; promote students' care about family, community, national, and global environmental issues; educate students correctly about the interrelationship between individuals, society, and nature; teach students the knowledge, methods, and skills to live harmoniously with the environment; nurture students' positive affections, attitudes, and values toward the environment; and guide students to choose lifestyles beneficial to the environment. The document also detailed the objectives, teaching contents, and activities to be adopted in EE at different levels (Ministry of Education of the People's Republic of China 2003). The *Primary and Middle School Environmental Education Implementation Guidelines* were issued in the same year to ensure an effective implementation of EE in primary and secondary schools. This makes EE an indispensable part of the primary and secondary curricula (Chen 2005). Yet, Lee et al. (2006) argue that EE in this phase tended to be "marginalized within the curriculum, mostly operating at the extra-curricular or informal curricular level" (p. 230).

Apart from the school curriculum, the Chinese government has also worked with nongovernmental organizations to promote EE in the post-1990s. For example, the State Education Commission, WWF-China, and British Petroleum codeveloped the Environmental Educators' Initiative (EEI) (1997–2007). In addition, in October 1998, the Chinese National Commission for UNESCO acted as the liaison organization between the Chinese government and UNESCO and supervised the introduction of the Project on Education for Environment, Population, and Sustainable Development in China (EPD-China Project) (1998 to present). One of the objectives of the project is to enhance the awareness and capability of educators in primary and secondary schools and colleges to promote education initiatives for environment, population, and sustainable development (Lee and Huang 2009). With reference to the Project expertise accumulated since 1998, and in light of the UNDESD framework, the *Guidelines for Integrating ESD in Teaching and Learning in Primary and Secondary Schools* were published in Beijing in December 2007 to guide teachers to integrate ESD notions into teaching and learning processes to effectively equip youths and children with the values and abilities required for sustainable development (Zhang 2010). The *Outline of China's National Plan for Medium and Long-term Educational Reform and Development* (*2010–2020*) was also issued in July 2010 to make ESD a thematic strategy for mainland China's long-term educational reform and development (Wang 2011).

To recap, the government is currently the central dominating factor in the development of ESD in mainland China, whether in school EE or in community-based projects or initiatives for sustainability education, whether in the pre- or post-1990s. A major difference between the two periods of time rests with the ways EE was perceived. Environmental education as one single, discrete subject in the school curriculum has transformed to an interdisciplinary subject, integrating multiple perspectives in solving environmental challenges since the 1990s, despite that the dominance of basic knowledge about the environment and a government-controlled curriculum has always been prevalent in mainland China. Contrary to the longer history of EE in mainland China, the documentary information which outlined the development of EE in Hong Kong and Taiwan showed a much later start of work on EE in these two regions. Major works on EE in these two regions were mainly initiated in the post-1990s.

7.3.2 Hong Kong

7.3.2.1 Pre-1990s Development

Since Hong Kong was constitutionally a colony of the United Kingdom until 1997, its curriculum development followed a typical British territory line (McClelland

1991). Before the 1990s, the British government influence on education was indirect, through encouragement and lobbying rather than edict (Stimpson 1997b). To enable the British government to protect its colonial interests, according to Morris and Marsh (1992), the curriculum development of Hong Kong in the colonial period since 1841 was "a compromise between a laissez-faire system and one that is centrally controlled" (p. 253). The EE curriculum in Hong Kong has been developed into a more open system which allows involvement of different sectors within and outside the society (Stimpson 1997b), given that the British influence is predominant in those sectors outside Hong Kong. This situation has persisted for more than 100 years until the approaching of 1997 when the transfer of sovereignty over Hong Kong from the United Kingdom to mainland China had to take place in 1997. Different from her mainland China counterpart, EE in Hong Kong in the post-1990s period was free from government intervention, in terms of its place in the school curriculum (whether it be a mandatory subject or being a single subject or integrated with other subjects in schools), its focus on scientific knowledge or public awareness and participation, and one that is centrally or less centrally controlled.

7.3.2.2 Post-1990s Development

Just before the 1997 handover of Hong Kong to mainland China, the *Guidelines on Environmental Education in Schools* (hereafter *Guidelines*) were issued in Hong Kong in 1992 (revised in 1999) in response to the publication of the *White Paper on Pollution in Hong Kong – A Time to Act* in June 1989, to promote EE at the school level. It is stated in the *Guidelines* that the ultimate aims of EE for schools in Hong Kong are "to promote in students a lifelong and forward-looking concern for the environment, and to prepare them for making well-informed, justifiable and practical decisions and taking action that would lead to the creation of sustainable environments in which they can live and work" (Curriculum Development Council 1999, pp. 4–5).

To achieve the abovementioned aims, three interrelated components, namely, education *about* the environment (i.e., acquiring knowledge about the environment), education *in* the environment (i.e., providing opportunities for learning in the environment), and education *for* the environment (i.e., developing an informed concern for the environment), are noted in the *Guidelines* as essential constituents for the implementation of EE. Though schools may have varying degrees of emphasis on the three components in implementation, it is suggested that all three elements be incorporated in their curriculum design (Curriculum Development Council 1999).

As with mainland China in her sustainability phase (starting from 1993), Hong Kong does not provide EE as an independent subject in its own right. Instead, a cross-curricular approach is adopted to integrate EE into the formal curriculum through a number of subjects. Principles are suggested in the *Guidelines* to help teachers design programs in both the formal and informal school curricula (Curriculum Development Council 1999, pp. 11–12), namely:

- Experiential learning
- A balanced viewpoint for all issues
- Emphasis on the formation of attitudes
- Individual contribution and participation

Under the school-based curriculum development initiative, Hong Kong schools are free to develop their own EE curricula. School-based learning activities such as visits, exhibitions, projects, fieldwork, film shows, competitions, waste paper collection, and "greening" the school garden are suggested in the *Guidelines* to promote students' awareness, sense of responsibility, and commitment to the environment (Curriculum Development Council 1999). Nevertheless, school practices are multifarious, and learning activities vary from school to school (see, e.g., Pui Ching Primary School 2010; St. Bonaventure Catholic Primary School 2010; Tung Wah Group of Hospitals Sin Chu Wan Primary School 2010). Table 7.1 below outlines some examples of the environmental learning activities conducted in these three schools.

As reported by Lee (1997), visits, competitions, tree planting, and community activities (e.g., cleaning and collecting rubbish) are activities commonly organized by Hong Kong primary schools to promote EE, while field trips, board displays and exhibitions, as well as talks and video shows are popular EE activities in secondary schools.

Apart from the *Guidelines*, with the aim of promoting EE at the primary and secondary schooling levels, the Student Environmental Protection Ambassador

Table 7.1 Examples of school-based EE activities in 2010

Pui Ching Primary School	Game booths on EE
	Video and photo gallery shows on the Hot Planet
	Competition on recycling
	Moon cake box recycling
	Environmental protection weeks
St. Bonaventure Catholic Primary School	Renewable energy projects (solar and wind power)
	Urban organic farming program
	Growing together activity
	Waste separation competition
	Organic waste disposal program
	Green the rooftop
Tung Wah Group of Hospitals Sin Chu Wan Primary School	Luncheon game booths, e.g., "How green are you?," "Trash to Treasure," etc.
	Composition competition on creativity in EE
	E-card promotion program
	Waste recycling promotion
	Greening the school activity

Scheme (SEPAS) and the Green School Award (HKGSA) were organized jointly by the Environmental Campaign Committee, Environmental Protection Department, and Education Bureau since 1995 and 2000, respectively (Environmental Campaign Committee 2011a, b). Moreover, schools are encouraged to involve students in activities organized by nongovernmental environmental organizations such as the Friends of the Earth (Hong Kong), the World Wide Fund for Nature (Hong Kong), the Conservancy Association, and the Green Power (Curriculum Development Council 1999).

The pre-1990s period in mainland China and Hong Kong revealed a fundamental difference in the forces behind the development of EE in these two places, with the former centrally controlled and knowledge focused. The gap between these two places seemed to have narrowed in the post-1990s when both places started to formalize the inclusion of EE in the school curriculum as an interdisciplinary area. Yet, fundamental differences still prevailed under the different political and social contexts of the two places. The way sustainability education works in Hong Kong can be summarized as a balanced effort between the government through school practices and community experts on EE. The nongovernmental environmental organizations in Hong Kong have long been deeply involved in educating the public on sustainable development and have initiated numerous community-wide projects on EE. Given the school-based initiatives in EE wherein schools can choose whether or not to thoroughly pursue EE, the work of these organizations is complementary and supplementary to formal school ESD. This diverges from the government-controlled and knowledge-focused type of EE in mainland China wherein nongovernmental environmental organizations usually follow closely the policies and initiatives of the central government, in contrast with the complementary, divergent, and community-initiated approach adopted by many nongovernmental environmental organizations in Hong Kong.

7.3.3 Taiwan

While mainland China and Hong Kong have undergone different political and social contexts which affected their development of ESD in schools in the pre-1990s, both places have witnessed the interdisciplinary curriculum development of EE in schools, with collaborative efforts from nongovernmental organizations in the post-1990s. In Taiwan, however, major works on EE will be examined since the 1990s.

7.3.3.1 Post-1990s Development

To keep pace with the global developments on sustainability, the National Council for Sustainable Development (NCSD) was established in Taiwan in 1997, aiming to "create a safe, healthy, comfortable, beautiful, and sustainable living environment;

build a pluralistic, harmonious, flourishing, vital, and vigorous society; and be a responsible citizen of the global village" (National Council for Sustainable Development Network 2004). In 1999, the Division of Environmental Protection Education at the Ministry of Education (MOE) proposed a master plan entitled *Stepping into the twenty-first century – an action strategy for sustainable development education of the MOE* with the purpose of enhancing the awareness and understanding of individuals of environmental problems and cultivating their environmental literacy, ethics, skills, and values (Su and Chang 2011). In May 2000, the Council drafted the *Agenda 21 of Taiwan*: *National Sustainable Development Strategy Guidelines* as the basis for the promotion of sustainable development. Additionally, the *Taiwan Sustainable Development Action Plan* was announced in 2002 as the guideline for the NCSD to monitor the progress and enforce the development of sustainable development through eight working groups, in which sustainable education was one of the groups included in the action plan (National Council for Sustainable Development Network 2004). Under the action plan, the MOE, the responsible agency for the sustainable education working group, is accountable for introducing and incorporating the concept of sustainable development into school education; coordinating the resources of governmental agencies, the private sector, business, and schools to implement ESD; and coordinating ESD at the international level (Su and Chang 2011).

At the curriculum level, Taiwan reformed its school curriculum to include more environmental elements in response to international dialogs (Wang 2009; Yueh and Barker 2011). The table below (Table 7.2) summarizes the components in the new curriculum, which was fully implemented in 2001.

Table 7.2 The development of the elementary school curriculum in Taiwan in 2010

11 School Subjects		7 Major Learning Areas		6 Important Issues
1. Civics and Ethics Health Education		1. Language Arts		1. Information technology education
2. Mandarin		2. Health and Physical Education		2. *EE*
3. Mathematics		3. Social Studies		3. Gender education
4. Social Studies	integrated into →	4. Arts and Humanities	Infused with	4. Career development education
5. Natural Science		5. Science and Technology		5. Human rights education
6. Physical Education		6. Mathematics		6. Home economics education
7. Music		7. Integrated Activities		
8. Fine Arts				
9. Group Activities				
10. Guidance Activities				
11. Native Art Teaching Activities				

In the new curriculum, the discipline-based elementary school subjects were integrated into seven major learning areas: language arts, health and physical education, social studies, arts and humanities, science and technology, mathematics, and integrative activities. Also, six important issues (including information technology education, EE, gender education, career development education, human rights education, and home economics education) were infused into all the learning areas (Yueh, et al. 2010; Ministry of Education 2003). This curriculum design, as with mainland China and Hong Kong in the post-1990s, implies the interdisciplinary and inter-subject nature of EE and infers the historical, cultural, political, and sociological implications of EE (Hung 2010). Since Taiwan adopts a bottom-up approach to implement the new curriculum, schools are free to make their own judgments on how and to what extent EE is to be integrated into the school-based curricula (Chen 2010). Such an encouragement of school-based curriculum development transferred educational power to schools and communities and fostered the uniqueness and autonomy of each school (Wang 2009). Nevertheless, there are still guidelines offered by the government to help schools infuse EE into different subjects such as Chinese language, music, physical education, and social studies (Environmental Protection Administration 2011).

According to the new curriculum, EE covers five main areas, namely, environmental awareness, environmental knowledge, environmental ethic values, environmental skills, and experiences of environmental action (Ministry of Education 2003; Hung 2010). Apart from knowledge building, EE in the new curriculum also cares about the development of the students' values and attitudes. It stresses the cultivation of students' appreciation of nature and aims to develop students' positive environmental attitudes and to equip them with problem-solving skills so as to prepare them to act sustainably. As students' environmental actions and experiences are emphasized in the new curriculum, outdoor/experiential teaching, problem-solving, and issue-based inquiry approaches are proposed to be adopted in the teaching and learning process (A River Runs Through thematic teaching website on Water, Ecology, and the Environment 2001). Also, it is suggested that teachers act as facilitators in EE and work together with students to solve environmental problems (Lai 2000).

To align with the decentralized and democratic nature of school-based curriculum development, a green school concept, by providing informal learning opportunities beyond formal school curriculum, is promoted through the Greenschool Partnership Project in Taiwan (GPPT), which emphasizes establishing school autonomy. Under the GPPT, partnerships were initiated and implemented by the Graduate Institute of Environmental Education of the National Taiwan Normal University (NTNU) and funded and promoted by the Ministry of Education in Taiwan in 1999 (Wang 2009). Besides the provision of green school concepts, the GPPT also develops examples of action plans and instructional materials to help schools turn into green schools (Taiwan Greenschool Partnership Network 2005). Member schools in the GPPT are self-directed to develop their own action plans for implementation. They are also encouraged to share their EE experiences with their green school partners and become self-evaluative in the whole process (Wang

2009). Since the GPPT adopts a holistic perspective and encourages the connection of school life with teaching and learning and administration, it is viewed as the software renovation project for ESD (Su and Chang 2011, p. 8). The Taiwan Sustainable Campus Program (TSCP), developed from the idea of the GPPT, offers funding for participatory schools to make revolution in hardware, such as in physical construction and utility operation (Wang 2009).

Apart from the GPPT and TSCP, a number of nongovernmental organizations in Taiwan are also engaged in EE. Through publications on environmental issues and conducting a variety of activities, such as seminars, exhibitions, training courses, and ecotours, they seek to enhance the public's knowledge and awareness on environmental issues and participation in environmental protection in the community (Green Citizens' Action Alliance 2013; Taiwan Ecological Education Association 2013; Taiwan Environmental Protection Union 2013; Wild at Heart Legal Defense Association, Taiwan 2013).

From the document analysis, the development of ESD in Taiwan, in addition to the curriculum initiatives proposed by the government through national curriculum reform led by the Ministry of Education, suggests a community-based focus through the green school concepts that have been built up under the GPPT, TSCP, and various nongovernmental environmental organizations. This is indicative of an involvement skewed toward community-based efforts and initiatives and proliferation of school-based sustainability education in Taiwan, in contrast with the government-directed ESD in mainland China and the balanced effort of the government and the nongovernmental organizations in Hong Kong. Insofar as the nature of sustainability education is concerned, Taiwan also witnesses a holistic approach emphasizing knowledge, attitudes, and problem-solving skills as fundamental bases for educating people to be environmentally literate in schools and in the community. An interdisciplinary and issue-based approach to school EE in which learning is student centered and participatory is outlined in the official documents.

In sum, by reviewing the historical development of ESD in the three Chinese communities, it is found that no matter how much the government or the community has involved in taking the initiatives in developing the EE/ESD curricula, the promulgation of *Agenda 21* in 1992 was a landmark of EE in the three communities. In mainland China, it moved EE from a predominant focus on environmental knowledge to a broader view of cultivating environmental awareness and public participation, whereas in Hong Kong and Taiwan, it led to an increasing concern of the public toward sustainable development and facilitated a more important role of EE in the school curricula.

7.4 Current ESD Practices and Challenges Facing the Practice of ESD in Schools

As suggested earlier, besides knowledge building, ESD concerns students' values and behavioral development. It suggests adopting interdisciplinary and holistic approaches to teaching the interrelationship between mankind and the environment. Teaching approaches are also proposed to be more active and more able to be directed toward inquiry. Are teachers ready to face these changes? Would they encounter difficulties when they undergo these changes? In this section, we examine how schools in these three places practice ESD at the implementation level from the perspectives of curriculum coherence, content, and pedagogical approaches.

7.4.1 Lack of Curriculum Coherence for EE in Schools

In mainland China, following the initiative of the State Education Commission that EE should be emphasized and promoted in certain primary and lower secondary subjects (State Education Commission 1994), environmental knowledge, in primary schooling, is infused into language subjects, mathematics, natural science, and social studies. At the secondary level, EE tends to be integrated into subjects such as physics, chemistry, biology, and geography (He and Zhang 2007). Chen (2005) pointed out that to ensure that the class can run smoothly and to enhance the teaching quality of the subjects, many schools in mainland China do not attend to the inclusion of content of EE in the curriculum planning and implementation stages. Subject teachers are not specifically required to teach environmental knowledge in their lessons. More importantly, teachers are not trained to integrate subject knowledge with EE. They do not understand the roles and effects of the subject that they teach on the implementation of EE (Chen 2005; Zhou 1995) and are weak in implementing EE through a whole-school or cross-curricular approach (Lee et al. 2006).

Akin to mainland China, teachers in Hong Kong were also found to hold firmly the subject-bound belief. They believe that teaching should aim at transmitting knowledge, which can be derived from the established academic disciplines (Yeung and Lam 2007). Though cross-curricular measures are suggested in the *Guidelines* for the promotion of sustainable development, Lee (1997) found that there is a relative underemphasis on the recommendation of adopting the whole-school approach suggested by the *Guidelines*. Nevertheless, some schools do try to practice cross-curricular education to promote sustainability, though they may sometimes find an overlap of content of one subject with another (Lee et al. 2009).

Schools in Taiwan in general are not committed to developing school-based EE curricula or infusing EE into the seven learning areas mentioned earlier. These learning areas account for about 80 % of the school timetable per week or per school year only. The other 20 %, or less, of curriculum time is allocated to the flexible curriculum, wherein the central government allows schools the autonomy to make their own decisions about teaching and school development. Schools can use the flexible curriculum component to develop their own school-based curricula to teach the six important issues, including EE, suggested in the reformed curriculum (Yueh et al. 2010). However, as the development of a school-based EE curriculum is not mandated, it is reported that schools would therefore more likely allocate the time in the flexible curriculum to supplementary learning of the examined subjects of English, Chinese, mathematics, or social studies. And even if the schools have designed a school-based EE curriculum, the curriculum is not lasting, and it may appear only in the first year of study (Yueh et al. 2010). Moreover, as subject teachers are trained specifically in their own discipline, many of them are reported to be unfamiliar with and to lack confidence about teaching EE. Given the heavy workload, packed curriculum, and pressure from parents for good examination results of their children, many teachers tend to focus on achieving the goals of their own teaching subjects instead of spending time on EE (Yueh and Barker 2011). Environmental education in Taiwan is hence commented on by Yueh and Barker (2011) as "unstructured" and "opportunistic" (p. 145).

According to Hargreaves (2006), when a curriculum is coherent, the various parts of the curriculum have a clear and explicit relationship with one another to make it a whole (p. 33). Students' learning activities will then be combined to form a coherent experience (Education Scotland 2014). Yet, the discussion in this section illustrates that there has been a limited collaboration, in varying degrees, among school subject curricula in implementing EE in the three Chinese communities. The learning experiences of students therefore become fragmented and are less relevant to students as goals of education such as the transfer of learning and the development of students' skills in thinking and reasoning are not supported (Marzano 1991; Perkins 1991; Jacob 1989).

7.4.2 *Focus on Transmitting Knowledge* About *the Environment*

Environmental education in mainland China focuses mainly on the transmission of environmental knowledge, which neglects the development of students' values and skills. It does not emphasize training students to be future citizens with correct environmental ethics and social responsibilities or to have problem-solving skills (Chen 2005). Hong Kong schools and teachers, according to Morris and Adamson (2010), often focus on academic rationalism that seeks to enlighten students with knowledge and concepts instead of serving as an agent for social reform, change, and criticism. Under pressure to prepare their students well for examinations, teachers find that teaching the content of textbooks is "safer" in terms of securing good academic results for the students (Tilbury 1997, p. 4); hence there is a lack of motivation and interest in implementing EE as a form of values education (Lee et al. 2006; Lee 2002; Ham and Sewing 1987). Citizenship components and political elements of education for the environment thus receive little coverage by Hong Kong teachers or are even absent from the curriculum (Wong 1994; Gerber 1990). No wonder it is reported in Fien's (2003) survey that students usually have knowledge *about* the environment rather than knowledge of how to work *for* the environment.

Similarly, Hsu (2004) noted that environmental educators in Taiwan overemphasize knowledge-based instruction. Although 66 % of teachers in Yueh and Barker's (2011) research indicated that action-oriented behavioral change was their expected student learning outcome in EE, 79 % believed that teaching knowledge about a wide variety of EE topics is paramount. It is argued that by just imparting knowledge of ecology and environmental problems, it is difficult to develop students' responsible environmental behavior.

It becomes apparent that the achievement goal in the Asian communities of mainland China, Hong Kong, and Taiwan has resulted in an overemphasis in schools on attaining good examination results of their students. Success in the examination system has long been believed to be the only key to social mobility in Confucian societies (Hui 2005). Therefore, schools in the three Chinese communities in which Confucianism is deeply rooted strive hard to train their students to attain good grades in public examinations. The pressure for schools to boost up the academic achievement levels of their students is in close connection with the expectations of most parents in these communities. The satisfaction or disappointment of these parents with their children is, according to Hui (2005), largely associated with the examination marks attained. In this regard, learning, teaching, and assessment in classrooms can be seen to a large extent as serving only the purpose of success in examinations.

In mainland China, EE is considered irrelevant to examinations and has often been ignored or marginalized (Lee et al. 2006; Zhu 1995). In assessing students' achievement, teachers in Hong Kong are dominated by traditional modes of assessment such as paper and pencil tests, since they are the most common assessment forms adopted in public examinations. Alternative modes of assessment which test for students' attitudes and behavior are less emphasized (Lee et al. 2009). Likewise, as EE is a non-examination learning area in Taiwan, it receives little attention from either teachers or students in practice. It is often perceived as a subset of other subjects and clear assessment procedures are not stated (Yueh and Barker 2011, p. 143). Hence, Yueh et al. (2010) argue that "examination, rather than governmental requirement, is the most influential theme in the implementation and emergence of environmental education in Taiwan" (p. 279).

This backwash effect of examination on school curricula and students' learning in mainland China, Hong Kong, and Taiwan is predominant; it does not only influence and minimize the attempts to move beyond "knowledge" goals for EE—"if it is not examined, it does not receive priority"—but also undermines the use of inquiry and student-focused pedagogical approaches in implementing EE.

7.4.3 Low Priority Given to the Inquiry Approach and Active Learning Strategies

When Hargreaves proposed the concept of *curriculum manageability* in 2006, he referred to it to as the amount of knowledge and skills to be put into the curriculum and the capacity of the teachers to relate the parts together in a balanced whole. Manageability is conceived as a teacher's use of the most appropriate pedagogies to make the curriculum content comprehensible to students within the allocated amount of time and resources. In this, even though the focus of EE in the three Chinese communities is on knowledge *about* the environment, learning outcomes could still be broadened beyond "knowledge," depending on the pedagogical approaches adopted by teachers. Yet, as noted in the preceding sections, teaching and learning strategies for EE are mainly teacher directed and expository (Chen 2005; Hsu 2004; Tilbury 1997).

In mainland China, personal experiences and direct participation of students in learning environmental issues are stressed in the *Syllabuses for Special Topics Education on Environmental Education for Primary and Middle School Students.* The document also states that schools should cultivate students' consciousness of and ability to reflect on the relationship between humans and their environment through self-inquiry (Ministry of Education of the People's Republic of China 2003). Though most schools organize extracurricular activities such as visits, camping, competitions, exhibitions, and talks to boost students' environmental engagement, Xu (1995) found that most of them are "occasional arrangements" and "unsystematic" (p. 90). Scholars have remarked that outdoor education, field inquiry, and hands-on activities are not sufficiently carried out due to the examination-oriented culture wherein study time is predominantly reserved for examined subjects. Didactic approaches, rather than the participatory decision-making, critical-thinking, and problem-solving practices promoted by UNESCO (2006), are frequently used and thus the strong practicality of EE cannot be satisfactorily revealed (Yang et al. 2010; He and Zhang 2007; Chen 2005).

Likewise, critical inquiry and active learning strategies were found to be of low priority in the teaching and learning of environmental issues in Hong Kong (Tilbury 1997; Wong and Stimpson 1994; Lee 1993). Wong and Stimpson (1994) argued that although many teachers adopt open approaches at the beginning of lessons, their teaching style becomes more restrictive as lessons proceed. Traditional and resource-based learning is perceived by teachers as most effective (Lee 1995). Moreover, pressure from parents also confines teachers to the textbook when teaching environmental issues (Lee et al. 2009). Thus, teaching strategies which involve pupils in exploring personal values and responses are infrequently used in Hong Kong classrooms (Man 1993). Furthermore, most of the teaching time spent on environmental issues revolves around "short-lived" activities (Lee et al. 2009, p. 175), such as recycling schemes, tree planting, and green consumerism, without a critical analysis of the underpinning rationale of these activities and the issues that brought forth these activities.

What is more, as noted by Beames and Brown (2005), though most children in Hong Kong might have a formal outdoor experience during their school life, its educational value and experiential nature are considered to be minimal due to the shortness and "super-protectiveness" of the outdoor programs usually offered (p. 77) and the "spoon-fed" nature of Hong Kong Chinese educational culture (p. 75). Students' experiences are believed to be largely driven by teachers with predetermined outcomes and hence could hardly be regarded as experiential (Beames and Brown 2005).

Teaching strategies did not change dramatically with the curriculum reform in Taiwan either. Though sometimes there is team or cooperative learning, lecturing, lecturing via media, and lecturing coupled with whole-class teacher-student discussion are still dominant in the lessons (Yueh et al. 2010). It is found in Liu's (2008) report that though group discussions, role-playing, and issue investigations are the preferred teaching strategies for EE, lecturing is perceived by both teachers and administrative staff as the simplest, most convenient, and most time-saving teaching method which can teach students about environmental knowledge systematically.

To sum up, these three Chinese communities are apparently not yet well prepared to implement ESD, and they share commonalities regarding the lack of curriculum coherence, overemphasis on knowledge acquisition, and lack of attention on EE, which is not an examination area, as well as a dominance of didactic teaching and lecturing.

7.5 The Step Forward

Since teachers' capacities are central to their approach and the content they deliver to students, to improve the situation mentioned above, measures should be taken to enhance the quality of teacher education programs on ESD. A paradigm shift in the EE curriculum design, teaching emphasis, and teaching approaches adopted in classrooms, a good knowledge of environmental problems, and a high commitment of teachers to ESD can all be seen as contributory factors leading toward a betterment of ESD in schools. Suggested below are the dimensions to improve the existing teacher education programs on EE, namely, instilling subject integration in the school curriculum to enhance curriculum coherence, preparing teachers to be facilitators of learning, and equipping teachers with student-focused inquiry-based teaching strategies.

7.5.1 Enhancing Curriculum Coherence Through Whole-School Collaboration

Document review and analysis has indicated that school curriculum lacks coherence and there is little collaboration among school subjects in implementing EE. To enhance whole-school responsibility of promoting ESD, sustainability issues should not be perceived as an add-on topic in the curriculum or be viewed too differently from other curriculum initiatives (Lee et al. 2009). Teachers in all subjects and positions should be obligated to instill the ideas of environmental sustainability into their teaching. This is an important concept that should be built into the minds of teachers of all subjects.

Apart from this change of mind-set, teacher education programs should help improve prospective teachers' environmental management skills and their ability to coordinate a whole-school or cross-curricular approach to ESD (Hargreaves 2006; Lee et al. 2006, 2009). Teachers should also be trained to attain better skills in the coordination and coherence among subjects. It is anticipated that when teachers are equipped with such skills, collaboration could then be carried out more easily and smoothly between subjects. The school-based curriculum would become more coherent with a clear rationale, and the various parts of the curriculum, such as learning sequences, content, activities and resources, time allocation, and assignments, would fit together to form a whole (Hargreaves 2006). At the same time, duplicate contents of different ESD subjects could also be restructured. A better teaching timetable would lead to a more effective teaching.

The development of a coherent curriculum for implementing EE in schools means that teachers of different subjects would need to be equipped with sufficiently good subject knowledge (Summers et al. 2000). Teacher education programs should be informative enough to equip teachers with sufficient know-how. Both knowledge of environmental issues and knowledge of how to act on these issues are essential (Hines et al. 1987). It is believed that a mastery of good environmental knowledge could well provide an easier start for developing a balanced and student-centered school-based curriculum on ESD. In addition, equipping teachers with better environmental knowledge can help emancipate teachers from the boundaries of their disciplines or professions and build up their confidence in teaching environmental issues, which would help teachers to work on curriculum integration and collaboration.

7.5.2 Preparing Teachers to Be Facilitators of Change Rather Than Transmitters of Knowledge

The fundamental problem facing EE in schools is the fact that it is a non-examination learning area; thus teachers pay less attention to designing its curriculum or simply just ignore it at the implementation level. Teacher education programs would therefore need to cultivate teachers' environmental attitudes and facilitate them to act environmentally. Only when the teachers themselves understand and agree with the importance of ESD will they then be willing to incorporate it into the curriculum.

Whereas principal and parental support provide the necessary conditions for a smooth process of implementing ESD (Lee 2009), teachers' environmental beliefs and awareness are seen as the leading factors in effective environmental teaching. According to Stimpson (1997a), "teachers as individuals need the understanding, skills, and commitment to environmentalize their teaching. Without this it is unlikely that [programs] will be effective in producing environmentally literate pupils as intended in the *Guidelines*" (p. 352). Stimpson's ideas echoed Lee's (1993) view that teachers would contribute more to EE if they were willing and able to adopt more effective teaching and learning strategies. Teachers would be more willing to overcome the barriers of teaching environmental issues and work for the betterment of the environment (Ko and Lee 2003) if they were urged to act as facilitators to bring about change in their students regarding environmental behaviors. Though such a participatory action approach (teacher as change facilitator) might not be received equally well across the three Chinese communities, teacher education programs on sustainable development evidently need to be moved toward facilitating teachers' awareness of their role as activators and curriculum developers in promoting ESD, building up their awareness and sensitivity to everyday environmental issues, and equipping them with the necessary skills to engage students in environmental activities (Taylor et al. 2007).

As the examination-oriented culture is a dominant factor urging teachers to focus more on education *about* the environment than on education *for* the environment, it is suggested that the range of types of assessment should be widened beyond paper and pen examinations to assess a full range of curriculum outcomes beyond knowledge outcomes, including environmental beliefs, attitudes, values, practices, processing, and decision-making. Scholars have suggested that public examinations should include more questions that are related to values in order to upgrade the teaching of values from an "ideal" to an "immediate practical task" in teachers' work plans (Lam 1990). They argue that there is also a need for greater emphasis on the assessment of those values and attitudes related to environmental concerns as well as on decision-making both in internal and external examinations (Lee 1993). Though these are somewhat related to changes at the policy level, it is believed that teacher education programs should help enhance teachers' aesthetic appreciation of the environment so as to make them more concerned about nature and thus bring this value to their environmental teaching. In this way, their strong emphasis on knowledge transmission is expected to be counterbalanced.

7.5.3 Equipping Teachers with Student-Focused Inquiry-Based Teaching Strategies

Teacher pedagogical knowledge is of vital importance in environmental teaching (Summers et al. 2000). Hence the ways that teachers make environmental knowledge accessible to students play a pivotal role in the effective teaching and learning about the environment (Summers et al. 1998). Jensen (2002) believes that the traditional teaching approaches (which overemphasize the delivery of knowledge and facts) adopted by teachers hinder students from actively contributing toward environmental protection. Research has revealed that inquiry teaching approaches and participatory field trips could better develop students' positive environmental attitudes and responsible environmental behavior than didactic approaches. These inquiry approaches do not focus on knowledge transmission, but open up more opportunities for students to reflect on their everyday practices and link them positively with the environment (Ou et al. 2010; Yeung 2002). Thus, to achieve education *in* and *for* the environment that is emphasized in ESD, there is a need to transform conventional teaching and assessment approaches. Teacher education programs should give insights to future teachers other than just preparing them to instill environmental knowledge in the younger generation. These programs should deal with sustainability issues using action-oriented and inquiry-based approaches that can involve more participation of future teachers in environmentally friendly behaviors and better develop their thinking, problem-solving, and action-taking skills regarding environmental problems, all of which can later be transferred to school contexts. Also, given the highly regulated and teacher-directed inquiry process in Hong Kong classrooms, teachers should be mindful of addressing closed and convergent questions that confine students' thinking when carrying out inquiry approaches (Yeung 2009).

It is also important that teacher education programs on ESD should possess strong relevance to school curricula and teaching. Stimpson (1994) argued that if teaching in the school needs to be informal, open, participatory, experiential, and thought inducing, then so does the education of teachers. In this regard, teacher education programs should be "less theoretical" and "more practical" (Lee et al. 2009, p. 193), with hands-on participation in workshops and visits to schools that have good reputations as green schools or schools with effective ESD programs. Through observing the actual practices of green schools, it is easier for teachers to enrich and broaden their perspectives on the planning of green programs and activities.

Offering teaching materials or support in teacher education programs can also help improve teachers' strategies for ESD. According to Chan (2000), existing environmental teaching kits are not readily accessible to schools, and teachers long for teaching kits in video format, with teachers' guides, worksheets, and slides. Teacher education programs should offer teachers comprehensive teaching kits or resources which comprise diversified and up-to-date teaching materials on the ever-changing environmental issues. Informing teachers about the means to acquire up-to-date resources is also important. It is anticipated that by providing teachers with teaching kits and other supporting materials, teachers' preparation time can be reduced, hence diminishing their workload. These materials can also give teachers insights into how to plan their lessons in an issue-inquiry way. In order to expand resource bases, teachers can be encouraged to share expertise and teaching materials among and between schools (Lee 1993). Taking into account the advancement of information communication technologies, the potential of e-learning and video-based community platforms should be fully utilized in teacher education programs to disseminate the most up-to-date resources developed on issue-inquiry and student-centered participatory approaches.

7.6 Conclusion

In the context of deteriorating environmental quality and the growing tendency to promote environmental sustainability in the community at large, greater emphases have been put on the development of ESD in schools all over the world, including in the Chinese communities. However, it is found that schools in the Chinese communities are not yet well prepared to keep up with the changes brought by ESD. School curricula are found to not have a coherent approach and to be without well-trained teaching teams that can integrate subject knowledge into EE and implement it through a cross-curricular approach (Yueh and Parker 2011; Lee 1997; Lee et al. 2006).

Despite some endeavors by the governments in the three Chinese communities to initiate changes at the curriculum level and school programs and activities to promote sustainable development, the lack of teachers' training on the integration of subject knowledge into EE, teachers' limited understanding of the roles and effects of EE, the emphasis on acquiring knowledge *about* the environment, and the low priority on inquiry and active learning strategies cause teachers to tend to retain their teaching focus on the causes and effects of environmental problems rather than spending time developing students' environmental values. Also, due to the packed curriculum and teaching schedule, time is tight for outdoor learning or inquiry. Collaborative teaching between subjects or across the whole school is rare due to the subject-bounded mind-set of teachers.

There are still many problems that need to be resolved by higher authorities at the policy level. Nevertheless, it is anticipated that by developing teacher education programs on ESD, which offer teachers better understanding of curriculum manageability and coherence, mastery of strategies to plan for better relationships between content and skills in the curriculum (Hargreaves 2006), better collaboration skills at the implementation level, increased comprehension and mastery of environmental knowledge and active student-focused teaching approaches, higher incentives regarding values education, etc., EE in the three Chinese communities could ultimately be more readily reformed into ESD.

References

Beames, S., & Brown, A. (2005). Outdoor education in Hong Kong: Past, present and future. *Journal of Adventure Education and Outdoor Learning, 5*(1), 69–82.

Belgrade Charter. (1975). Connect. *UNESCO-UNEP Environmental Education Newsletter, 1*(1), 1–2.

Chan, K. (2000). Use of environmental teaching kits in Hong Kong. *The Environmentalist, 20*, 113–121.

Chen, X. P. (2005). Analysis of environmental education history and its content and characteristics in China's primary and middle schools (我國中小學環境教育的歷史演進及其內容與特點分析). *Journal of Zhejiang Education Institute* (浙江教育學院學報), *3*, 42–47. [In Chinese].

Chen, P. S. (2010). Comment and analysis on the policy implementation of grade 1-9 program (九年一貫課程政策執行評析). *Journal of Education Resources and Research* (教育資料與研), *92*, 47–74. [In Chinese].

Cobern, W. (1998). Science and a social constructivist view of science education. In W. Cobern (Ed.), *Socio cultural perspectives on science education: An international dialogue* (pp. 7–23). Dordrecht: Kluwer Academic.

Curriculum and Development Council. (1999). *Guidelines on environmental education in schools (draft)*. Hong Kong: Government Printer.

Dillon, J. (2014). Environmental education. In N. G. Lederman & S. K. Abell (Eds.), *Handbook of research on science education* (Vol. II, pp. 497–514). New York: Routledge.

Duschl, R., & Osborne, J. (2002). Supporting and promoting argumentation discourse in science education. *Studies in Science Education, 38*, 39–72.

Education Scotland Foghlam Alba. (2014). *Principles of curriculum design*. Retrieved from http://www.educationscotland.gov.uk/thecurriculum/howisthecurriculumorganised/principles/index.asp

Environmental Campaign Committee. (2011a, August 8). *Hong Kong Green School Award*. Retrieved from http://www.ecc.org.hk/english/core/core_detail_list.php?catid=29

Environmental Campaign Committee. (2011b, November 14). *Student environmental protection ambassador scheme*. Retrieved from http://www.ecc.org.hk/english/core/core_detail.php?id=211

Environmental Protection Administration. (2011). *Elementary and junior high schools' handbook on the development of environmental education curriculum* (國民中小學環境教育課程編製指導手冊). Taipei: Author. [In Chinese].

Fien, J. (2003). Young people and the environment: Results and implications of a study of environmental knowledge, beliefs and behavior in Asia and the Pacific. In P. Hills & C. S. Man (Eds.), *New directions in environmental education* (pp. 15–32). Hong Kong: Centre for Urban Planning and Sustainable Development, Hong Kong University.

Gerber, H. (1990). Asian geographical educators' perceptions and practice. In P. Stimpson & T. Y. L. Kwan (Eds.), *Teaching geography in and about Asian Pacific countries* (pp. 17–32). Hong Kong: IGU Commission on Geographical Education/University of Hong Kong.

Gough, A. (2002). Mutualism: A different agenda for environmental and science education. *International Journal of Science Education, 24*(11), 1201–1215.

Green Citizens' Action Alliance. (2013). Retrieved from http://www.gcaa.org.tw/

Ham, S. H., & Sewing, D. R. (1987). Barriers to environmental education. *Journal of Environmental Education, 19*(2), 17–24.

Hargreaves, D. H. (2006). Coherence and manageability: Reflections on the national curriculum and cross-curricular provision. *The Curriculum Journal, 2*(1), 33–41.

Hart, P. (2007). Environmental education. In S. K. Abell & N. G. Lederman (Eds.), *Handbook of research on science education* (pp. 689–726). Mahwah: Lawrence Erlbaum Associates, Inc., Publishers.

He, W., & Zhang, B. (2007). Comparative analysis of environmental education in primary and secondary schools in the Mainland China and Hong Kong (中國大陸和香港地區中小學環境教育對比分析). *Education of Geography* (地理教育), *5*, 8–9. [In Chinese].

Wild at Heart Legal Defense Association, Taiwan. (2013). Retrieved from http://zh.wildatheart.org.tw/environment

Hestness, E., McGinnis, J. R., Riedinger, K., & Marbach-Ad, G. (2011). A study of teacher candidates' experiences investigating global climate change within an elementary science methods course. *Journal of Science Teacher Education, 22*(4), 351–369.

Hines, J. M., Hungerford, H. R., & Tomera, A. N. (1987). Analysis and synthesis of research on responsible environmental behaviour: A meta-analysis. *Journal of Environmental Education, 18*(2), 1–8.

Hsu, S.-J. (2004). The effects of an environmental education program on responsible environmental behaviour and associated environmental literacy variables in Taiwanese college students. *The Journal of Environmental Education, 35*(2), 37–48.

Hui, L. (2005). Chinese cultural schema of education: Implications for communication between Chinese students and Australian education. *Issues in Educational Research, 15*(1), 17–36.

Hung, R. (2010). A critical examination of the legitimacy of the seven important issues of grade 1-9 curricula (九年一貫課程七大議題正當性之批判性&#x6AA2視). *Journal of Educational Research and Development* (教育研究與發展期刊), *6*(2), 33–57. [In Chinese].

Jacobs, H. H. (1989). *Interdisciplinary curriculum: Design and implementation*. Alexandria: Association for Supervision and Curriculum Development.

Jensen, B. B. (2002). Knowledge, action and pro-environmental behaviour. *Environmental Education Research, 8*(3), 325–334.

Ko, A. C.-c., & Lee, J. C.-k. (2003). Teachers' perceptions of teaching environmental issues within the science curriculum: A Hong Kong perspective. *Journal of Science Education and Technology, 12*(3), 187–204.

Lai, Y. F. (2000). Introduction and case studies of environmental education [in Chinese]. In *Yunlin county 88th school year environmental and ecological education training ,manual for elementary school teachers* (pp. 69–72).

Lam, C. C. (1990). Values teaching in school geography in Hong Kong. In P. Stimpson & T. Y. L. Kwan (Eds.), *Teaching geography in and about Asian Pacific countries* (pp. 55–62). Hong Kong: IGU Commission on Geographical Education/University of Hong Kong.

Leal Filho, W., & Behrens, R. (2003). Trends in education for sustainability. In P. Hills & C. S. Man (Eds.), *New directions in environmental education* (pp. 1–14). Hong Kong: Centre for Urban Planning and Sustainable Development, Hong Kong University.

Lee, C. K. J. (1993). Geography teaching in England and Hong Kong: Contributions towards environmental education. *Geographic and Environmental Education, 2*(1), 25–40.

Lee, J. C.-k. (1995). Environmental education in Hong Kong secondary schools: State of the art. *New Horizons in Education, 36*, 63–70.

Lee, J. C. K. (1997). Environmental education in schools in Hong Kong. *Environmental Education Research, 3*(3), 359–371.

Lee, J. C. K. (2002). Approach to environmental education for sustainable development in schools: Challenges and opportunities. *Hong Kong Teachers' Centre Journal, 1*, 131–137.

Lee, J. C.-K. (2009). Green primary schools in Hong Kong. In J. C.-K. Lee & M. Williams (Eds.), *Schooling for sustainable development in Chinese communities: Experience with younger children* (pp. 195–212). Dordrecht: Springer.

Lee, J. C.-K., & Huang, Y. (2009). Education for sustainable development projects and curriculum reform in china: The EEI and the EPD. In J. C.-K. Lee & M. Williams (Eds.), *Schooling for sustainable development in Chinese communities: Experience with younger children* (pp. 115–135). Dordrecht: Springer.

Lee, J. C.-K., & Tilbury, D. (1998). Changing environments – the challenge for environmental education in China. *Geography, 83*(3), 227–236.

Lee, J. C.-K., Lin, P., & Kwan, T. Y.-L. (2006). Environmental and geographical education for sustainability in Hong Kong and China: Trends and issues. In J. C.-K. Lee & M. Williams (Eds.), *Environmental and geographical education for sustainability: Cultural context* (pp. 229–245). New York: Nova Science Publishers.

Lee, J. C.-K., Lam, W.-P., & Williams, M. (2009a). The education for sustainable development project in Hong Kong. In J. C.-K. Lee & M. Williams (Eds.), *Schooling for sustainable development in Chinese communities: Experience with younger children* (pp. 157–176). Dordrecht: Springer.

Lee, J. C.-K., Williams, M., & Lam, W.-P. (2009b). Case studies from the education for sustainable development project in Hong Kong. In J. C.-K. Lee & M. Williams (Eds.), *Schooling for sustainable development in Chinese communities: Experience with younger children* (pp. 177–193). Dordrecht: Springer.

Liu, C-N. (2008). *Current situation and difficulties of environmental education teaching in elementary schools in Yunlin County* (Unpublished master's thesis) (雲林&#x7E23國小實施環境教育教學現況與困境之研究). Taiwan: National University of Tainan. [In Chinese].

Man, S. W. (1993, September). *Impacts of recent approaches to value education on environment – A critical appraisal*. Paper presented at the annual conference of the educational research association, Singapore.

Marzano, R. J. (1991). Fostering thinking across the curriculum through knowledge restructuring. *Journal of Reading, 34*(7), 518–525.

McClelland, J. A. G. (1991). Curriculum development in Hong Kong. In C. Marsh & P. Morris (Eds.), *Curriculum development in East Asia* (pp. 106–128). London: The Falmer Press.

McKeown, R., & Hopkins, C. (2003). EE≠ESD: Defusing the worry. *Environmental Education Research, 9*(1), 117–128.

Meadows, D. (1990). *Harvesting one hundredfold*. Kenya: United Nations Environment Programme.

Ministry of Education. (2003). *General guidelines of grade 1–9 curriculum of elementary and junior high school education* (國民中小學九年一貫課程綱要). Taipei: Author. [In Chinese].

Ministry of Education of the People's Republic of China. (2003). *Syllabuses for special topics education on environmental education for primary and middle school students* [in Chinese]. Retrieved from http://www.moe.gov.cn/publicfiles/business/htmlfiles/moe/s3320/201001/81832.html. [In Chinese].

Morris, P., & Adamson, B. (2010). *Curriculum, schooling and society in Hong Kong*. Hong Kong: Hong Kong University Press.

Morris, P., & Marsh, C. (1992). Curriculum patterns and issues in East Asia: A comparative survey of seven East Asian societies. *Journal of Education Policy, 7*(3), 251–266.

Moseley, C., Desjean-Perrotta, B., & Crim, C. (2014). Programmatic implementation of environmental education in an elementary educator preparation program: A case study. In J. C.-K. Lee & R. Efird (Eds.), *Schooling for sustainable development across the Pacific, Schooling for sustainable development volume 5* (pp. 243–259). Dordrecht: Springer.

National Council for Sustainable Development Network. (2004). *The Taiwan declaration on sustainable development*. Retrieved from http://sta.epa.gov.tw/NSDN/en/development/index.htm

Ou, T-Y., Weng, J-H., & Huang, C-K. (2010). The effect of educational program on the cognition and attitude of college students on environmental protection (環境教育課程對&#x5C08科學生環境保護認知與態度之影響). *Research and Development in Science Education Quarterly* (科學教育研究與發展季刊), *57*, 1–20. [In Chinese].

Parliamentary Commissioner for the Environment (PCE), New Zealand. (2004). *See change: Learning and education for sustainability*. Wellington: PCE.

Perkins, D. N. (1991). Educating for insight. *Educational Leadership, 49*(2), 4–8.

Pui Ching Primary School. (2010). *Environmental activities* [in Chinese]. Retrieved from http://photo.pcps.edu.hk/gallery/gcampus09_02

St. Bonaventure Catholic Primary School. (2010). *St. Bonaventure Catholic Primary School: Hong Kong Awards for Environmental Excellence – sectoral awards* (聖文德天主教小學:香港環境卓越大&#x596C—界別卓越&#x596C), Retrieved from http://www.hkaee.org.hk/questcms/userfile/File/StBonaventureCatholicPrimarySchool_20100520.pdf. [In Chinese].

State Education Commission. (1994). *A collection of educational documents on nine-year compulsory education (primary) [in Chinese]*. Beijing: Beijing Normal University Press Department of Elementary Education.

Stimpson, P. (1994). Environmental education and classroom practice: Some implications for teacher education. *Geographic and Environmental Education, 3*(2), 22–34.

Stimpson, P. G. (1997a). Environmental challenge and curricular responses in Hong Kong. *Environmental Education Research, 3*(3), 345–357.

Stimpson, P. (1997b). Environmental education in Hong Kong and Guangzhou: One purpose, two systems? *Compare: A Journal of Comparative and International Education, 27*(1), 63–74.

Su, H-J., & Chang, T-C. (2011). The status of education for sustainable development in Taiwan. Retrieved from http://140.135.10.180/esdtaiwan_english/index.php

Summers, M., Kruger, C., & Mant, J. (1998). Teaching electricity effectively in primary schools: A case study. *International Journal of Science Education, 20*(2), 153–172.

Summers, M., Kruger, C., Childs, A., & Mant, J. (2000). Primary school teachers' understanding of environmental issues: An interview study. *Environmental Education Research, 6*(4), 293–312.

Taiwan Ecological Education Association. (2013). Retrieved from http://www.lovenature.org.tw/

Taiwan Environmental Protection Union. (2013). Retrieved from http://www.tepu.org.tw/

Taiwan Greenschool Partnership Network. (2005). *Taiwan greenschool partnership program.* Retrieved from http://www.greenschool.moe.edu.tw/eng/about.htm

Taylor, N., Doff, T., Jenkins, K., & Kennelly, J. (2007). Environmental knowledge and attitudes among a cohort of pre-service primary school teachers in Fiji. *International Research in Geographical and Environmental Education, 16*(4), 367–379.

Tilbury, D. (1995). Environmental education for sustainability: Defining the new focus of environmental education in the 1990s. *Environmental Education Research, 1*(2), 195–212.

Tilbury, D. (1997). Environmental education: A head, heart and hand approach to learning about environmental problems. *New Horizons in Education, 38*, 1–9.

Tilbury, D., & Wortman, D. (2006). Geography and sustainability the future of school geography? In J. Lidstone & M. Willians (Eds.), *Geographical education in a changing world: Past experience, current trends and future challenges* (pp. 195–211). Dordrecht: Springer.

Tuncer, G., Tekkaya, C., Sungur, S., Cakiroglu, J., Ertepinar, H., & Kaplowitz, M. (2009). Assessing pre-service teachers' environmental literacy in Turkey as a mean to develop teacher education programs. *International Journal of Educational Development, 29*, 426–436.

Tung Wah Group of Hospitals Sin Chi Wan Primary School. (2010). *Activities of environmental week in Tung Wah Group of Hospitals Sin Chu Wan Primary School* (東華三院&#x51BC次雲小學環保週活動), Retrieved from http://www.scwps.edu.hk/popup/20100426/20100426.pdf. [In Chinese].

UNESCO. (1978). *Intergovernmental conference on environmental education—Tbilisi (USSR)—final report*. Retrieved from http://unesdoc.unesco.org/images/0003/000327/032763eo.pdf

UNESCO. (1992). *UN conference on environment and development: Agenda 21*. Switzerland: UNESCO.

UNESCO. (2006). *Framework for the UNDESD international implementation scheme*. Paris: UNESCO.

UNESCO-EPD. (1997). *Declaration of Thessaloniki (UNESCO publication no. EPD-97/CONF.401/CLD.2)*. Paris: UNESCO-EPD.

Wang, S.-M. (2009). The greenschool project in Taiwan. In J. C.-K. Lee & M. Williams (Eds.), *Schooling for sustainable development in Chinese communities: Experience with younger children* (pp. 213–231). Dordrecht: Springer.

Wang, Q. (2011). Characteristics of ESD-promoting strategies in China's basic education. *Journal of Education for Sustainable Development, 5*(2), 215–223.

Wasmer, C. (2005). Towards sustainability: Environmental education in China – A German strategy for Chinese schools? *Duisburger Arbeitspapiere zur Ostasienwirtschaft, No. 73*. Retrieved from https://www.econstor.eu/dspace/bitstream/10419/23138/1/AP73.pdf

A River Runs Through thematic teaching website on Water, Ecology, and the Environment. (2001). *Preliminary discussion of environmental education* (環境教育初探). Retrieved from http://contest.ks.edu.tw/~river/teach/theory/theory5.htm. [In Chinese].

Wong, F. (1994). The political dimensions of environmental education in Hong Kong. *International Research in Geographical and Environmental Education, 3*(1), 4–18.

Wong, E., & Stimpson, P. (1994). Teaching styles of Hong Kong in environmental education in secondary schools. *Research in Education, 52*, 1–12.

Xiao, F. (2002, December). "Green schools" in China. *Asia-Pacific forum on science learning and teaching, 3*(2). Retrieved from http://www.ied.edu.hk/apfslt/v3_issue2/xiaof/abstract.htm

Xu, H. (1995). Some considerations on school environmental education in China. *International Research in Geographical and Environmental Education, 4*(1), 90–91.

Yang, G., Lam, C.-C., & Wong, N.-Y. (2010). Developing an instrument for identifying secondary teachers' beliefs about education for sustainable development in China. *The Journal of Environmental Education, 4*(4), 195–207.

Yavetz, B., Goldman, D., & Pe'er, S. (2014). How do preservice teachers perceive 'environment' and its relevance to their area of teaching? *Environmental Education Research, 20*(3), 354–371.

Yeung, S. P. M. (2002). Teaching approaches and the development of responsible environmental behavior: The case of Hong Kong. *Ethics, Place and Environment, 5*(3), 239–269.

Yeung, S. Y. S. (2009). Is student-centered pedagogy impossible in Hong Kong? The case of inquiry in classrooms. *Asia Pacific Education Review, 10*, 377–386.

Yeung, S. S.-Y., & Lam, C.-C. (2007). Teachers' conception of curriculum integration: A problem hindering its implementation in Hong Kong. *Education Journal, 35*(2), 109–144.

Yueh, M.-C. M., & Barker, M. (2011). Framework thinking, subject thinking and "Taiwan-ness" in environmental education. *Australian Journal of Environmental Education, 27*(1), 134–148.

Yueh, M.-C. M., Cowie, B., Barker, M., & Jones, A. (2010). What influences the emergence of a new subject in schools? The case of environmental education. *International Journal of Environmental and Science Education, 5*(3), 265–285.

Zhang, Y. (2004). Special issue on "green school" – thoughts on issues concerning "green schools.". *Chinese Education and Society, 37*(3), 64–70.

Zhang, T. (2010). From environment to sustainable development: China's strategies for ESD in basic education. *International Review of Education, 56*, 329–341.

Zhou, L. (1995). Creating favourable conditions for environmental education – A case study from Xuejun high school, Hangzhou. *International Research in Geographical and Environmental Education, 4*(1), 96–98.

Zhu, H. (1995). Education and examinations – a major constraint hindering environmental education in the People's Republic of China. *International Research in Geographical and Environmental Education, 4*(1), 106–107.

Chapter 8
Hong Kong Students' Decision-Making About Ecological and Health Issues

Yeung Chung Lee

8.1 Introduction

This chapter presents four studies that examine Hong Kong students' reasoning for decision-making about socioscientific issues (SSIs) related to health and ecology. These four studies collectively represent a progressive endeavor to elicit students' reasoning and decision-making on SSIs and help them to reflect on their reasoning through specially designed strategies that include cross contextual or cross-cultural exchanges. The participants in these studies were mainly senior secondary students in Hong Kong and other locations. The relative merits of these strategies for improving students' reasoning are discussed, drawing on the insights gained from these studies.

8.2 SSIs as Part of Science Education

Socioscientific issues (SSIs) have been used as a learning context for both science and educational courses. The SSI movement has grown out of the science-technology-society (STS) or science, technology, society, and environment (STSE) approaches, which emphasize the interactions among science, technology, and society. It is strongly linked with the "Science for All" movement and the curricular goal of developing students' scientific literacy. Recently, the traditional STS movement has split into two main movements. The first is SSI education,

Y.C. Lee (✉)
Department of Science and Environmental Studies, The Hong Kong Institute of Education, Room 36, 1/F, Block D3, 10 Lo Ping Road, Tai Po, Hong Kong
e-mail: yclee@ied.edu.hk

L.L. Liang et al. (eds.), *Chinese Science Education in the 21st Century: Policy, Practice, and Research*, Contemporary Trends and Issues in Science Education 45, DOI 10.1007/978-94-017-9864-8_8

which focuses on the development of multi-perspective thinking, critical reasoning, argumentation, and value judgment through decision-making on SSIs. In the Hong Kong science curricula at senior secondary levels (e.g., CDC and HKEAA 2007a), these attributes are manifested as STSE connections. The second movement emphasizes the relationship among science, technology, and engineering and the crosscutting concepts that connect these three disciplinary areas (NRC 2012). This approach is intended to bring engineering practices to the fore and to bridge science and technology. Engineering is seen as a process in which scientific knowledge is applied to achieve technological ends through design and systematic testing, including experimentation (NRC 2012). This new emphasis is in response to the twenty-first-century need to develop human capacity in science, technology, and engineering. However, this view of engineering is contentious as it implies technology and engineering are applied sciences, which oversimplifies the relationships between technology and science (Van Eijck and Claxton 2008). This section briefly reviews studies on the SSI movement in the context of STS, with a particular focus on the reasoning involved in decision-making about SSIs among young people.

Decision-making on SSIs is widely associated with the ability to informally reason (e.g., Sadler 2004; Sadler and Zeidler 2005; Wu and Tsai 2007), which includes reasoning about cause-effect relationships from multiple perspectives, weighing the pros and cons of decision (Means and Voss 1996; Zohar and Nemet 2002; Ranyard et al. 1997), moral and ethical reasoning (e.g., Zeidler and Keefer 2003; Zeidler and Sadler 2008), and reasoning based on values (Grace 2002, 2009). Millar and Osborne (1998) argued that in considering these kinds of issues, students should be taught to "distinguish between technical issues (what is possible) and ethical issues (what ought to be done)" (p. 2022). Research has identified several factors that influence students' reasoning about SSIs, including prior beliefs, personal consequences (Sadler and Zeidler 2004), the ideas of probability and risk (Levinson et al. 2011; Millar and Osborne 1998), personal values (Bell and Lederman 2003), the capability of moral and ethical reasoning (e.g., Zeidler and Keefer 2003; Zeidler and Sadler 2008), argumentation skills (Foong and Daniel 2013), and cultural standpoints (Braund et al. 2007), all of which are intricately related (NRC 2005). In sum, these factors are integrated into higher-order cognitive skills essential for making rational decisions, which include multi-perspective thinking, assessment of causal relationships, moral reasoning, risk assessment, and evaluation of values.

Research on students' ability to engage in informal reasoning on SSIs has yielded mixed findings, reflecting the complexity of this kind of reasoning compared with that on pure scientific issues (Braund et al. 2007). On the positive side, middle school to senior secondary students were found to be able to approach SSIs from a reasonable number of perspectives in the context of nuclear energy usage, including social, economic, scientific, and ecological perspectives (e.g., Patronis et al. 1999; Wu and Tsai 2007), a clear demonstration of multi-perspective thinking. It was further found that college students used an even wider range of perspectives, including medical, ethical, economic, religious, political, and scientific perspectives, to consider the issue of stem-cell research. Middle school to college students are able to engage in value- or moral-based reasoning on a wide range of themes, including biotechnology (Zeidler and Keefer 2003; Zeidler and Sadler 2008), biological conservation (Grace 2002, 2009), wetland environmental management (Jimenez-Aleixandre and Pereiro-Muñoz 2002), and medical research (Halverson et al. 2009). On the less positive side, middle school students were found to be fairly limited in their ability to employ multi-perspective reasoning to address environmental issues (Hogan 2002; Kortland 1996) and to integrate scientific and other information in making personal decisions (Zeidler et al. 2002). Another related concern is that scientific perspectives or science content knowledge appear to be less commonly relied upon than ethical perspectives in weighing different decision alternatives and justifying decisions on SSIs (Grooms et al. 2014; Halverson et al. 2009). Personal values were shown to play a much more important role in influencing university students' decision-making than scientific understanding (Bell and Lederman 2003). This finding led Oulton et al. (2004) to argue that students should be provided with opportunities to separate arguments from their own values and beliefs in discussing SSIs and Nielsen (2013) to conclude that decision-making about SSIs should not focus only on grounding the decision in scientific evidence but rather on how students consider a multitude of factors and their relative importance. All of these results point to the need to integrate scientific and nonscientific perspectives in SSI education without compromising the importance of either perspective.

Teacher facilitation may be useful in guiding secondary students on decision-making about issues through the use of argumentation (Patronis et al. 1999). Various frameworks for scaffolding the decision-making process have been proposed to promote informal reasoning ability for informed decision-making. These decision-making frameworks generally involve students generating options or decision alternatives with the support of justifications based on appropriate criteria, followed by a review of the decision making process with a view to making further improvements (e.g., Keefer 2003; Ratcliffe 1997; Tal et al. 2001; Zeidler et al. 2009). A more generic process was proposed by Svenson (1992, 1996) based on the "differentiation and consolidation theory," which mirrors how people make decisions in more general contexts. This framework consists of three progressive phases. The first phase involves the recognition of the problem and the identification of decision alternatives. The second involves the differentiation of alternatives by either holistic differentiation or systematic process differentiation in which the pros and cons are weighed. The third stage comprises post-decision consolidation, where differentiation continues to take place to ensure that the chosen alternative is the most appropriate.

Within these decision-making frameworks, various strategies for enhancing students' informal reasoning and decision-making ability have been extensively studied, with evidence of varying degrees of success (e.g., Kolsto 2001; Sadler 2004; Wu and Tsai 2007). These include the use of argumentation to promote understanding and decision-making (Patronis et al. 1999; Simon et al. 2006; Simon and Maloney 2007); the use of meta-cognitive strategies such as reflective thinking to integrate multiple perspectives (e.g., Zeidler et al. 2002); confronting students with opposing arguments through class discussions, role-plays, and debates to clarify their thoughts (Simonneaux 2001); and group discussion to enhance collaborative or peer-group decision-making (Grace 2009). However, the implementation of these strategies in the classroom is not without difficulty. Teachers may lack the skills to use argumentation, role-play, and even discussion. The effective use of such strategies also needs considerable preparation and training on the part of both teachers and students (Oulton et al. 2004). More importantly, multi-perspective reasoning and value judgment do not occur in a vacuum, which means that teachers must be able to recognize the importance of the sociocultural dimension of decision-making and to take students' sociocultural contexts into consideration in designing relevant pedagogy. Although the influences of students' experiences arising from the particular context of their decision-making are well recognized (e.g., Aikenhead and Jegede 1999; Haidt 2008; Lee 2001; Lynch 2001; Kozoll and Osborne 2004), the influence of contextual and sociocultural factors and sociocultural exchanges on decision-making about SSIs has been under explored until very recently (e.g., Lee and Grace 2012; Lee et al. 2009; Zeidler et al. 2013).

8.3 Studies of Hong Kong Students' Reasoning and Decision-Making About SSIs

In this chapter, two aspects of Hong Kong students' decision-making are examined that are drawn from four independent studies. The first aspect is the way in which young people in Hong Kong address selected SSIs, their decision-making about such issues, and their reasoning behind their decisions. In some of these studies, students' decision-making was also discussed from a comparative perspective, drawing insights from evidence from cross contextual and cross-cultural comparisons that involved mainland Chinese and European students. The second aspect concerns how and to what extent different teaching strategies help to promote Hong Kong students' reasoning and decision-making about SSIs. This aspect also considers the role of sociocultural contexts in shaping students' reasoning on SSIs and how inter-contextual or intercultural exchanges can facilitate multi-perspective reasoning and decision-making, particularly about regional and global SSIs. The SSIs discussed in these four studies were related to health and animal conservation in the local, regional, or global contexts. Researchers have advocated the integration of health and environmental issues with science education to promote scientific literacy in light of the increased importance of these issues (Lee 2012; Roth 2014; Zeidler et al. 2014). All of these studies involved secondary students aged 13–17. In the following sections, the context of Hong Kong is briefly introduced as a background to the studies. The two aspects of decision-making about SSIs are then discussed, drawing on the findings of the four studies. Due to the different contexts of the four case studies, there are limitations in cross case comparison and hence synthesis of potential findings. Despite this, a discussion of the patterns including the unity and variations among Hong Kong students in reasoning and decision-making about SSIs across these cases will be contemplated wherever possible in order to inform future studies. This is followed by an evaluation of the impact of different teaching strategies on their reasoning and decision-making processes and outcomes. To conclude the chapter, some implications of the findings for SSI education are discussed.

8.4 Context of Hong Kong

Hong Kong was a British colony until 1997. After China regained sovereignty of Hong Kong, Hong Kong became a Special Administrative Region (SAR) governed by a constitution set up according to the Basic Law. The Basic Law, the mini-constitution of the Hong Kong SAR, gives high autonomy to the local government, with policy making based largely on consultation and consensus (Chiu et al. 1998). Freedom of speech and of the press is honored. Hong Kong citizens are accustomed to expressing their views on social issues. Opposition to unpopular policies is often strong, fuelled by the increased pursuit by political parties and nongovernment organizations of political, economic, or environmental causes. The development of independent and critical thinking has been emphasized in recent curriculum reforms, culminating in the implementation of a new Liberal Studies curriculum at the senior secondary level that aims to help students to become independent thinkers capable of viewing issues from multiple perspectives and to acquire positive values in an interdisciplinary context (CDC and HKEAA 2007b). In terms of science education, the current trend is to integrate scientific inquiry, STSE connections, and the nature of science into the various science subjects (e.g., CDC and HKEAA 2007a).

8.5 Hong Kong Students' Reasoning and Decision-Making About SSIs

In this section, the findings on Hong Kong students' reasoning and decision-making about four SSIs—the banning of smoking in restaurants, bat conservation, the control of whaling, and the implementation of the central slaughtering of chickens—are discussed in turn.

8.5.1 Issue 1: Banning of Smoking in Restaurants

The Hong Kong Government proposed a bill in 2005 to ban smoking in all restaurants. Although the rationale was to protect the public from the risks of passive smoking, the bill was met with strong opposition from the restaurant industry. As public opinion was divided over the issue, the issue was used to connect 15- to 16-year-old students' understanding of biological concepts with their reasoning and decision-making through a specially designed classroom discourse (Lee 2007). The classroom discourse focused first on the development of the students' conceptual understanding of smoking and passive smoking and the potential risks to bodily health through the teacher's exposition and modeling activities to show how smoke enters the breathing system and deposits tar in the lungs. The risks of smoking were substantiated by five pieces of scientific evidence presented to the students in the form of statistical data or summary statements of empirical findings. The first contained statistics about the death rate due to lung cancer and the quantity of cigarettes smoked daily that were collated by the American Cancer Society (Salber 1968). The second dataset was obtained by the Hong Kong Council on Smoking and Health (Hedley et al. 2001a) on the outcomes of passive smoking on nonsmoking catering workers in Hong Kong in terms of levels of cotinine, a metabolite of nicotine, in their urine. Questions were set by the teacher to guide the students' analysis and interpretation of these two datasets. Examples of the questions are: What are the possible explanations for the patterns shown by the statistics on lung cancer and the quantity of cigarettes smoked daily? How sure are you that smoking is a major cause of lung cancer? Why did nonsmoking workers working in restaurants that allowed smoking have higher cotinine levels than workers in other industries? What would you expect your urine cotinine levels to be if you ate out frequently in restaurants that allow smoking? Does the evidence available justify a ban on smoking in restaurants? (Lee 2007, p. 173).

The third piece of evidence was a survey that clearly showed that more people would choose to dine out if the law were passed (Hedley et al. 2001b). The fourth established, by empirical tests, that it is difficult for cigarette smoke to disperse in a restaurant even when air-conditioning is operating and that the level of ventilation required to significantly reduce the smoke causes discomfort to people (Hong Kong Council on Smoking and Health 2005). The last piece of evidence was a report by government experts that stated that about 150 nonsmoking restaurant workers die from illnesses related to exposure to secondhand smoke annually. The students were asked to consider in small groups the arguments for and against a complete ban on smoking in restaurants before making a final decision on the issue. The students were presented with guiding questions to facilitate their decision-making, such as: What are the arguments for and against the proposed ban? To what extent are the arguments based on scientific evidence? Are you in favor of banning smoking in all restaurants? To what extent is your decision based on scientific evidence or personal values, including morals, personal rights and freedoms, and altruism? (Lee 2007, p. 174).

The findings showed that the students were able to identify the relevant arguments for and against a complete ban on smoking in restaurants. The majority of the students were supportive of the ban. However, data from student interviews showed that some of the students had strong reservations against a complete ban. Some of the counterarguments provided by these students are noteworthy (Lee 2007, p. 175) and can be summarized as follows:

- There is only limited evidence showing that passive smoking is dangerous, as the study on cotinine levels involved only a small number of subjects.
- Most secondhand smoke is dispersed into the environment and thus should not cause serious harm.
- Secondhand smoke is comparatively less harmful than car exhaust pollutants, which are the major source of air pollution in Hong Kong.
- A smoking ban should be postponed until Hong Kong's ailing economy recovers.
- A complete ban on smoking in restaurants may lead to social instability because a large number of smokers are heavily addicted to smoking. There should be a transition period in which all restaurants are allowed to designate a dining area for smokers so that people have time to adapt to the change.

Apart from the first counterargument, the rest are based essentially on common-sense reasoning, lack empirical support, and may even contradict the evidence provided. Furthermore, the advocacy of a transition period in the last argument indicates that these students tended to value social stability more than personal or public health. Together, these factors contributed to their adoption of a more conservative stance on the issue.

8.5.2 Issue 2: Bat Conservation

This study involved a complete class of 31 Secondary Four students aged 15–16. The discourse focused on a local SSI about the intrusion of a colony of bats into a village house. The issue was introduced to the students in the form of a videotape produced by a local nongovernment animal conservation agency. The students were asked to make and record their personal decision on and justification for a solution to the issue. The students were then engaged in a group decision-making activity guided by a framework that involved identifying the stakeholders and their possible views, reading written materials about bats and their ecology, suggesting decision alternatives and discussing their pros and cons, and making final decisions supported by justifications (Fig. 8.1).

The students' initial and final decisions were categorized into five main types of actions (Lee and Grace 2010, p. 161).

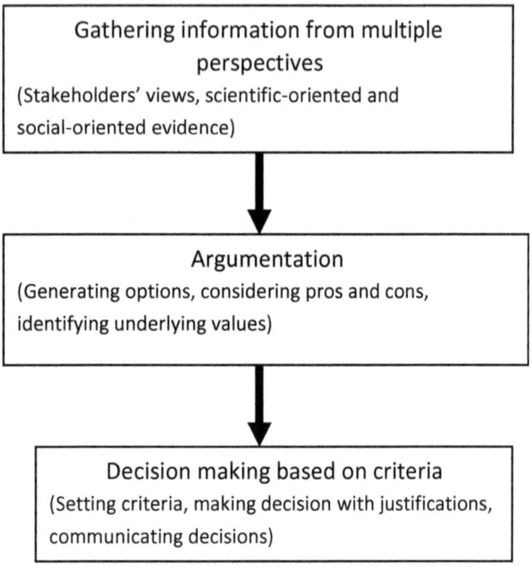

Fig. 8.1 The decision-making framework used in Issue 2 (Lee and Grace 2010, p. 157)

(1) Driving the bats away from the house with no regard for their fate
(2) Leaving experts to decide what to do
(3) Protecting and restoring bat habitats
(4) Allowing the bats to stay
(5) Making use of the bats for specific purposes

Before the group activity, the initial decision that predominated was to force the bats out of the house (42 %), which reflected an emotive type of reasoning (Sadler and Zeidler 2005) and a rather negative attitude toward bats. The second most dominant view was to leave the matter to the discretion of experts (36 %). However, after the activity, 84 % of the students had shifted to a more tolerant attitude toward bats, probably due to their increased understanding of the ecological importance of bats (Lee and Grace 2010). The change in the students' stance on the issue is consistent with another research finding that a better conceptual understanding of bats resulted in less negative attitudes toward these animals (Prokop and Tunnicliffe 2008). Although leaving the matter to the experts was still a common choice after the activity, the students tended to develop their own views based on reasoning from both scientific and social perspectives.

From the justifications provided by the students for their decisions, four types of values were inferred: "no clear value" (e.g., "Let's ask the experts and see whether we should drive the bats away or take other action"), "anthropocentric" (e.g., "The bats disrupt normal living"), "biocentric" (e.g., "Both bats and humans have the right to live") and "eclectic," or a combination of the anthropocentric and biocentric views (e.g., "We can find a new habitat for the bats so that the villagers' lives will not be affected any more"). There was a tendency for the students to shift between the initial decision and the final decision from a purely anthropocentric view to a biocentric view, although to different extents. A fairly large proportion of the students (42 %) adopted an eclectic view to accommodate the needs of both humans and bats. Overall, 77 % of the students changed their values after the activity (Lee and Grace 2010).

8.5.3 Issue 3: Control of Whaling

Unlike the previous two issues, the control of whaling is a global issue that is rather remote to Hong Kong, as whaling is not practiced locally and whales are rarely observed in local waters. The intention in studying Hong Kong students' decision-making on the control of whaling was to understand how they reason through an issue that is of little direct relevance to their everyday life but has an impact on global ecology and biodiversity (Lee et al. 2009). The study was intended to elicit students' reasoning chiefly from the scientific perspective. A cross-cultural dimension was added to reflect how the students' reasoning is compared with that of their counterparts in England and Sweden, two European countries that are culturally different from Hong Kong. The study examined in particular how cultural differences may affect the way in which science students engage in conservation decision-making. The focus was to understand how the students approached the task, what values they drew upon, and how they interacted with each other regarding their decisions and justifications.

A science class of 16- to 17-year-olds in the three locations engaged in a decision-making discussion (lasting about an hour) about whaling. The method shared some similarities with that employed for the second study, except that the final group decisions were videotaped and shown to their counterparts in the other two locations. The students also recorded their personal final decision after the group discussion. Part of the findings is reported in Grace et al. (2015).

The decisions of the students before and after the activity were classified into six categories (Grace et al. 2015).

(1) Whaling should continue for humanitarian reasons: accept whaling if it is a must to fulfill human needs, such as the survival of aboriginal people who rely on whale meat as their main food source.
(2) Whaling should be regulated: impose restrictions on whaling, such as some sort of quota system.
(3) Whaling should continue but only using humane methods: kill whales in a humane way only.
(4) Whaling should continue for commercial gain: it is a normal hunting activity much like fishing.
(5) Whaling should continue for scientific research.
(6) Whaling should be banned: disapproval of any form of whaling under any circumstances.

These categories are not mutually exclusive, except for number 6. Although not many students were completely against whaling, more of the English students took this stance than the Swedish and Hong Kong students before the activity. The regulation of whaling for the protection of whales was accorded a high priority in all three locations. The majority of the students approved of whaling of both types, and most of the students, irrespective of their location, supported whaling for humanitarian reasons but with some sort of regulation or control to conserve whale populations. After the activity, none of the students maintained an absolute anti-whaling stance, and more students opted for categories 1, 3, 4, and 5 as their final decision, suggesting that the activity had moderated their anti-whaling views.

The pro- and anti-whaling justifications provided by the students before and after the activity were categorized. The reasons that the students gave for accepting whaling were divided into the following categories:

(1) Humans' right to survive.
(2) Whales are a legitimate source of food like other animals.
(3) Humane methods can be used.
(4) Respect for indigenous cultures/communities.
(5) The value of scientific research.
(6) Restrictions to maintain sustainable populations.

The students also gave reasons for proposing that whaling was unacceptable to support their anti-whaling stance or conditional approval for whaling. Six main reasons were identified:

(1) Whales might become endangered or extinct.
(2) Aesthetic reasons.
(3) Whaling is inhumane/cruel.
(4) No need for scientific whaling.
(5) Developed countries do not need whale meat.
(6) Whales have the right to survive; it is unethical/immoral to kill whales.

After the activity, more students were sympathetic with indigenous communities who have traditionally relied on whales as their staple food. This indicates that although the students strove to prevent whales from becoming endangered, they tended to accept whaling when human lives and community welfare were at stake. The Hong Kong group seemed to be the most tolerant of whaling and tended to accept whaling under wider circumstances, including whaling for commercial benefit or research purposes. Conversely, more of the English students disapproved of whaling for these purposes than in the other two groups. The Swedish group came somewhere in between the English and the Hong Kong groups in this regard.

Concern about whales becoming endangered was a common justification against whaling in all three locations. This concern became more prominent among the Hong Kong and English students after the activity but became less important to the Swedish group. All three groups, but especially the Swedish students, became more concerned about the cruelty of whaling after the activity. The English students were the only group who justified their decisions against whaling with aesthetic reasons (e.g., whales are magnificent or beautiful), although this was a fairly small proportion of students.

The increased tolerance for whaling after the activity, particularly among the Hong Kong students, suggests that both anthropocentric and biocentric or ethical values guided the reasoning of most of the students, a phenomenon that apparently transcended cultural boundaries. This dualism led students to come to a compromised view in their final decision, whereby they accepted whaling but only with certain caveats. One final point to note is that although the use of scientific principles and concepts such as biodiversity, conservation, and ecosystems were prevalent, especially among the Hong Kong students, scientific data and evidence were relatively underused across the three locations to justify the students' decisions for or against whaling both before and after the activity, even though data such as whale population figures were provided for the students' reference.

8.5.4 Issue 4: Central Slaughtering of Chickens

The central slaughtering of chickens has often been suggested as a protective measure to counter the spread of avian flu caused by the H5N1 virus, which originated in Hong Kong and southeastern China, instead of the existing practice of slaughtering chickens in situ in wet market retail outlets. From the scientific perspective, central slaughtering could protect citizens from avian flu by minimizing avian-human contact (Hong Kong SAR Government 2006). The issue has generated opposition from various stakeholders, including wet market vendors, the transport industry, and Chinese citizens who have traditionally bought freshly killed chicken for consumption and for use as a special offering on ceremonial occasions.

A local Secondary Two class (13- to 14-year-olds) of 40 students participated in the study (Lee and Grace 2012). To understand the influence of contextual variables on the students' reasoning and decision-making, another class of secondary students at the same level in Guangzhou, a major city in southeastern mainland China, was engaged in the study. Guangzhou is similar to Hong Kong in that its population is predominantly Chinese and it is also at risk from bird flu. However, the political system of Guangzhou is different from that of Hong Kong, which has implications for the flow of information and negotiation by the public on SSIs. Moreover, the sociocultural context is also different, as Hong Kong has been influenced by the West to a greater degree than Guangzhou as a result of its colonial status until 1997.

The design of the study essentially followed the decision-making framework employed in the third study, except that the students were given an opportunity to reflect on their groups' final decision after sharing their decisions with the other groups in the same location or the other location through videotaped oral presentations. The design of this revised framework (Fig. 8.2) was based on Svenson's differentiation and consolidation theory (Svenson 1992, 1996). The whole activity

Group discussion (Differentiation)

Identifying stakeholders and their possible views (Step 1)
Identifying and collecting necessary information or evidence to inform arguments (Step 2)
Generating options for resolving the issue (Step 3)
Considering the pros and cons of options; identifying values underlying options (Step 4)
Formulating criteria for evaluating options (Step 5)
Making decisions with justifications (Step 6)

Sharing and reflecting on the group decision (further differentiation or post-decision consolidation)

Sharing among students in the same location (Step 7)
Sharing with students in a different location (Step 8)

Revising decisions

Fig. 8.2 Decision-making framework used in Issue 4, encouraging outcome-relevant involvement

took place over three lessons, each lasting about an hour and a half. The lessons were spread over about 2 weeks, and any change in decision between the pretest and the posttest was presumed to be a result of the prescribed activity.

The findings showed that the Hong Kong students were capable of multi-perspective thinking, as demonstrated by the number of perspectives underlying their justifications before and after the activity (Lee and Grace 2012). These included science and health, economic, sociocultural, consumer choice, practicality, and environmental hygiene perspectives. There was an increasing trend of students justifying their decision from the science/health and environmental perspectives after the activity. This trend contrasts with that displayed by their counterparts in Guangzhou, who attached increased importance to economic, cultural, and consumer perspectives after the activity. A more in-depth analysis of the criteria adopted by the students in making the decision, and the justifications underlying these criteria, showed that both the Hong Kong and Guangzhou students adopted multiple values. However, for the Hong Kong students, concerns over human health were of greater importance, whereas the students in Guangzhou placed a significantly greater emphasis on protecting the interests of consumers (Lee and Grace 2012).

After the two rounds of post-sharing reflection, there was a general tendency for all of the students to become more receptive to alternative viewpoints, resulting in the adoption of a mixed approach to the issue: supporting central slaughtering while allowing the sale of live poultry to a certain degree. During the reflection process, the students from both locations were able to suggest a wide range of possible factors that might affect their views on the issue, including sociocultural, political, values, epidemiological, economic, educational, and practical factors.

8.6 Discussion and Educational Implications

8.6.1 Patterns of Students' Reasoning

It is difficult to generalize from the four studies with respect to Hong Kong students' reasoning and decision-making about SSIs given the heterogeneity of the issues studied. However, certain patterns of students' reasoning emerge from these studies that are worthy of further investigation and have implications for SSI education in Hong Kong and perhaps elsewhere. First, although Hong Kong students' reasoning and decision-making tended to be context dependent, they were generally able to adopt reasonably wide perspectives in considering SSIs, even before going through any teaching intervention, but certain perspectives tended to dominate over others depending on the specific context of the issue. For instance, before the intervention, emotive-intuitive reasoning tended to dominate on the issue of bat intrusion, whereas the scientific/health perspective was emphasized to a greater extent than the other perspectives in the context of personal and public health, as exemplified in the studies on smoking and central slaughtering. In the "bat" case, the students' resorting to emotive-intuitive reasoning was probably a result of the threat they perceived or a natural fear evoked as they encountered wildlife (Hermann and Menzel 2013; Prokop and Tunnicliffe 2010). The economic perspective seemed to be manifested more prominently on the issue of whaling. These emphases suggest that Hong Kong students tend to give a higher priority to people's immediate concerns than ecological well-being or wildlife welfare. Among these concerns, personal health was accorded a higher priority particularly when health is at stake; otherwise concerns such as economic factors assume importance. These patterns reflect an anthropocentric view rather than a biocentric view. It is premature at this stage to hypothesize any cultural factor or value that underlies these views, but it seems reasonable to infer that Hong Kong students emphasize personal interests and economic development more than altruistic values which could be applied to other living beings. This kind of utilitarian value may be attributed to the highly urbanized and densely populated environment in which Hong Kong students reside, which might have distanced them from nature and wildlife. In this kind of environment, personal well-being is particularly vulnerable to contagious diseases as the history of epidemiology has indicated; hence, disease prevention is of utmost importance as publicized by the local government ever since the outbreak of severe acute respiratory syndrome (SARS) in Hong Kong in 2003. Moreover, it comes as no surprise that in a socioeconomic milieu such as Hong Kong, which lacks natural resources but is well known for its economic vibrancy, students would attach greater importance to socioeconomic well-being in evaluating different decision alternatives in SSIs.

Second, there was insufficient emphasis, at least among a certain percentage of Hong Kong students, on the use of scientific data or evidence as a basis to inform their decision-making. This weakness was clearly exhibited by the students who argued against the banning of smoking in restaurants using the commonsense justification that restaurant owners would lose business if smoking in restaurants were prohibited, despite evidence to the contrary (Lee 2007). This lack of focus on scientific data in the reasoning process was also shown in the issue of whaling. Students across cultural contexts seem to habitually engage in causal reasoning by relying on general scientific or ecological principles rather than context-specific data or evidence to evaluate different decision alternatives. These observations reaffirm the research finding that students give little consideration to the nature of science in considering decision alternatives (Bell and Lederman 2003) although science content knowledge was used in a general sense. This finding highlights the need to develop evidence-based reasoning among students, with due consideration of the nature of science, to develop their ability to reason and make decisions about controversial socioscientific issues at a more advanced level (Kolsto 2001).

The third pattern of note is that the students' perspectives appeared to be widened through interactions with peers in group discussions and even further through sharing and post-sharing reflection and discussion. Students' decisions are liable to change after such activities, provided that they are convinced by others' justifications. This seems to contradict previous findings that people tend to stick to their guns once a decision has been made and consolidate their opinion thereafter to minimize regret (e.g., Bäck et al. 2011; Svenson 1996; Shamoun and Svenson 2002; Svenson et al. 2009). The impact of the students' reflection and cross-cultural sharing on their decision-making is exemplified by the following interview excerpts.

> *Students in different regions received different information. This probably led them to different decisions. If we can listen to the opinions of other students we will understand the issue better and hence make better decisions.* (*A Hong Kong student discussing Issue 3*)
>
> *What they* [*the students in Guangzhou*] *value is citizens' interests and preferences, whereas we treasure people's health more.* (*A Hong Kong student on Issue 4*)
>
> *I think this kind of sharing between the two schools is good, because it allows us to understand each other's values.* (*A Hong Kong student on Issue 4*)

8.6.2 Importance of Decision-Making Frameworks

Besides revealing the aforementioned patterns of reasoning, this review of the four studies involving Hong Kong students shows that a decision-making framework that guides students through multi-perspective thinking based on stakeholders' views, argumentation involving the weighing of pros and cons, and the framing criteria for making judgments facilitates students' decision-making. The following extract from the interview transcript vividly demonstrates the positive impact of such a framework:

With these step-wise guidelines, we could not only think about various options, but also delve into the arguments and counter-arguments for each option. In this way, we could make more accurate decisions. If we were asked to draw a mind map, we would probably draw it only according to our initial impression of the issue. Our opinions would not be objective enough. (*A Hong Kong student on Issue 3*)

In addition to a basic decision-making framework to guide students toward informed decision-making, the studies reviewed noted the impact of the social and sociocultural dimension on decision-making about SSIs. This impact is created by interpersonal exchanges and collective decision-making within the same cultural group and by cross contextual or cross-cultural sharing among different cultural groups. These exchanges enhance multi-perspective thinking by exposing students to different value systems based on contextual or sociocultural foundations. The studies on the issues of whaling and central slaughtering clearly demonstrated how students' reasoning on SSIs of regional and global concern can be extended through meta-cognitive reflection on the overt or covert influences of social and cultural contexts on personal or group decision-making.

8.6.3 The Roles of Cross Contextual and Cross-Cultural Exchanges

Cross contextual and cross-cultural exchanges further enhance multi-perspective reasoning and meta-cognitive reflection on more deeply seated sociocultural issues and influence an individual's viewpoint. Teachers should, however, adopt this approach with the understanding that the final decision made may not necessarily be the most appropriate decision. This is demonstrated by the outcomes of the students' discussions on the issues of whaling and central slaughtering. In terms of the former issue, the Hong Kong students tended to adopt a consensual view to balance the interests of all stakeholders, which led to the acceptance of whaling for both economic gain and scientific purposes. In terms of the latter issue, the students adopted an "additive" approach by incorporating different decision alternatives. The following interview excerpt illustrates this practical stance:

We think we can find a balance so that nobody will be disadvantaged. (*A Hong Kong student on Issue 3*)

However, such mixed or composite decisions may not be the most appropriate, especially from a public health viewpoint, nor are they necessarily feasible to implement (Lee and Grace 2012). As such, teachers need to be sensitive to the overall impact of contextual or cultural considerations and, if necessary, draw students' attention to legitimate concerns grounded firmly in scientific evidence and universal values. They should not lose sight of the complexity of global or regional issues and hence the need to develop students' capacity to thinking

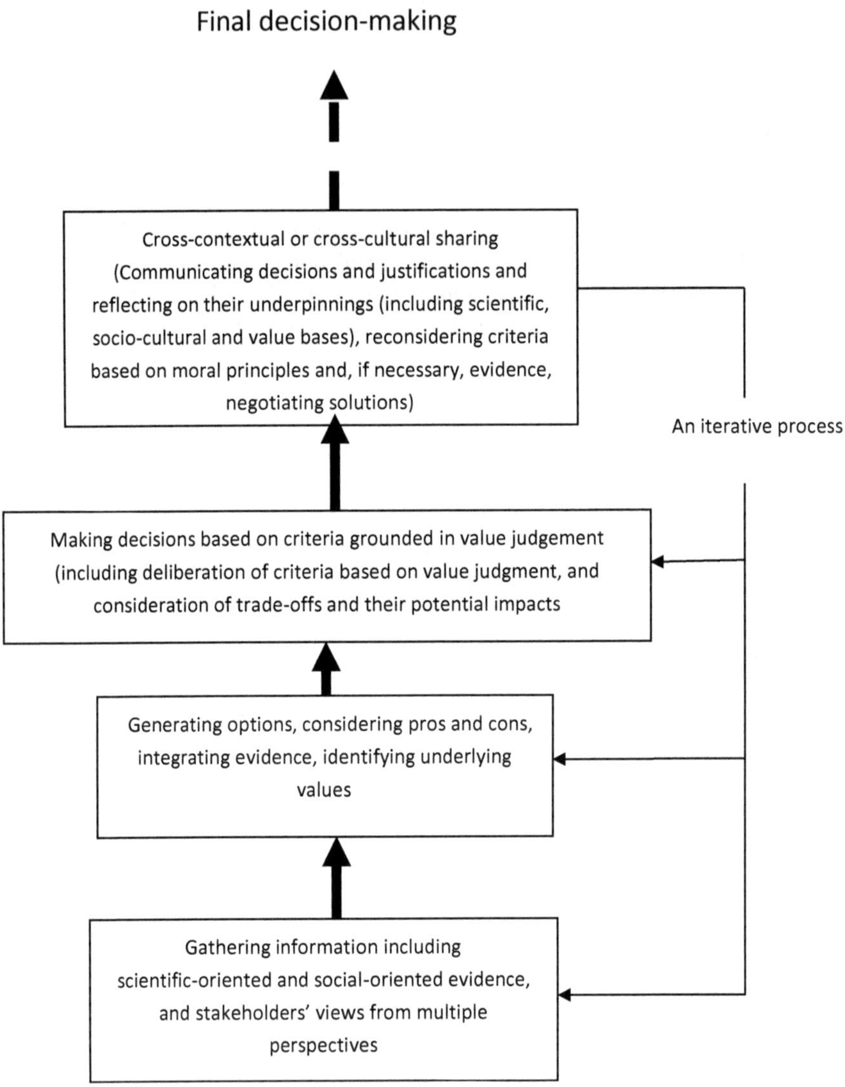

Fig. 8.3 A tentative model for cross contextual or cross-cultural decision-making about socioscientific issues

pluralistically and make decisions democratically in resolving ethical disagreements stemming from cultural differences (Zeidler et al. 2005). To achieve these goals, cross-cultural SSI education models based on the decision-making framework tested in these studies need to be conceptualized to provide teachers with the necessary theoretical and pedagogical frameworks to enhance students' reasoning about SSIs (Lee and Grace 2012). These models should help teachers capitalize on the benefits of cross-cultural sharing while staying alert to indiscriminate compromise or simple eclecticism and emphasizing conflict resolution based on sound principles and values as an effective strategy for problem solving. Figure 8.3 presents a tentative framework that could serve as a starting point for further deliberation in this direction.

Obviously, a major challenge of moving forward is to develop concrete guidelines to help students consider trade-offs among multiple perspectives. These guidelines should help students judge the relative value of different, often conflicting, principles or perspectives. It may entail making necessary sacrifices on the part of certain stakeholders. However, it is difficult to set hard and fast rules for resolving socioscientific issues across the board as many of these issues tend to be unique and dependent on contexts and cultures. However, some kind of framework could be proposed with important guidelines embedded to serve as a starting point for further deliberation by science educators. Figure 8.3 presents a possible framework distilled from the foregone case studies.

This framework outlines several stages through which students may be involved in coming to a final decision about an issue. The first stage is to gather information including stakeholders' views from multiple perspectives, while collecting and evaluating both scientific and nonscientific evidences that support those views. Based on the evaluation of these information and multi-perspective views, options can be generated, with their pros and cons carefully deciphered. It is an imperative to reveal the value that underpins each option so that these options can be subject to value judgment in the subsequent stage. Before making a decision as to which option is preferred, the criteria for making the choice have to be deliberated and decided upon. This entails two steps. The first one is to make judgment by consensus as to which value is more justifiable, and the second is to consider possible impacts of necessary trade-offs. For the first one, students have to address questions such as: What are the core values of society? Which value is overriding? For the second one, they need to ask: What are the impacts of the decision on society? Will the impact of the decision on particular stakeholders be so adverse as to affect their well-being? A case in point is the sole dependence of certain indigenous groups on whales as their food source; hence, a complete ban on whaling will be detrimental to their livelihood.

A useful strategy of leading students to come to reasonable trade-offs is to engage students in "reversing perspectives," a strategy recommended by Johnson and Johnson (1988, p. 58) in their model of developing "critical thinking through structured controversy." This strategy places students in a position opposite to their own stance and requires them to strengthen that position with additional evidence so that they can understand more fully the arguments of the opposing side. This helps to minimize biases and develop a more complete picture of the arguments from multiple perspectives, leading to a more balanced decision. The last stage is to subject the group decision to cross-cultural sharing. Students communicate their decisions and justifications to their peers in other cultures, reconsidering their criteria for decision-making if necessary. This could be followed by further negotiation until a consensus is reached. This stage is no easy task as it may bring cultural variation in values sharply into focus. Space is limited in this paper to consider in detail the nature of values and the theories about value clarification and judgment. Suffice it to say that, as noted by Aspin (2000), values represent generalizations of a population's inclination toward certain objects or situations, which are "settled at the level of the culture of a community" (Aspin 2000, p. 27). Hence, any negotiation of values invoked by socioscientific issues among cultures entails value clarification and judgment at a higher level. Moral principles may have to be drawn upon at this level of discussion. Students could argue or debate over which principles are universalistic and overriding, such as promoting sustainability in a global sense, enhancing the general welfare of mankind and other living beings, and ensuring equity for all cultures. Although the different stages in this framework are presented in a linear fashion, the process is iterative in that students may need to revisit previous stages to collect more evidence and consolidate or revise their views as their perspectives become widened, entailing progressively higher levels of critical thinking.

In sum, this review should be able to illuminate future efforts to promote SSI education and suggests that such efforts should be directed toward understanding and appreciating the nature of science and scientific inquiry rather than applying general scientific concepts to unique contexts in a stereotypical manner. SSI education should also integrate scientific and nonscientific evidence, formulate informed criteria for differentiating decision alternatives, and evaluate these criteria based on sound value judgment grounded in meta-cognitive reflection. To stimulate meta-cognitive reflection among students, alternative pedagogies for cross contextual or cross-cultural sharing need to be developed. This kind of sharing may range from an exchange of views through written communication or videotaped oral presentations to direct dialogue through videoconferencing or face-to-face discussions, as circumstances allow.

With the increase in the number of controversial issues—such as climate change, habitat destruction, new pandemics, and the depletion of fossil fuels—that require concerted regional or international efforts to solve, it makes sense for students from different localities to establish some level of mutual understanding so that in the future they can communicate and collaborate with each other to tackle these issues more effectively and in more sustainable ways.

References

Aikenhead, G. S., & Jegede, O. J. (1999). Cross-cultural science education: A cognitive explanation of a cultural phenomenon. *Journal of Research in Science Teaching, 36*(3), 269–287.

Aspin, D. (2000). A clarification of some key terms in values discussions. In M. Leicester, C. Modgil, & S. Modgil (Eds.), *Moral education and pluralism* (pp. 16–31). London: RoutledgeFalmer.

Bäck, E. A., Esaiasson, P., Gilljam, M., Svenson, O., & Lindholm, T. (2011). Post-decision consolidation in large group decision-making. *Scandinavian Journal of Psychology*. doi:10.1111/j.1467-9450.2011.00878.x.

Bell, R. L., & Lederman, N. G. (2003). Understandings of the nature of science and decision making on science and technology based issues. *Science Education, 87*, 352–377.

Braund, M., Lubben, F., Scholtz, Z., Sadeck, M., & Hodges, M. (2007). Comparing the effect of scientific and socio-scientific argumentation tasks: Lessons from South Africa. *School Science Review, 88*(324), 67–76.

Chiu, C. H., Ting, K. F., Tso, G. K. F., & Cai, H. (1998). A comparison of occupational values between capitalist Hong Kong and socialist Guangzhou. *Economic Development and Cultural Change, 46*(4), 749–770.

Curriculum Development Council, & Hong Kong Examinations and Assessment Authority. (2007a). *Biology: Curriculum and assessment guide (secondary 4–6)*. Hong Kong: Curriculum Development Council.

Curriculum Development Council, & Hong Kong Examinations and Assessment Authority. (2007b). *Liberal studies: Curriculum and assessment guide (secondary 4–6)*. Hong Kong: Curriculum Development Council.

Foong, C.-C., & Daniel, E. G. S. (2013). Students' argumentation skills across two socio-scientific issues in a confucian classroom: Is transfer possible? *International Journal of Science Education, 35*(14), 2331–2355.

Grace. (2002). The science and values that young people draw upon to make decisions about biological conservation issues. *International Journal of Science Education, 24*(11), 1157–1169.

Grace. (2009). Developing high quality decision-making discussions about biological conservation in a normal classroom setting. *International Journal of Science Education, 31*(4), 551–570.

Grace, M., Lee, Y. C., Asshoff, R., & Wallin, A. (2015). Student decision making about a globally familiar socioscientific issue: The value of sharing and comparing views with international counterparts. *International Journal of Science Education, 37*(11), 1855–1874.

Grooms, J., Sampson, V., & Golden, B. (2014). Comparing the effectiveness of verification and inquiry laboratories in supporting undergraduate science students in constructing arguments around socioscientific issues. *International Journal of Science Education, 36*(9), 1412–1433.

Haidt, J. (2008). The emotional dog and its rational tail: A social intuitionist approach to moral judgment. In J. E. Adler & L. J. Rips (Eds.), *Reasoning: Studies of human inference and its foundations* (pp. 1024–1052). Cambridge: Cambridge University Press.

Halverson, K. L., Siegel, M. A., & Freyermuth, S. K. (2009). Lenses for framing decisions: Undergraduates' decision making about stem cell research. *International Journal of Science Education, 31*(9), 1249–1268.

Hedley, A. J., McGhee, S. M., Chan, B., & Lam, T. H. (2001a). *The impact of smoke-free policies on the patronage of restaurants in Hong Kong 2001, Report No. 9*. Hong Kong: Hong Kong Council on Smoking and Health.

Hedley, A. J., McGhee, S. M., Repace, J., Wong, T. W., Yu, M. Y. S., Chan, A. Y. W., Lam, T. H., Lo, P. C. K., Tsang, M., Wong, L. C., Chan, A. L. N., Ng, E. S. L., & Janghorbani, M. (2001b). *Passive smoking and risks for heart disease and cancer in Hong Kong catering workers 2001, Report No. 8*. Hong Kong: Hong Kong Council on Smoking and Health.

Hermann, N., & Menzel, S. (2013). Threat perception and attitudes of adolescents towards re-introduced wild animals: A qualitative study of young learners from affected regions in Germany. *International Journal of Science Education, 35*(18), 3062–3094.

Hogan, K. (2002). Small groups' ecological reasoning while making an environmental management decision. *Journal of Research in Science Teaching, 39*, 341–368.

Hong Kong Council on Smoking and Health. (2005). *Smoking (public health) (amendment) bills committee 2005: Position paper of the Hong Kong Council on Smoking and Health*. Hong Kong: Hong Kong Council on Smoking and Health.

Hong Kong SAR Government. (2006). *LCQ2: Poultry slaughtering plant (press release)*. Hong Kong: Hong Kong Information Services Department.

Jimenez-Aleixandre, M.-P., & Pereiro-Muñoz, C. (2002). Knowledge producers or knowledge consumers? Argumentation and decision making about environmental management. *International Journal of Science Education, 24*(11), 1171–1190.

Johnson, D. W., & Johnson, R. T. (1988). Critical thinking through structured controversy. *Educational Leadership, 45*(8), 58–64.

Keefer, M. (2003). Moral reasoning and case-based approaches to ethical instruction in science. In D. L. Zeidler (Ed.), *The role of moral reasoning on socioscientific issues and discourse in science education* (pp. 241–259). Dordrecht: Kluwer Academic Publishers.

Kolstø, S. D. (2001). To trust or not to trust – pupils' ways of judging information encountered in a socio-scientific issue. *International Journal of Science Education, 23*(9), 877–901.

Kortland, K. (1996). An STS case study about students' decision making on the waste issue. *Science Education, 80*, 673–689.

Kozoll, R. H., & Osborne, M. D. (2004). Finding meaning in science: Lifeworld, identity, and self. *Science Education, 88*, 157–181.

Lee, O. (2001). Preface – culture and language in science education: What do we know and what do we need to know? *Journal of Research in Science Teaching, 38*(5), 499–501.

Lee, Y. C. (2012). Socio-scientific issues in health contexts: Treading a rugged terrain. *International Journal of Science Education, 34*(3–4), 459–483.

Lee, Y. C. (2007). Developing decision-making skills for socio-scientific issues. *Journal of Biological Education, 41*(4), 170–177.

Lee, Y. C., & Grace, M. (2010). Students' reasoning processes in making decisions about an authentic, local socio-scientific issue: Bat conservation. *Journal of Biological Education, 44*(4), 156–165.

Lee, Y. C., & Grace, M. (2012). Students' reasoning and decision making about a socio-scientific issue: A cross-context comparison. *Science Education, 96*(5), 787–807.

Lee, Y. C., Grace, M., Wallin, A., & Byrne, J. (2009). *Cultural aspects of decision making about a controversial socio-scientific issue: Whaling*. Paper presented at the conference of European Science Education Research Association (ESERA), 31 Aug–4 Sept, 2009, Istanbul.

Levinson, R., Kent, P., Pratt, D., Kapadia, R., & Yogui, C. (2011). Developing a pedagogy of risk in socio-scientific issues. *Journal of Biological Education, 45*(3), 136.

Lynch, S. (2001). Conclusion – science for all is not equal to "One size fits all": Linguistic and cultural diversity and science education reform. *Journal of Research in Science Teaching, 38* (5), 622–627.

Means, L. M., & Voss, J. F. (1996). Who reasons well? Two studies of informal reasoning among children of different grade, ability, and knowledge levels. *Cognition and Instruction, 14*(2), 139–178.

Millar, R., & Osborne, J. (Eds.). (1998). *Beyond 2000: Science education for the future*. London: King's College.

National Research Council [NRC]. (2005). *Decision making for the environment: Social and behaviouralbehavioral science research priorities*. Washington, DC: National Academies Press.

National Research Council [NRC]. (2012). *A framework for K-12 science education: Practices, crosscutting concepts, and core ideas*. Washington, DC: The National Academies Press.

Nielsen, J. A. (2013). Delusions about evidence: On why scientific evidence should not be the main concern in socioscientific decision making. *Canadian Journal of Science, Mathematics, and Technology Education, 13*(4), 373–385.

Oulton, C., Dillon, J., & Grace, M. M. (2004). Reconceptualizing the teaching of controversial issues. *International Journal of Science Education, 26*(4), 411–423.

Patronis, T., Potari, D., & Spiliotopoulou, V. (1999). Students' argumentation in decision-making on a socio-scientific issue: Implications for teaching. *International Journal of Science Education, 21*(7), 745–754.

Prokop, P., & Tunnicliffe, S. D. (2008). "Disgusting" animals: Primary school children's attitudes and myths of bats and spiders. *Eurasia Journal of Mathematics, Science and Technology Education, 4*(2), 87–97.

Prokop, P., & Tunnicliffe, S. D. (2010). Effects of having pets at home on children's attitudes toward popular and unpopular animals. *Anthorozoö, 23*(1), 21–35.

Ranyard, R., Crozier, W. R., & Svenson, O. (1997). Introduction to part II. In R. Ranyard et al. (Eds.), *Decision making: Cognitive models and explanations* (pp. 37–39). London: Routledge.

Ratcliffe, M. (1997). Student decision-making about socio-scientific issues within the science curriculum. *International Journal of Science Education, 19*(2), 167–182.
Roth, W.-M. (2014). Personal health – personalized science: A new driver for science education? *International Journal of Science Education, 36*(9), 1434–1456.
Sadler, T. D. (2004). Informal reasoning regarding socioscientific issues: A critical review of research. *Journal of Research in Science Teaching, 41*(5), 513–536.
Sadler, T. D., & Zeidler, D. L. (2004). Student conceptualizations of the nature of science in response to a socioscientific issue. *International Journal of Science Education, 26*(4), 387–409.
Sadler, T. D., & Zeidler, D. L. (2005). The significance of content knowledge for informal reasoning regarding socioscientific issues: Applying genetics knowledge to genetic engineering issues. *Science Education, 89*(1), 71–93.
Salber, E. J. (1968). *Facts about smoking and health.* Chicago: Science Research Associates.
Shamoun, S., & Svenson, O. (2002). Value conflict and post-decision consolidation. *Scandinavian Journal of Psychology, 43*, 325–333.
Simon, S., & Maloney, J. (2007). Activities for promoting small-group discussion and argumentation. *School Science Review, 88*(324), 49–57.
Simon, S., Erduran, S., & Osborne, J. (2006). Learning to teach argumentation: Research and development in the science classroom. *International Journal of Science Education, 28*(2–3), 235–260.
Simonneaux, L. (2001). Role-play or debate to promote students' argumentation and justification on an issue in animal transgenesis. *International Journal of Science Education, 23*(9), 903–927.
Svenson, O. (1992). Differentiation and consolidation theory of human decision making: A frame of reference for the study of pre- and post-decision processes. *Acta Psychologica, 80*, 143–168.
Svenson, O. (1996). Decision making and the search for fundamental psychological regularities: What can be learned from a process perspective? *Organizational Behaviour and Human Decision Processes, 65*, 252–267.
Svenson, O., Salo, I., & Lindholm, T. (2009). Post-decision consolidation and distortion of facts. *Judgment and Decision Making, 4*(5), 397–407.
Tal, R. T., Dori, Y. J., Keiny, S., & Zoller, U. (2001). Assessing conceptual change of teachers involved in STES education and curriculum development; STEMS project approach. *International Journal of Science Education, 23*, 247–262.
Van Eijck, M., & Claxton, N. X. (2008). Rethinking the notion of technology in education: Techno-epistemology as a feature inherent to human praxis. *Science Education, 93*, 218–232.
Wu, Y.-T., & Tsai, C.-C. (2007). High school students' informal reasoning on a socioscientific issue: Qualitative and quantitative analyses. *International Journal of Science Education, 29*(9), 1163–1187.
Zeidler, D. L., & Keefer, M. (2003). The role of moral reasoning and the status of socioscientific issues in science education. In D. L. Zeidler (Ed.), *The role of moral reasoning on socioscientific issues and discourse in science education* (pp. 7–38). Dordrecht: Kluwer Academic Publishers.
Zeidler, D. L., & Sadler, T. D. (2008). Social and ethical issues in science education: A prelude to action. *Science and Education, 17*, 799–803.
Zeidler, D. L., Walker, K. A., Ackett, W. A., & Simmons, M. L. (2002). Tangled up in views: Beliefs in the nature of science and responses to socioscientific dilemmas. *Science Education, 86*, 343–367.
Zeidler, D. L., Sadler, T. D., Simmons, M. L., & Howes, E. V. (2005). Beyond STS: A research-based framework for socioscientific issues education. *Science Education, 89*, 357–377.
Zeidler, D. L., Sadler, T. D., Applebaum, S., & Callahan, B. E. (2009). Advancing reflective judgementjudgment through socioscientific issues. *Journal of Research in Science Teaching, 46*(1), 74–101.
Zeidler, D. L., Herman, B. C., Ruzek, M., Linder, A., & Lin, S.-S. (2013). Cross-cultural epistemological orientations to socioscientific issues. *Journal of Research in Science Teaching, 50*(3), 251–283.

Zeidler, D. L., Zeyer, A., & Dillon, J. (2014). Science/environment/health – Towards a reconceptualization of three critical and inter-linked areas of education. *International Journal of Science Education, 36*(9), 1409–1411.

Zohar, A., & Nemet, F. (2002). Fostering students' knowledge and argumentation skills through dilemmas in human genetics. *Journal of Research in Science Teaching, 39*(1), 35–62.

Part IV
Assessment

Gavin W. Fulmer

Editor's Introduction: Part IV

The chapters in this section focus on assessment for science education in mainland China. As Lee (2001) noted some 15 years ago, there is a developing trend toward educational exchange across East and West: over time the West is moving toward tightening of educational policy more like the East's, while the East is moving to a loosening of educational policy. Assessment is, in particular, one of the areas where this bilateral exchange is evident. This shift has continued and is particularly visible in assessment policy, as Western countries move more toward test-based accountability and national standards or curriculum (Herman and Dietel 2005; Mattei 2012) while some Asian countries are moving to looser and more broad-spectrum assessment systems (e.g., South Korea Ministry of Education 2011). China, in particular among East Asian societies, has a famously long history of uniform, national examination systems that began over 1,000 years ago during the Western Zhou dynasty, and that continued and expanded into subsequent dynasties (see Han and Yang 2001, for a thorough review). The nature of this examination-oriented education changed from the early 1900s, but examinations remained central to the education system into modern times, such as examination-based promotion from primary to secondary education. However, since the 1990s, there have been a series of shifts to reduce the centrality of examinations in the system and to "reduce stress among pupils and the demoralization associated with failure" (Han and Yang 2001, p. 9). Even so, such wide-scale changes are slow-going, and reduction of attention to national examinations is accompanied by shifts to examinations at the provincial level.

In recognition of the historical importance and ongoing change, the chapters in this section provide a view of the current status of assessment in science education within mainland China. The authors provide a broad range of views on assessment, from the level of national and provincial policy down to the level of individual classroom teachers. Reading across the chapters, readers will also see a flow from the more system-focused to a teacher-focused set of studies. Looking across the chapters for key connections, readers will also note the importance of understanding the relationship between assessments and mandated learning outcomes—an important element to consider across contexts, but particularly so in the Chinese education system, which relies on a high-stakes system (Kennedy 2007).

As can be seen, when taken together, the chapters show notable similarities with educational assessment research in the West. There are commonalities in the importance of understanding alignment (e.g., Liu and Fulmer 2008; Porter 2002) and in the study of teachers' beliefs and views regarding assessment (e.g., Brown et al. 2011a). In addition, the chapters show how China's science education system has attended to the international discussion of assessment reconceptualized to emphasize formative in addition to summative purposes of assessment (e.g., Black and Wiliam 1998; Taras 2005), as well as efforts to implement classroom assessment practices not only *of* learning but also *for* learning (e.g., Black and Wiliam 2009).

Yet, there are also striking differences. For example, the assessment-related chapters presented here do not focus explicitly on the students' knowledge and understanding. This belies its importance within the field, as evidenced by papers in a recent special issue of the *Journal of Research in Science Teaching*. Papers explored key topics such as how well large-scale standardized tests capture non-majority students' knowledge or understanding of science concepts (Noble et al. 2012), and the general importance of specifying a cognitive basis for an assessment instrument and an assessment tool (Opfer et al. 2012). This emphasis at the student level is also reflected in work in the *Second International Handbook of Science Education*, which included chapters on assessing affective outcomes in students (e.g., Kerr and Murphy 2012; Tytler and Osborne 2012). As another example, an important general area of research is on the relationships between assessments and instruction, such as studies on the value of instructional sensitivity in assessment tools (Ruiz-Primo et al. 2012). More broadly, the international field is also debating the problem of over-assessment as a result of the deluge of assessments *for* and *of* learning, to the point that scholars bemoan the over-assessed state of *as* learning (e.g., Torrance 2007).

While the chapters in this section are only a limited sample of all science education assessment research in mainland China, it does show a focus more at the system and the teacher levels, whereas current research in the West also pays significant attention to the student level. Thus, there seems to be great opportunity for collaborative research among Western and Chinese scholars using this student-centered perspective on cognitive foundations of assessment and the validity and interpretation of large-scale assessments.

Chapter Introductions

Chapter 9 provides key information on the current state of assessment policies within China, including both the assessment reforms and current policies and programs for teacher education with respect to assessment. A particularly interesting point in this chapter is how the reforms for assessment have increased attention to the role of formative assessment. That is, that assessment reforms have encouraged a view of assessment as a method for student development, not only for ranking and placement. Much as Lee (2001) noted, this indicates a loosening of educational assessment policy. It also reflects other shifts in East Asian education systems toward educational goals that emphasize innovation, creativity, and twenty-first-century skills (e.g., Singapore Ministry of Education 2010).

Chapter 10 addresses the alignment of standardized tests with its associated curriculum materials, examining the extent to which the tests emphasize similar subject matter knowledge or cognitive demands for students. Chen and colleagues address the ways that standardized tests, particularly ones that have potentially high stakes for students and teachers, reflect the intended educational outcomes they are purported to measure. In that way it has broad connections to the rich literature on educational alignment outside of China (Fulmer 2011; Liu and Fulmer 2008; Porter 2002). As with previous work on alignment within educational settings (Liu et al. 2009), this chapter identifies areas where there is misalignment between the test and a curriculum—with the test emphasizing higher-level cognitive skills (according to a revised Bloom's taxonomy) than the curriculum indicates. Much like the previous research that the chapter extends, the findings described in Chap. 10 underscore the need to monitor alignment of tests and standards/curriculum, because such alignment is itself a part of the validity argument for such testing systems (Beck 2007). In addition, in situations of notable misalignment, teachers and students will often focus on the structure and format of the test itself while ignoring the intended curriculum materials, especially under high-stakes conditions. Therefore, depending on the nature of the educational assessment, the test itself can serve as an impetus for high-quality instruction—if tests emphasize higher-order cognitive demands such as analyzing, evaluating, and creating, as shown in previous study in China and other East Asian countries (e.g., Liu et al. 2009)—but such a demanding examination system can also bring along stress and anxiety for students (e.g., Han and Yang 2001).

Chapter 11 examines teachers' conceptions of assessment using a mixed method study, providing insight into how teachers in China view assessment and how this may relate to policy and to their instruction. This topic is one of considerable interest internationally, with research in many countries and educational jurisdictions including Australia (Brown et al. 2011a), New Zealand (Brown 2004; Fletcher et al. 2012), Singapore (Fulmer et al. 2013; Koh and Luke 2009; Tan 2012), the UK (James and Pedder 2006; Winterbottom et al. 2008), and the USA (Adams and Hsu 1998; Aydeniz and Southerland 2012). While there has been previous research on Chinese teachers' conceptions of assessment, focusing on Hong Kong and the

Shenzhen area of China (Brown et al. 2009, 2011b), Chap. 11 goes well beyond this research by combining qualitative and quantitative approaches. Like Tan's (2012) phenomenographic study in Singapore, Chap. 11 emphasizes the complex and occasionally conflicting relationships among conceptions that teachers hold and demonstrates the great importance of context in how teachers consider their conceptions and what this can mean for their instruction. This particular chapter also connects this general line of work on teachers' conceptions of assessment to *science* teachers' conceptions of assessment, an area with infrequent explicit attention in the literature (for exceptions, see Aydeniz and Southerland 2012; Cowie 2012). Hopefully, it will provide a promising stimulus for future work both within mainland China and internationally on science teachers' conceptions of assessment.

References

Adams, T. L., & Hsu, J.-W. Y. (1998). Classroom assessment: Teachers' conceptions and practices in mathematics. *School Science and Mathematics, 98*(4), 174–180. doi:10.1111/j.1949-8594.1998.tb17413.x.

Aydeniz, M., & Southerland, S. A. (2012). A national survey of middle and high school science teachers' responses to standardized testing: Is science being devalued in schools? *Journal of Science Teacher Education, 23*(3), 233–257.

Beck, M. D. (2007). Review and other views: "Alignment" as a psychometric issue. *Applied Measurement in Education, 20*(1), 127–135. doi:10.1207/s15324818ame2001_7.

Black, P., & Wiliam, D. (1998). Assessment and classroom learning. *Assessment in Education, 5*, 7–74.

Black, P., & Wiliam, D. (2009). Developing the theory of formative assessment. *Educational Assessment, Evaluation & Accountability, 21*, 5–31. doi:10.1007/s11092-008-9068-5.

Brown, G. T. L. (2004). Teachers' conceptions of assessment: Implications for policy and professional development. *Assessment in Education: Principles, Policy & Practice, 11*(3), 301–318. doi:10.1080/0969594042000304609.

Brown, G. T. L., Kennedy, K. J., Fok, P. K., Chan, J. K. S., & Yu, W. M. (2009). Assessment for student improvement: Understanding Hong Kong teachers' conceptions and practices of assessment. *Assessment in Education: Principles, Policy & Practice, 16*(3), 347–363. doi:10.1080/09695940903319737.

Brown, G. T. L., Hui, S. K. F., Yu, F. W. M., & Kennedy, K. J. (2011a). Teachers' conceptions of assessment in Chinese contexts: A tripartite model of accountability, improvement, and irrelevance. *International Journal of Educational Research, 50*(5/6), 307–320. doi:10.1016/j.ijer.2011.10.003.

Brown, G. T. L., Lake, R., & Matters, G. (2011b). Queensland teachers' conceptions of assessment: The impact of policy priorities on teacher attitudes. *Teaching & Teacher Education, 27*(1), 210–220. doi:10.1016/j.tate.2010.08.003.

Cowie, B. (2012). Focusing on the classroom: Assessment for learning. In B. J. Fraser, K. G. Tobin, & C. J. McRobbie (Eds.), *Second international handbook of science education* (pp. 679–690). Dordrecht: Springer.

Fletcher, R., Meyer, L., Anderson, H., Johnston, P., & Rees, M. (2012). Faculty and students conceptions of assessment in higher education. *Higher Education, 64*(1), 119–133. doi:10.1007/s10734-011-9484-1.

Fulmer, G. W. (2011). Estimating critical values for strength of alignment among curriculum, assessments, and instruction. *Journal of Educational and Behavioral Statistics, 36*(3), 381–402.

Fulmer, G. W., Lee, C. H. I., & Tan, K. H. K. (2013). *Integrating research on contextual factors and teachers' views and practices of assessment*. Paper presented at the annual meeting of AEA-Europe, Paris, France.

Han, M., & Yang, X. (2001). Educational assessment in China: Lessons from history and future prospects. *Assessment in Education: Principles, Policy & Practice, 8*(1), 5–10. doi:10.1080/09695940120033216.

Herman, J. L., & Dietel, R. (2005). A primer on accountability. *American School Board Journal, 192*(12), 26–28.

James, M., & Pedder, D. (2006). Beyond method: Assessment and learning practices and values. *The Curriculum Journal, 17*(2), 109–138. doi:10.1080/09585170600792712.

Kennedy, K. J. (2007). *Barriers to innovative school practice: A socio-cultural framework for understanding assessment practices in Asia*. Paper presented at the Redesigning Pedagogy: Culture, Understanding and Practice Conference, Singapore.

Kerr, K., & Murphy, C. (2012). Children's attitudes to primary science. In B. J. Fraser, K. G. Tobin, & C. J. McRobbie (Eds.), *Second international handbook of science education* (pp. 627–650). Dordrecht: Springer.

Koh, K. H., & Luke, A. (2009). Authentic and conventional assessment in Singapore schools: An empirical study of teacher assignments and student work. *Assessment in Education: Principles, Policy & Practice, 16*(3), 291–318.

Lee, J. (2001). School reform initiatives as balancing acts: Policy variation and educational convergence among Japan, Korea, England and the United States. *Education Policy Analysis Archives, 9*(13), 1–9.

Liu, X., & Fulmer, G. W. (2008). Alignment between the science curriculum and assessment in selected NY State regents exams. *Journal of Science Education & Technology, 17*(4), 373–383. doi:10.1007/s10956-008-9107-5.

Liu, X., Zhang, B., Liang, L. L., Fulmer, G. W., Kim, B., & Yuan, H. (2009). Alignment between the physics content standard and the standardized test: A comparison among the United States-New York State, Singapore, and China-Jiangsu. *Science Education, 93*(5), 777–797.

Mattei, P. (2012). Market accountability in schools: Policy reforms in England, Germany, France and Italy. *Oxford Review of Education, 38*(3), 247–266. doi:10.1080/03054985.2012.689694.

Noble, T., Suarez, C., Rosebery, A., O'Connor, M. C., Warren, B., & Hudicourt Barnes, J. (2012). "I never thought of it as freezing": How students answer questions on large-scale science tests and what they know about science. *Journal of Research in Science Teaching, 49*(6), 778–803. doi:10.1002/tea.21026.

Opfer, J. E., Nehm, R. H., & Ha, M. (2012). Cognitive foundations for science assessment design: Knowing what students know about evolution. *Journal of Research in Science Teaching, 49*(6), 744–777. doi:10.1002/tea.21028.

Porter, A. C. (2002). Measuring the content of instruction: Uses in research and practice. *Educational Researcher, 31*(7), 3–14.

Ruiz-Primo, M. A., Li, M., Wills, K., Giamellaro, M., Lan, M.-C., Mason, H., & Sands, D. (2012). Developing and evaluating instructionally sensitive assessments in science. *Journal of Research in Science Teaching, 49*(6), 691–712. doi:10.1002/tea.21030.

Singapore Ministry of Education. (2010). *MOE to enhance learning of 21st century competencies and strengthen art, music and physical education*. Singapore: Singapore Ministry of Education.

South Korea Ministry of Education. (2011). *Curriculum innovation for elementary and secondary schools based on practicing teachers*. Seoul: National Curriculum Planning Division.

Tan, K. H. K. (2012). Variation in teachers' conceptions of alternative assessment in Singapore primary schools. *Educational Research for Policy & Practice, 12*, 21–41. doi:10.1007/s10671-012-9130-4.

Taras, M. (2005). Assessment – Summative and formative – Some theoretical reflections. *British Journal of Educational Studies, 53*, 466–478.

Torrance, H. (2007). Assessment as learning? How the use of explicit learning objectives, assessment criteria and feedback in post-secondary education and training can come to dominate learning. *Assessment in Education, 14*, 281–294.

Tytler, R., & Osborne, J. (2012). Student attitudes and aspirations towards science. In B. J. Fraser, K. G. Tobin, & C. J. McRobbie (Eds.), *Second international handbook of science education* (pp. 597–626). Dordrecht: Springer.

Winterbottom, M., Brindley, S., Taber, K. S., Fisher, L. G., Finney, J., & Riga, F. (2008). Conceptions of assessment: Trainee teachers' practice and values. *The Curriculum Journal, 19*(3), 193–213. doi:10.1080/09585170802357504.

Chapter 9
Assessing Science Learning in Schools: Current Policy and Practices

Cuidian Feng and Lingbiao Gao

9.1 Context of School Science Assessment

China is known as the "kingdom of examinations." Various tests and exams are set for students during their school time. With the trend of "exam-oriented education" getting more powerful, only pen-and-paper tests were used in all kinds and all levels of exams in science after the early 1990s. These tests and exams used to be achievement-oriented, elitist, and bureaucratic (Gao 1998). The power of examination also made it "a baton conducting teachers, students, and the teaching-learning process" which led to an exam-oriented style of teaching and learning in all schools throughout China. Teaching and learning focused sharply on drilling students with exam techniques in order to get higher marks (Gao and Watkins 2002). In this way, assessment became an obstacle for improving the quality of learning and teaching. It stood on the opposite side of research findings in education and worldwide trends of science education and curriculum reforms in the past decades (Gao 2002).

To keep pace with the world, China moved to reform its national curriculum for basic education, as well as the school assessment system, and changed its orientation toward personal quality development. According to the *Guidelines for Curriculum Reform in Basic Education* (*Trial*) (hereafter referred to as *Guidelines* in this chapter) published by the Chinese Ministry of Education (MOE) in 2001, the aims of this reform pursue to change the curriculum:

C. Feng (✉)
School of Teachers Education, Taizhou University, Linhai, Zhejiang 317000, China
e-mail: fengcuidian@163.com

L. Gao
South China Normal University, 55 Zhongshan Dadao, Tianhe, Guangzhou 510631, China

L.L. Liang et al. (eds.), *Chinese Science Education in the 21st Century: Policy, Practice, and Research*, Contemporary Trends and Issues in Science Education 45, DOI 10.1007/978-94-017-9864-8_9

- From a knowledge-delivery center to fostering students' all-around development, including knowledge, processing skills and methods, emotions, attitudes, and values
- From a subject-based and decollated structure to a well-balanced, comprehensive, and properly flexible structure that able to meet the needs of all students
- From course contents that were overloaded and overly difficult to ones that have appropriate amounts of material and degree of difficulty and which can enhance the relations between course contents and students' lives as well as keep pace with progress in science and technology
- From too much emphasis on rote learning to higher emphasis on meaningful learning
- From testing for ranking and selecting students to assessing mainly for facilitating students' learning and development
- From an overly centralized and unified system to a more diverse and flexible system which allows schools and local governments to share the responsibility in curriculum development and management (MOE 2001a)

New policies and techniques of assessment were adopted in order to:

> build up a new assessment system aimed at facilitating students' all-around development. It will not only assess students' learning achievement, but also discover and develop students' potential in a variety of ways, identify their needs in progress, and help them to develop their self-understanding and self-confidence. Assessment needs to play its roles in educating students and facilitating their development. (MOE 2001a, Guideline No.14)

A new wave of assessment reform has spread over China since the release of the above *Guidelines* in 2001. In 2002, the Ministry of Education released another document named *Circular of the Ministry of Education on Promoting Reforms on School Evaluation and Examination System* (hereafter referred to as *Circular*) to elaborate further the aims and philosophy of the assessment reform:

(a) The reform should follow the education policy of the Chinese Communist Party, which aims at assessing and facilitating students' all-around development morally, intellectually, physically, and aesthetically.
(b) School evaluation system, including both student assessment and teacher evaluation, is not only for administration purposes, but more importantly, for the development of students, teachers, and schools. The most important purpose of this reform is to change traditional school evaluation into a facilitative process for the development of students, teachers, and schools in terms of quality of learning, teaching, and educating.
(c) The contents of assessment should be diverse in order to cover all aspects of student development rather than subject knowledge only. Great attention needs to be paid to keeping a good balance between the uniform curriculum standards and personal difference in students' capacity and personality.
(d) Improvements on assessment techniques and instruments are encouraged in order that pen-and-paper tests are not the only tools used in assessment. Summative and formative tests, quantitative and qualitative techniques, and

intrinsic and extrinsic assessments all should be included in the new system. Intrinsic assessments and self-assessments are especially encouraged. Students and teachers are no longer treated only as objects to be evaluated but also as subjects to assess their own progress.

(e) Attention should be paid not only to the results of assessment but also to changes and progress that have occurred in the process of assessment.

(f) The importance of the roles of students, teachers, and schools in the process of evaluation should be noted. It is expected that assessment will become a process of interaction among students, teachers, schools, education officers, and administration sections as well as parents (MOE 2002).

The *Circular* also drafted a framework of the objectives of the new student assessment system. Assessment objectives were classified into two groups: (1) "general development objectives" in six major domains of student development as shown in Table 9.1 and (2) "subject-based learning objectives" in three dimensions of learning outcomes, which were described in a series of national curriculum standards for all school subjects.

The *Circular* also mandated a portfolio named "records of progress" that to be established for every student to promote ongoing assessment focusing on the process of learning. The portfolio is supposed to collect qualitative information, i.e., the students' self-records of their learning and schooling, peer evaluation results, the students' best work, students' performance in community service, awards received by students in any competition, teachers' observations and comments, comments from parents, etc. It should give an all-around description and deep understanding of the students' development in the process of learning. Furthermore, it will promote the students' reflection about their learning process and encourage students to play a more active role in self-assessment.

At the end of each term and school year, a qualitative assessment focused on the "general literacy and capacity" of students, as shown in Table 9.1, is conducted. Reports including students' achievement test scores at the end of the term or year, their self-reports on learning, peer evaluations, and teachers' comments are put together and sent to students' parents to give an all-around assessment on the progress of students. Final reports of a student's "general literacy and capacity" at the end of junior or senior secondary stages also play a role in the acceptance of students into senior high schools or universities. In the junior secondary stage, general scores are given to students based on the results of assessment of their "general literacy and capacity." This score is 10 % of the final score for senior secondary school entrance. Another 90 % of the score comes from the results of public examinations. In the senior secondary stage, information provided by the report on "general literacy and capacity" becomes one of the basic requirements for college acceptance. Tests and examinations are still the major approach for summative assessments at midterm (in the senior secondary stage, the end of a module), at the end of term, and at the end of school year. A rating scale such as "excellent, good, or satisfactory" or "pass or fail" is recommended to grade students' achievement instead of the popular percentage marking scale in primary schools. Ranking

Table 9.1 The general development objectives of students in six major domains

Domains	Indicators
Ethics and morality	Love our motherland, our people, and our socialist system. Be fond of work. Respect for laws. Be honest. Be public spirited and concerned with the community and environment
Civic literacy	Have self-confidence, self-esteem, self-reliance, self-discipline, and diligence. Be responsible for personal behavior and the society. Be active in activities for public welfare
Learning ability	Have aspiration and interest in learning. Be able to learn in a variety of ways and learn at higher level. Be accustomed to thinking reflectively on the process and outcome of one's own learning. Be able to analyze and solve problems by applying the knowledge, skills, and experience learned in different areas. Develop abilities of inquiry and creativity
Capacity of communication and cooperation	Be able to set and complete a task with others. Be respectful and receptive to the views and conditions of others. Be able to evaluate and regulate one's own behavior. Be cooperative and able to communicate and interact with others in a variety of ways
Participation in physical activities and conditions of health	Love to participate in sports. Regularly take part in physical exercises. Practice and be skillful in sports. Keep the body strong and healthy and live in a healthy way
Awareness of beauty	Be able to enjoy and appreciate the beauty of life, nature, art, music, and science. Have a good and positive aesthetic taste. Be active in a variety of art and music activities, and be able to perform in variety of ways

Based on the Circular (MOE 2002)

students according to their scores on the tests is not suggested in primary school stage. The primary graduation examination and the junior secondary school entrance examination are eliminated. Primary graduates are distributed randomly to nearby junior secondary schools in their communities. At the end of the 9-year compulsory education, the junior secondary graduation examination and the senior secondary entrance examination are merged into one regional public examination. The university matriculation examinations are decentralized and become public examinations at the provincial level, though they are still named as the national university matriculation examinations.

In sum, the MOE *Circular* drew a blueprint of the expected assessment system, which aims at facilitating the all-around development of students (and so was named "developmental assessment system"). This new system is characterized by (a) a diversity of its acting personnel and (b) a diversity of assessing methods and techniques. It seems to agree with the concepts of "assessment for learning" (Black and Wiliam 1998) and "assessment as learning" (Eral 2003) that have acknowledged increased attention and importance internationally. In the past decades, it has been argued that assessment has to move from "assessment of learning" to

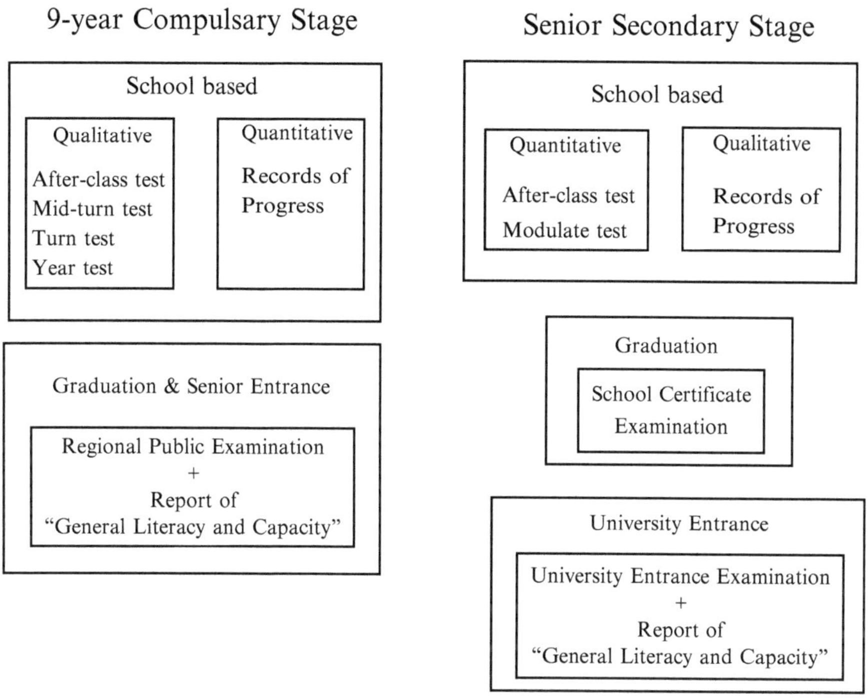

Fig. 9.1 School assessment system in China

"assessment for learning," where assessment procedures and practices are developed to support learning rather than undermine learning (Gipps 1999; Shepard 2000; Stiggins 2002). In addition, student-involved assessment, i.e., assessment as learning, has received more emphasis and recognition as a core component where assessment can support learning, (Eral 2006; Noonan and Duncan 2005; Torrance 2007). Figure 9.1 gives a brief summary of the new school assessment system in China today (Gao 2013).

9.2 Distinctive Features of the Current Assessment Reform

As mentioned above, China has reconstructed its national curriculum and released a reform to change its assessment system accordingly. In order to have a deep understanding of the new system of learning assessment, it is necessary to understand how this reform is distinct from the former assessment system. These distinctions also served as indicators for the review and inspection of the success of the reform.

Firstly, the aims of assessing student learning are different. Traditionally, student assessment in China is aimed at inspecting and evaluating student learning in terms of students' performance on summative pen-and-paper tests. The new policy advocates that student assessment will inform student learning and development by identifying their progress and diagnosing their problems in the process of learning. Not only are the progress and problems in cognitive field learning assessed but also those related to student learning and development in all-around aspects.

Secondly, the focus on the contents of assessment is different. Traditional assessments focus only on knowledge, especially textbook knowledge. Under the new policy, objectives of assessment cover three dimensions, as suggested in the *Circular*: (a) knowledge, (b) processing skills and methods, and (c) emotion, attitudes, and value (MOE 2002). This suggests that not only will scientific knowledge be assessed, but students' understanding of scientific inquiry; inquiry skills and methods; capacity for applying scientific knowledge, skills, and methods in scientific inquiry and everyday problem solving; and understanding of the underlying philosophy of scientific inquiry will all be included in learning assessment. In addition, students' emotions about, attitudes toward, and value of science are also parts of assessment content. This enlarges the scope and strengthens the role of learning assessment. Gao (2004) recognized this as the highlight of the assessment reform and the most significant distinction between the new and the traditional assessment systems.

Details of the contents of learning assessment are further defined in the national curriculum standards developed by MOE. They are subject dependent and varied with school stages. At the primary level, the main areas of assessment in science include (MOE 2001b):

(a) Children's knowledge and understanding of the basic scientific concepts in the living world, the physical world, the earth, and the space, as defined in the curriculum standards
(b) Children's skills and abilities in "doing" scientific activities
(c) Children's awareness and understanding of scientific inquiry and learning science
(d) Children's interest in and attitudes toward science and learning science
(e) Children's affection for science and the enterprise of science

At the junior secondary stage, the subject courses, physics, chemistry, biology, and geography, are separate in the entire country except for Zhejiang province, which has an integrated science course. At the senior secondary stage, all subject courses are separate all over the country. The details of knowledge contents of learning assessment are different from subject to subject. However, the focus areas of learning assessment are similar for all subjects and at both the junior and senior stages. According to the junior secondary school curriculum standards of science, physics, chemistry, biology, and geography (MOE 2001c, d, e, f, g) and the senior secondary school curriculum standards of physics, chemistry, biology, and geography (MOE 2003a, b, c, d), learning assessment focuses of on the following areas:

(a) Students' understanding of the scientific inquiry process and its key steps: challenging and questioning, assuming, planning, experimenting and collecting evidence, analyzing and concluding, reporting and communicating, examining, and reviewing; the capacity for completing learning activities relevant to one or several of the above steps
(b) Students' understanding of the scientific knowledge defined by the curriculum standards, especially the big ideas that underlie the scientific concepts
(c) Students' skills in learning and doing science, such as skills in observing, measuring, experimenting, operating instruments and tools, and data collecting and analyzing
(d) Students' ability to apply scientific knowledge and solve problems
(e) Students' interests in and attitudes to science and learning
(f) Students' understanding about the nature of science and scientific enterprise
(g) Students' understanding and views on STS issues

Thirdly, with the changes in the focus areas in assessment, the strategies of assessment need to change accordingly. The new policy encourages a variety of means of assessment rather than relying on pen-and-paper tests only. At the primary school stage, it strongly emphasizes that strategies of learning assessment in science should (MOE 2001b):

(a) Develop carefully to meet the aims of the new curriculum
(b) Focus sharply on the indicators defined by the curriculum standards
(c) Handle the level of difficulty properly to meet the level of student learning
(d) Develop carefully and apply properly the new techniques and instruments for assessment

It should be noted that a diversity of qualitative techniques are introduced in the primary science curriculum standards. These include classroom observation, interviews, after-class learning tasks, DIY (do-it-yourself) activities, project learning activities, student portfolios of progress, and performance tests. Self-assessment and peer assessment of students are encouraged. It is suggested that teachers should pay attention to their reflective language and behaviors while communicating or interacting with students since these are specific ways of assessment that happen subconsciously. Furthermore, summative pen-and-paper tests are not compulsory for all students; local educational authorities can decide independently whether or not to conduct summative tests at the end of a period of learning. This means that quantitative and summative examination is no longer the major approach to learning assessment in primary schools. Thus, qualitative and process-oriented assessment techniques become the mainstream of learning assessment strategies in primary school science.

At secondary school stage, there is little difference in assessment strategies adopted by different subjects and at different periods. According to the national secondary school science curriculum standards documents (MOE 2001a, b, c, d, e, f, g, h), the strategies of learning assessment in secondary science will:

(a) Serve to diagnose and improve student learning.

(b) Involve multiple personnel in the assessment. Not only teachers, external experts, and administrators but also the parents and students could all act as assessors.
(c) Limit the scope of knowledge in assessment and control of the level of difficulty properly.
(d) Develop and apply a variety of flexible techniques and instruments in learning assessment.

It can be seen from the national curriculum standards that summative pen-and-paper tests are still emphasized in the secondary school stage. However, all the curriculum standards call for a change in designing test items to shift the focus from memorizing knowledge to understanding the scientific concepts and the underlying big scientific ideas. The use of open-ended items is encouraged in summative tests, although they might cause difficulty in marking. In addition, the importance of internal assessment is recognized due to the overemphasis of external assessments in past decades. This is different from the policy at the primary stage where summative pen-and-paper tests are not encouraged.

The use of performance tests is encouraged in the secondary school stage. Operational skills in observation and experiments, including skills in handling tools and instruments, identifying and focusing on objects, collecting data and evidence, and properly recording the observed information and data measured, have become the major content of performance tests. Students' self-assessment and peer assessment are also encouraged in performance tests.

The record of progress is introduced as one of the most important instruments in learning assessment for improving students' learning autonomy. It acts as not only an instrument of assessment but also a platform for interaction and communication among students, teachers, and parents and a good way for students to review their own learning. The strategy applied here is similar to that in the primary stage. However, the techniques employed in different stages must be different to fit the level of maturity and development of students.

Qualitative techniques, such as observations and interviews, as already described at primary stage, are also introduced in order to make a change in assessment.

9.3 Internal Science Assessment in Practice

As described in Fig. 9.1, internal assessment in science includes processing assessments in and after class and the summative assessments after a certain period of learning, i.e., after a module, half-term, a term, or a school year. Primary graduation tests are still in practice in most primary schools in China. According to the *Circular*, these tests ought to be school-based. However, in reality, most of the primary graduation tests are run by district/county educational authorities to compare the quality of schools (Gao 2013), so these tests are only for school accountability purposes rather than for student accountability. Thus, such graduation tests and examinations are excluded here and will be addressed in Sect. 9.4.

9.3.1 Portfolio Assessment in Practice

Portfolio assessment was first introduced by the Ministry of Education in the *Circular* (MOE 2002). Every student in China has to work with a "record of progress" when schooling. It was considered a form of authentic assessment in China. Authentic assessments offer multiple indicators of student progress and encourage students to take an active role in their own learning and to demonstrate what they know in ways that encompass their personal learning styles. It increases students' ownership of course content, provides a first step in researching, and offers more opportunities for writing, discussion, and the use of technology. Independent learning and creative problem solving are also encouraged. A move toward more authentic application tasks and outcomes thus improves learning and teaching, and students have greater clarity about their obligations. Teachers can thus come to believe that assessment results are both meaningful and useful for improving instruction (Wiggins 1993).

As a popular instrument, the portfolio requires students and teachers to document students' growth and changes by selecting evidence from their teaching and learning practices. This will help students to become more self-regulated and gain personal control and independence in their learning. They will be able to use a wide variety of learning styles to demonstrate their learning. They will also be able to develop a greater understanding of their particular learning style when they self-evaluate and reflect on the evidence they have selected for inclusion in the portfolio to demonstrate competence.

Neither the central government nor the local educational authorities give a clear definition of the construction and contents of this "record of progress." As a result, a diversity of "portfolios" have been developed by teachers and educators in the past decade. A review of the portfolio used in practice shows that there are three main kinds of portfolios (Zhao 2012):

1. *Demonstrative portfolios*. These are self-selected collections of students' best work chosen in order to demonstrate their achievements to teachers, parents, and fellow students. Students also explain the criteria and reasons for the selections. Teachers, parents, and other students give their comments on both the works demonstrated and the criteria for selection. This type of portfolio could work as a medium for student-teacher-parent communication. It could also be a good instrument to help students to learn how to assess themselves by setting assessment criteria and selecting their best work.
2. *Descriptive portfolios*. This is a systematic collection of a variety of records in the process of students' learning. It includes the teachers' evaluations, observations, and comments on students' performance in the learning process, the

students' achievement in tests or examinations, and the students' own works and any other records the students or their teachers regard as significant to collect. Both the students and their teachers have the right to add things into the portfolio. This portfolio is very important for students (or their teachers) to review their learning processes, identify their progress, and diagnose their problems.

3. *Summative portfolios*. This is a collection of a student's work selected by teachers and educators to report the achievement of that student to parents and the society. Preset standards are necessary and need standardization to maintain fairness for all students. The content of this type of progress record normally covers the six major domains of student development as shown in Table 9.1.

Since most of the portfolios are not subject-based, they facilitate students' all-around development, including their learning in science. In the National Forum on Assessment of the General Literacy of High School Students run in Xian in October 2010, teachers from schools in different provinces summed up the advantages of portfolio assessment:

1. Records of progress are a very rich resource of information about students. They give detailed descriptions of the all-around aspects of students in their process of maturity and development.
2. Records of progress can act as media promoting interactions between teachers, students, and their parents and can therefore benefit students' development.
3. Records of progress help teachers to learn the individual differences among students and the features of each student. Teachers can then give each student more appropriate guidance according to the characteristics shown in the portfolio.
4. Records of progress provide platforms for students to submit the work that they themselves are satisfied with and fond of. This enables students to see their own progress and experience the joy of success and progress and, in turn, improves students' attitudes toward and emotion about learning.
5. Records of progress highlight the active role of students in the process of assessment. They provide chances for students to review their own learning.
6. Records of progress collect a vast amount of qualitative and quantitative evidence of student development for summative assessment. This provides an effective way to integrate summative and formative assessments and to integrate teaching, learning, and assessment.

Many teachers enjoy their experience in portfolio assessment process and believe that the records of progress facilitate student progress in the right direction. Mr. Li Weifeng, a high school physics teacher at the Dongzhimen Middle School in Beijing, said[1]:

---

[1] Extracted and translated from Li Weifeng: *A speech on the National Forum on Assessment of the General Literacy of High School Students*. Unpublished manuscripts of the forum, Xian, China, October 2010.

> ... Student portfolios record the details of students' progress in the past three years. ... I was very impressed and excited as I read their portfolios page by page. I saw that my students grew up day after day. They are happy to learn and to enquire. They are enthusiastic and hardworking. They are highly concerned with our society. They are kind, honest, wise, and intelligent. They are the future of China. ...
>
> ... I am very surprised by and proud of my students when I review their progress. I am sure that they would not be able to progress so well if we focused only on their knowledge learning and achievement and if we are concerned only about their performance on the public examinations.
>
> What makes them progress so well? ...... Records-of-progress do not focus narrowly on subject learning and exam marks. Rather, the teachers' comments, the comments from their peers, the experience of thinking reflectively, and self-evaluation: all of these lead them to progress actively in the right direction.

Similar to teachers, many students enjoy their experience in portfolio assessment process and view the portfolio as a record of life. A student in the First Middle School of Yinchuan, the capital city of Ningxia Autonomous Area in Western China, wrote down her feeling on the record of progress:

> When I reviewed my experience in the past three years, I found this small thing [the portfolio] had changed me. I became more self-confident and more focused on learning. Furthermore, I saw a new me. ...It is a condensation of my life. Every detail in the portfolio is like a drop of water reflecting my life and it is so beautiful. I am very impressed by my own portfolio. I am so proud and happy with my experience.[2]

Some of the parents were also happy with their kids' experience, which is reflected actively in the portfolio. A parent wrote to her daughter in the portfolio:

> Half a semester has passed since you entered senior high school. We are very happy to see that you are progressing with joy. Meanwhile, we are impressed with your collection [in the portfolio]. Remember, what you have collected is not only a resume; further, it is the record of your progress and growth. It will enrich your life. Your learning is excellent in that you learn actively with self-consciousness. We hope that you will keep on learning in this way. This will help you progress well. Go ahead. Your future is bright.[3]

However, this is only one side of the picture. On the other side, problems remain unsolved in portfolio assessment. Firstly, conducting portfolio assessment is a heavy workload for teachers. This is partially due to the big class size in China. A typical class in primary schools has 40–50 students and in secondary schools 50–60. If a teacher spends 10 min per week to read and give feedback to one student, 500–600 min needs to be spent on portfolio assessment per week. This is a very heavy workload for the teacher. Many teachers are enthusiastic in the beginning but then begin to feel tired of doing portfolio assessment. Secondly, due to the very strong impact of public examinations, quite a large number of teachers still consider exam marks to be the most important result of schooling. They view

---

[2] Extracted and translated from the portfolio of Miss LYX, a portfolio shown at the National Forum on Assessment of the General Literacy of High School Students, Xian, China, October 2010.

[3] Extracted and translated from Zhao Xueqin: *A speech at the National Forum on Assessment of the General Literacy of High School Students*. Unpublished manuscripts of the forum, Xian, China, October 2010.

portfolios as less significant and not worth spending a lot of energy on. Similar problems trouble the students also. Many students, especially those with lower motivation to learn, feel that it is too tiresome to do portfolio assessment. They feel bored in collecting their own works and are not happy to show them to others, especially to parents. How to encourage teachers and students to be actively involved in portfolio assessment becomes a big problem.

These problems trouble educators and teachers and stand in the way of implementing portfolio assessment in schools. In fact, only a small number of teachers and schools, mainly in the primary stage, are involved actively in using records of progress. In most schools, especially at the secondary stage, portfolio assessment becomes a mere formality.

9.3.2 Performance Assessment in Practice

One popular type of performance assessment in China refers to a practical learning task, such as a science experiment, an inquiry learning activity, or a real or analogous problem-solving task. Through observing, recording, and analyzing the examinee's behaviors, teachers can assess not only operational skills, cognitive level, and thinking skills but also the desire and capacity for learning, cooperation, and communication. Students can also assess themselves and their peers. The history of performance assessment in China is rather long; however, due to the very strong influence of high-stakes public examinations which use only pen-and-paper tests, performance assessment has been neglected for many years. The *Circular* reemphasizes the importance and value of performance assessment, so it has become popular again, especially in primary and junior secondary stages.

Summarized from the Chinese literature in recent years, a typical performance test includes three key stages:

(a) *Preparation*. This stage focuses on setting a task with its learning objectives and criteria for assessment. The task and the corresponding objectives and criteria might be assigned by the curriculum or a textbook. In other cases, the task is assigned by the teacher, and the teacher sets the objectives and criteria for learning and assessing beforehand.
(b) *Assessment*. This stage includes the student's self-assessment, peer assessment, and teacher's assessment. The students' self-assessment and peer assessment focus on reviewing and rethinking their approaches and behaviors to find out what they achieved, the mistakes they made and the problems they met in the process, and their feelings and ideas about the activity, learning, and science. The teacher's assessment focuses on students' performance by observation and communication. As an example, Xu and Yan (2008) describe the details of assessing students' performance in an inquiry activity in chemistry. In order to help students to conduct self-assessment, the teacher negotiated with them and then finally constructed a rubric consisting of the criteria shown in Table 9.2.

Table 9.2 A rubric for student self-assessment in an inquiry activity

Title of the task:		
Name:	Gender:	Class:
Points to review		Your description
What is your general feeling about this experiment?		
Did you have your own assumptions before you started to conduct the experiment? If yes, did you make a plan to improve your assumption?		
Did you have problems in the process of experimenting? What are the problems? How did you solve the problems?		
How did you work as a team in the experiment?		
Did you find you needed to do anything to improve your cooperation?		
Did you stop during the process of experimenting? Why did you stop?		
Did you know how to use the instruments in the experiment? Did you operate them properly?		
Did you ask questions to the teacher when you were not sure about something during the experiment? What were the questions?		
Can you give a comment on this activity? Give reasons to support your comment		
If you need to rank your performance as excellent, good, fair, pass, or fail, what is your rank?		

Translated from Xu and Yan (2008)

Students were invited to write down their feelings and experience with the help of the rubric. Here, a qualitative approach was adopted.

Another rubric was designed for student's peer assessment, which is shown in Table 9.3.

Teachers are still the most important people in assessment since they can assess students' performance more objectively than outside people can. They prepare the criteria of assessment and share it with all students involved in the activity. A rubric of this assessment has been shown in Table 9.4. Students' capacity and skillfulness in designing and conducting an experiment are the focuses of this rubric. It evaluates process skills and results of the activity, identifies the student's capacity, and diagnoses problems the student has in the activity.

(c) *Marking and reporting*. Sometimes, it is necessary to rank or to mark the students' performance. Normally, all the results of students' self-assessments, peer assessments, and teacher assessments will contribute to the final result. However, a descriptive report is always more important than the rank/mark. The indicators shown in Table 9.4 also give an outline describing students' performance.

Most of the teachers accepted the ideas of performance assessment. However, due to the lack of tools and technical assistance, many of them do not know how to conduct performance assessment effectively in practice. This encouraged the

Table 9.3 A rubric for student peer assessment in an inquiry activity

Title of the task:		
Name:	Gender:	Class:
Points to review		Your comment
Was the student actively involved in searching for information relating to the activity?		
Did the student make his/her own assumption before the activity?		
What was the information the student collected? Where did he/she collect that information?		
What kind of work did the student complete?		
Did the student contribute any ideas or suggestions?		
Did the student help other group members when they had problems? If yes, what were the problems?		
Did the student cooperate actively with other group members to fulfill any task? If yes, what was the task?		
Did the student listen to others?		
Can the student express his/her ideas neat and clear while communicating with others?		
What is the result/conclusion of that activity?		
Please give your rank to this activity from one of the five following: excellent, good, fair, pass, and fail		

Translated from Xu and Yan (2008)

Chinese educators and teachers to develop a variety of techniques for performance assessment in recent years, including worksheets and rubrics for science inquiry activities, rubrics for observations, peer assessments and self-assessments, and team competitions.

Table 9.5 presents an example of a marking scale used to assess primary third grade pupils' skills in handling a thermometer (Ling 2009). In this activity, students are required to heat the cold water in a beaker and use a thermometer to measure the temperature of the water while heating. The group leaders are appointed by the teacher and are marked by the teacher beforehand. They also learn marking skills and then act as assessors to mark other members of the group. The group leaders rotate so that each member will have an opportunity to act as a group leader.

A worksheet can also be used as a tool to assess students' performance in this activity. Many researchers encourage the application of worksheets and have developed a large number of worksheets for different activities (Luo 2006; Zhao and Pan 2010; Cai 2012). Table 9.6 gives an example of a worksheet for an experiment on the relations among three variables: electric current, voltage, and resistance (Luo 2006). Students conduct the experiment following the guide in the worksheet step by step, writing down their answers to the questions presented on the worksheet. They report on the process and results of their performance in the activity, and then the teacher can assess the student on their knowledge and experimentation skills based on their written responses and the results of their experiment.

Table 9.4 A rubric for teacher assessment in an inquiry activity

Title of the task:					
Name:	Gender:	Class:			
	Qualitative description			Grade	
Indicators	Excellent	Good	Satisfactory	Pass	Fail
1. Was the student able to put forward proper questions as the focus of the experiment?					
2. Was the student able to put forward hypotheses for the experiment?					
3. Was the student able to provide a reasonable rationale to support the hypothesis?					
4. Was the plan for the experiment reasonably designed?					
5. Were the instruments for the experiment selected properly?					
6. Were the students able to set up the instruments quickly and correctly?					
7. Were the rules properly followed in the operation of the balance in weighing chemical reagents?					
8. Did the student take care to save the chemical reagents?					
9. Was the experiment table clean?					
10. Were the instruments clean?					
11. Were the results of the experiment correct?					
12. Was the way of analyzing reasonable?					
13. How was the attitude of the student in cooperating with others?					
14. How well did the student express her/himself orally?					
15. How actively was the student engaged in the experiment?					
16. Your overall evaluation of the activity					

Translated from Xu and Yan (2008)

Some assessment rubrics refer to scientific inquiry activities (project learning) generally. Jian et al. (2005) gave an example of this kind of rubric as shown in Table 9.7.

After 10 years of practice, many researchers and teachers agree that performance assessment can:

(a) Collect more and deeper information about students' learning
(b) Help understand students' science literacy in all-around aspects
(c) Vary with the level of students and their learning environment
(d) Improve the validity of student assessment
(e) Make science experiments and activities more attractive
(f) Promote students' active and creative involvement in learning

Table 9.5 A scale for student peer assessment of skills in using a thermometer. Grade 3, class, group, assessor (group leader)

Elements	[a]GM1	GM2	GM3	GM4
1. Grasp the upper end of thermometer and put it into the water. The lower end of the thermometer should not touch the bottom or the wall of the beaker				
2. Keep your eyes at the same level as the liquid surface inside the thermometer. The thermometer should not leave the water while reading the temperature				
3. Read and record the temperature of hot water correctly				
4. Read and record the temperature of cold water correctly				

Translated from Ling (2009)
[a]*GM* Group member

Table 9.6 Sample worksheet for a physics experiment

Exploring the relations among electric current, voltage, and resistance
I. The following materials are on your table: three fixed-value resistors, one variable resistor, one ammeter, one voltmeter, two batteries, one switch, and several pieces of wire. Please check these items. Please raise your hand if there is anything missing
II. The question for you to explore, what are the relations among the following variables: electric current, power, voltage, and resistance?
III. Please perform the following tasks:
1. Make a hypothesis about the relations among electric current, voltage, and resistance based on your knowledge and experience
2. Design an experiment to test your hypothesis (complete the electric circuit on the picture of the components. Explain your procedure
3. Conduct the experiment based on your plan. Record the data in the table you have preset
4. Analyze the data and draw your conclusions from the data

Translated from Luo (2006)

(g) Facilitate students' development of their inquiry abilities (Wei 2007)

Similar to portfolio assessment, performance assessment also has shortcomings. The major problems are as follows:

(a) It is time consuming to conduct performance assessment. A good way to solve this problem is to invite students to play the role of assessor. When the assessment focuses on some simple objectives, it is easier for students to handle the criterion of assessment. However, it is not easy for students to handle complex tasks.
(b) Most of the rubrics in the literature are too complicated. The criteria for assessment are not well defined. This increases the difficulty of implementing performance assessments. In addition, the rubrics developed by different people are different even when they focus on the same issue. On one hand, this enables the assessment to fit closely to a certain group of students and their level of

Table 9.7 A rubric for assessing scientific enquiry activities

Elements	Indicators	Marking scale			
		D (fail)	C (accept)	B (good)	A (excellent)
Observing and questioning	Student is able to ask question based on context	Unable to ask question	Ask vague question	Ask surface question	Ask valid scientific question
Predicting and hypothesizing	Student is able to make prediction and hypothesis	Unable to make prediction	Predict the results but incorrectly	Predict the results nearly correctly	Predict the results correctly
Planning and designing experiment	Student is able to plan and design experiment accordingly	Unable to design the experiment	Design a part of the experiment	Design the experiment but not feasible	Design a feasible experiment
Experimenting and collecting data	Student is able to implement the experiment and collect the necessary data	Unable to conduct the experiment	Conduct the experiment with errors and mistakes	Operate correctly but cannot record the data	Operate correctly and record the data exactly
Analyzing and concluding	Student is able to analyze the data and draw conclusions	Unable to analyze the data	Analyze a part of data without conclusions	Analyze data correctly and draw superficial conclusions	Analyze the data and draw in-depth conclusions
Arguing and evaluating results	Student is able to make argument and evaluate the process and results	Cannot evaluate the results	Compare results with the expectation without further consideration	Compare results with the expectation and discover/raise new questions	Compare results with the expectation, discover/raise new questions based on comparison, and attempt to answer the new question to some extent
Discussing and applying	Student is able to 1. Discuss actively with his/her own opinion 2. Write report with convincingness	Achieve none of the 1, 2, 3, 4, 5	Achieve one of the 1, 2, 3, 4, 5	Achieve two of the 1, 2, 3, 4, 5	Achieve three or more than three of the 1, 2, 3, 4, 5

(continued)

Table 9.7 (continued)

Elements	Indicators	Marking scale			
		D (fail)	C (accept)	B (good)	A (excellent)
	3. Question others' report scientifically				
	4. Cooperate well with others				
	5. Work creatively				

Translated from Jian et al. (2005)

learning, but on the other hand, different groups who using different assessments cannot be compared. For this reason, applying performance assessments is limited to larger-scale public examinations.

9.3.3 Summative Assessment in Practice

The situation of summative assessment in primary schools is different from that in secondary schools. In the primary stage, since the summative test in science is not compulsory, the situations vary from place to place. For example, in Guangdong province, no summative test or another type of assessment is set for primary school science. Teachers must rank their students at the end of each term and school year. However, since there are only two ranks, pass or fail, set for reporting students' achievement in their term/annual report, almost all students can get a pass except those absent from the science classes. This means that summative assessment does not in fact exist in primary schools in Guangdong. Zhejiang province is an example a different situation. In Zhejiang, most city bureaus give a pen-and-paper test at the end of each term. Since the pen-and-paper test can only test some of the surface knowledge of science at the primary stage, the final score of a student will consist of three parts: (a) students' records of their pen-and-paper test at the end of the school term and year, (b) students' capacity for processing skills and their understanding of scientific methods, and (c) students' emotions about and attitudes toward learning and inquiry. The last two parts are based on the formative/process assessment. Table 9.8 shows a sample marking scheme in a primary school (Ling 2009).

In both the junior and senior secondary stages, formal summative tests are administered at the end of each term and year. End-of-module tests and midterm tests are also administered in every school. Pen-and-paper tests are still the most popular method of summative assessment. Changes have mostly occurred in the techniques for developing test items. Increasing numbers of open-ended items testing students' abilities of comprehension and application are being developed and used in the pen-and-paper tests. Multiple-choice items that only can test trivial

Table 9.8 A summative marking scheme in primary science

	Student code:		Class:	
Dimensions	Grade	Comments	Assessment	Score
Science concepts			End-of-unit test	
			End-of-semester test	
Process and methods			Experimentation skills	
			Checklist of science learning	
			Best student work sample	
Emotion, attitudes, and values			Teacher observation	
			Peer and self-assessment	
			Monthly group-wide competition	
Final score				

Note:
1. Grading scale: a score between 90 and 100 will be given an A (excellent)
A score between 75 and 89 will be given a B (good)
A score between 60 and 74 will be given a C (acceptable)
A score below 60 will be given a D (needs improvement)
2. The weight of the two scores in scientific concepts is 50:50; the weight of the three scores in process and methods is 20:40:40; the weight of the three scores in emotion, attitude, and value is 40:30:30; and the weight of the three-dimensional scores in the final score is 40:30:30

knowledge are declining in tests and exams. In some schools, performance tests are also included in the summative tests. The tasks of performance tests could be (a) an experiment, (b) an academic paper, (c) a DIY product, (d) an oral presentation or demonstration, etc. (Gao 2011). A final score considering both the results of summative tests and process assessment is given to students to show their achievement in learning. Table 9.9 gives an example of the composition of a final score in senior physics (Zhang 2005).

From Table 9.9, one can see that summative assessment includes both pen-and-paper tests and performance tests. There are two kinds of pen-and-paper tests: one is the traditional closed-book test and the other is an open-book test. In the open-book test, students may be asked to write an academic paper on an issue relating to the knowledge learned, draw a concept map related to the concepts learned, present an argument on a STS issue, etc. Most of the performance tests ask students to conduct an experiment they have done in the term. Most of the unit tests in the process assessment use a pen-and-paper test to assess students' knowledge. Qualitative techniques are used in other parts of process assessment. These include students' self and peer assessments and teacher's assessment. It should be noted that attendance is also a factor in the assessment. Students must attend the course and accumulate enough class hours in order to get credit. Otherwise, they cannot get credit in that course even if they get a very high mark on the final test.

It should also be noted that not all the secondary schools assess student learning in the way shown in Table 9.7. In fact, many schools continue to use closed-book pen-and-paper tests as the only method of assessing students' learning in science, which is nearly the same as the situation before the new curriculum reform.

Table 9.9 A term report of student achievement in physics

	Elements and contents		Results of assessment				
Process assessment	Unit tests	Unit 1	Mark		Rank		Comments
		Unit 2	Mark		Rank		Comments
		Unit 3	Mark		Rank		Comments
		Unit 4	Mark		Rank		Comments
	Attitude and method	Learning attitude		Excellent	Good	Acceptable	Needs to improve
		Learning habit					
		Learning method					
	Cooperation and communication	Respect for others		Excellent	Good	Acceptable	Needs to improve
		Easy to get along with others					
		Express correctly					
	Peer assessment						
				Signature of group leader			Date
	Self-assessment						
					Signature of student		Date
	Teacher assessment						
		Final rank (pass/fail)			Signature of teacher		Date
Summative assessment	Pen-and-paper tests: Close	Mark		Rank		Comments	
	Pen-and-paper tests: Open	Mark		Rank		Comments	
	Performance test	Mark		Rank		Comments	
Total # of hours required for attendance					Total # of hours actually attended		
			Signature of teacher				Date

9.4 External Examinations in Science

As shown in Fig. 9.1, no official public examination is mandated in science at the primary stage. At the end of the junior secondary stage, a public examination is required as the graduation exam as well as the senior secondary school entrance exam. This exam is now a criterion-referenced exam named the junior secondary school certificate examination (JSSCE). There are two public examinations for senior secondary students: the high school certificate examination (HSCE), held in grades 11 and 12, and the university matriculation examinations (UMEs). The UME is the most important and high-stakes examination in China.

9.4.1 Features and Issues Related to Junior and Senior Secondary School Certificate Examinations (JSSCEs and HSCEs)

Science is one of the subjects of the JSSCE. The exam paper for science consists of three parts, physics, chemistry, and biology, which relate to the three separate science subjects in junior secondary schools. According to the Ministry of Education, JSSCE should use a variety of methods to assess students' levels and capacities comprehensively and include the results of processing assessment. A five-point or seven-point rating scale has been introduced as a substitution for the traditional percentage scale (MOE 2008). Since the JSSCE is organized by city or regional bureau of education, the overall situation in the country is too complicated to give a detailed description. However, a brief review on the JSSCE web page[4] shows that almost all senior high schools admit students solely based on their exam marks. In science, pen-and-paper tests are still the most important method of the JSSCE,[5] and the percentage score is still the only scale for marking students' performances. This suggests that in the past decade, there has been no significant change in external science exams in the junior secondary stage. The design of exam items, however, now seems to be more ability-oriented (Liao and Yuan 2010).

At senior secondary stage, the HSCE is run at the provincial level. Wu (2012) reviewed the HSCE in 19 provinces and found that science is one of the testing subjects in all the HSCE. In some provinces, science appears as a single testing subject, and in some provinces, it appears as three separate subjects. In some provinces, performance tests that focus on experimentation skills are now included in the exams (see Table 9.10). However, she found that the HSCEs are immature and underdeveloped in terms of their purposes, schemes, objectives, methods, and interpretations. She identified two major problems in the HSCE:

1. The nature of the HSCE is unclear. It should be a criterion-referenced test, but the existing form is a mixture of criterion- and norm-referenced examinations. The role of the HSCE is to confirm that the student meets the curriculum standards upon graduation. However, the current HSCE focuses partly on selection and overlaps with the national university matriculation examinations.

[4] http://gz.zhongkao.com/

[5] A few cities and provinces include listening and oral exams in the examinations for foreign languages. A few cities and provinces adopt open-book exams as a method in exam of political science.

Table 9.10 Methods used by different provinces in different subjects of HSCE

Subject	Student[a]	Method[b]	Provinces
Physics, chemistry, biology	All	Ext. test	5 provinces
	S & L		2 provinces
	All	Ext. test + experiment. 1	6 provinces
	S & L		1 province
	All	Ext. test + exp. 2	13 provinces
	S & L		1 province
	All	Ext. test + exp. 3	1 province

Notes:
Exp. 1 = external test of experiment, run by the educational authority
Exp. 2 = internal test of experiment, run by the school
Exp. 3 = internal test of experiment, run by the school + daily performance
[a]All = all students, S & L = students majoring in social science or liberal arts
[b]Ext. test = external pen-and-paper test

2. The criteria/standards of the HSCE tests are missing in the schemes of almost all provinces or are at least not open to the public.

These two serious problems might cause the HSCE to become invalid, unreliable, and unnecessary. Wu (2012) suggests that it is very urgent that the Chinese education authorities, at both national and provincial levels, clearly define the nature and roles of the HSCE in order to develop and publish the criteria/standards for the HSCE.

9.4.2 *Features and Issues Related to University Matriculation Examinations (UMEs)*

There have been a few changes in the UME. First, the UME has been decentralized. Only one UME scheme, developed by the National Education Examination Authority (NEEA), was used in the country before the national curriculum reform. The test was also developed solely by the NEEA and was the same everywhere in China. This was unfair, since the levels of social and economic development and the quality of education are different from one region to another. With the implementation of the new national curriculum, the administration of the UME has been moved down to the provincial level, and UME exam schemes are proposed by provincial educational examination authorities; the tests are developed by a group of university experts appointed by provincial education examination authorities.

Second, test items have been improved. Traditional pen-and-paper tests can assess only students' knowledge memory and lower-level thinking skills. A large number of experts and teachers have criticized this. However, the pen-and-paper test is still the only method used for the science portion of the UME. This has caused more people involved in these studies to focus on improving the quality of pen-and-paper test items in order to test students' concepts understanding and higher-level thinking skills. Currently, a variety of new items simulating the real scientific research situations and problems have been created. Students are requested to design a scheme for an experiment for inquiry or to solve a set questions. A deep understanding of knowledge, higher thinking, and problem-solving skills are needed to answer these items. This makes the UME highly challenging to most students.

There are other small changes too. However, these improvements have not significantly changed the basis of the UME.

9.5 Summary and Discussion

School science assessment in China has been characterized as examination-dominated, achievement-oriented, and selective in nature. It has been criticized as an obstacle for the cultivation of students' scientific literacy. When the MOE launched the innovation to develop a curriculum that is more student-centered, ability-focused, and aims at developing the all-around person for schools in China, assessment reform became urgent. It started soon after the national curriculum reform at the beginning of the new century. However, there are big gaps between the policy and practice.

On the policy level, a review of the government documents shows that the MOE has tried to build a new assessment system that aims at facilitating the all-around development of students. In this new system, the acting personnel are diverse: not only teachers and external experts can act as assessors, but also students can do so. The focuses of learning assessment extend from scientific knowledge to the all-around development of scientific literacy in three dimensions: scientific knowledge and skills, ability of scientific inquiry and creativity, and emotions about and attitudes toward science and the value of science. Self-assessment and peer assessment are encouraged. A variety of ideas, methods, and techniques have been introduced. The orientation of these changes agrees with the concepts of "assessment for learning" and "assessment as learning."

On the practical level, the ideas of process assessment, the introduction of the records of progress, and other qualitative techniques have become the most distinctive changes in science learning assessment, especially in primary stage when the wash-back effects of the high-stakes public examinations are not so strong. In addition, a variety of performance test techniques have been developed to be used in both formative and summative assessment. These new ideas, methods, and techniques have extended the scope of students, teachers, and, especially, parents on students' all-around development. They also combine students' learning and assessment into one process. Student self-evaluation and peer evaluation techniques are being widely used in the process of learning, which encourage students to review and learn from their own experiences of learning.

In secondary schools, especially in senior secondary schools, formative assessment and records of progress have not been as popular as expected. Summative tests are still the most prevalent mode of student assessment. The introduction and application of performance tests seem to be the most important change. This might be due to the fact that secondary school students have to face three public examinations, the JSSCE, HSCE, and UME. Only a few changes have occurred in the past decade in those public examinations, and those changes have concentrated on improving the quality of test items. Different kinds of new test items have been developed and tried in order to test not only students' knowledge memory and surface understanding but also their high-order thinking skills and problem-solving abilities. These have made the public examinations increasingly difficult and have also seemed to push students to work harder and focus more specifically only on the academic aspects of development. This is the opposite of curriculum reform's desired effect of cultivating students in all-around aspects. It also creates powerful wash-back effects which pull the students and teachers back to the traditional way of learning.

Why are the practical situations of assessment reform different from the original expectation? Gao (2013) reviewed the challenges in student assessment reform in the past decade:

> Firstly, the policy is top-down and the interpretations of this policy from different levels are confusing. For example, different experts and administration agencies at different levels interpret the concept "developmental assessment system" differently. Secondly, the qualitative assessment techniques are young and not well developed. More importantly, public examinations in China are still one of the most important ways to achieve social fairness and flexibility. All these strengthen the traditional routine of assessment. So in practice, the implementation of the assessment reform has not gone as smoothly as expected. Significant problems exist, which has made the reform progress slowly, and to some extent, drift off its original direction and become the so-called "bottleneck" of China's education innovation. (p. 455)

Student assessment in school science learning has undergone a series of changes toward a more student-centered, learning-facilitated, and development-oriented in China. However, assessment of learning is still the most dominant approach in schools since it fits the needs of social fairness and flexibility. There is still a long way to make the dreams of "assessment for learning" and "assessment as learning" come true in science learning assessment. It depends not only on the efforts of educators and teachers but also on changes in Chinese society.

References

Black, P. J., & Wiliam, D. (1998). *Inside the black box: Raising standards through classroom assessment*. London: King's College.

Cai, W. Y. (2012). The classroom assessment based on worksheet in process of science inquiry instruction (科学探究教学中基于工作单的课堂评价). *Contemporary Education Science, 2*, 25–28 (In Chinese).

Eral, L. M. (2003). *Assessment as learning: Using classroom assessment to maximize student learning*. Thousand Oaks: Cowin.

Eral, L. M. (2006). *Rethinking classroom assessment with purpose in mind. Manitoba Education, Citizenship, and Youth*. Canada: MECY.

Gao, L. B. (1998). Cultural context of school science teaching and learning in China. *Science Education, 82*(1), 1–14.

Gao, L. B. (2002). Issues in student assessment in the new national curriculum (新课程的评价问题). *Global Education, 31*(6), 52–56 (In Chinese).

Gao, L. B. (2004). The concept of processing evaluation and its function (过程性评价的功能和理念). *Journal of South China Normal University (Social Science Edition), 6*, 102–106 (In Chinese).

Gao, L. B. (2011). *Rethinking student assessment reform theoretically* (学生评价改革的理论思考). A lecturer on the national workshop for senior science teachers, 12/12/2011. Guangzhou: South China Normal University. (In Chinese).

Gao, L. B. (2013). Reforms in student assessment in Mainland China. In E. H. F. Law & C. Z. LI (Eds.), *Curriculum innovations in changing societies: Chinese perspectives from Hong Kong, Taiwan and Mainland China* (pp. 449–472). Rotterdam: Sense Publishers.

Gao, L. B., & Watkins, D. A. (2002). Conceptions of teaching held by school science teachers in P.R. China: Identification and cross-cultural comparisons. *International Journal of Science Education, 24*(1), 61–79.

Gipps, C. (1999). Social-cultural aspects of assessment. *Review of Research in Education, 24*, 355–392.

Jian, Y. G., Xian, H. Z., & Jin, P. (2005). The construction and practice of the assessment system of scientific inquiry teaching (科学探究教学评价体系的构建与实践). *Curriculum, Teaching Material and Method, Beijing China, 25*(12), 60–64 (In Chinese).

Liao, B. Q., & Yuan, L. M. (2010). The analysis of characteristics of science inquiry ability assessment in physics Senior-high School Entrance Examination in 2009 (2009年中考物理试题对探究能力考查的特点探析). *Basic Education Curriculum, 3*, 71–74 (In Chinese).

Ling, B. (2009). Ideas and practice of primary science learning assessment (小学科学课教学评价的思考与实践). *Science Class (Primary Edition), 10*, 20–21 (In Chinese).

Luo, G. Z. (2006). Study on performance assessment of scientific inquiry abilities (对科学探究能力表现性评价的研究). *Curriculum, Teaching Material and Methods, 26*(8), 66–69 (In Chinese).

Luo, G. Z. (2007a). Research of gender difference of performance evaluation in scientific exploration (科学探究表现性评价的性别差异研究). *Theory and Practice of Education, 27*(5), 41–42 (In Chinese).

Luo, G. Z. (2007b). Study on effectiveness of assessment methods of science inquiry ability based on paper and-pencil test (基于纸笔测验的科学探究能力评价的有效性研究). *Shanghai Education Research, 10*, 51–54 (In Chinese).

Luo, G. Z. (2007c). The assessment methods of science inquiry abilities (科学探究能力的评价方法). *Education Science, 23*(3), 7–10 (In Chinese).

Ministry of Education (MOE). (2001a). *Guidelines for curriculum reform in basic Education (trial)* (基础教育课程改革纲要(试行)). (Official document, No.17) (In Chinese).

Ministry of Education (MOE). (2001b). *Primary science curriculum standards for compulsory education (3–6) (trial) (科学课程标准(3–6年级)(实验稿))*. Beijing: Beijing Normal University Press (In Chinese).

Ministry of Education (MOE). (2001c). *Junior secondary science curriculum standards for compulsory education (7–9) (trial)* (科学课程标准(7–9年级)(实验稿)). Beijing: Beijing Normal University Press (In Chinese).

Ministry of Education (MOE). (2001d). *Physics curriculum standards for compulsory education (trial)* (义务教育物理课程标准(实验稿)). Beijing: Beijing Normal University Press (In Chinese).

Ministry of Education (MOE). (2001e). *Chemistry curriculum standards for compulsory education (trial)* (义务教育化学课程标准(实验稿)). Beijing: Beijing Normal University Press (In Chinese).

Ministry of Education (MOE). (2001f). *Biology curriculum standards for compulsory education (trial)* (义务教育生物课程标准(实验稿)). Beijing: Beijing Normal University Press (In Chinese).

Ministry of Education (MOE). (2001g). *Geography curriculum standards for compulsory education (trial) (义务教育地理课程标准(实验稿))*. Beijing: Beijing Normal University Press (In Chinese).

Ministry of Education (MOE). (2001h). *Guidelines for kindergarten education (trial)* (幼儿园教育指导纲要(试行)). No. 20, 2001. (In Chinese).

Ministry of Education (MOE). (2002). *Circular of the ministry of education on promoting reforms on school evaluation and examination system* (关于积极推进中小学评价与考试制度改革的通知). (*Official document*, No. 26). (In Chinese).

Ministry of Education (MOE). (2003a). *Physics curriculum standards for senior secondary schools (trial)* (普通高中物理课程标准(实验稿)). Beijing: People's Education Press (In Chinese).

Ministry of Education (MOE). (2003b). *Chemistry curriculum standards for senior secondary schools (trial) (普通高中化学课程标准(实验稿))*. Beijing: People's Education Press (In Chinese).

Ministry of Education (MOE). (2003c). *Biology curriculum standards for senior secondary schools (trial)* (普通高中生物课程标准(实验稿)). Beijing: People's Education Press (In Chinese).

Ministry of Education (MOE). (2003d). *Geography curriculum standards for senior secondary schools (trial)* (普通高中地理课程标准(实验稿)). Beijing: People's Education Press (In Chinese).

Ministry of Education (MOE). (2008). *Suggestions on promoting and improving the reform of senior secondary school entrance examinations* (关于深入推进和进一步完善中考改革的意见). No. 6, 2008. (In Chinese).

Noonan, B., & Duncan, R. (2005). Peer and self-assessment in high school. *Practical Assessment, Research & Evaluation, 10*(17), 1–8.

Shepard, L. (2000). The role of assessment in a learning culture. *Educational Researcher, 29*(7), 4–14.

Stiggins, R. J. (2002). Assessment crisis: The absence of assessment for learning. *Phi Delta Kappan, 83*, 758–765.

Torrance, H. (2007). Assessment as learning? How the use of explicit learning objectives, assessment criteria and feedback in post-secondary education and training can come to dominate learning. *Assessment in Education, 14*(3), 281–294.

Wei, F. S. (2007). Performance assessment in process of primary science instruction (小学科学教学中的表现性评价). *Basic Education Research, 5*, 39–41. Nanning China, In Chinese.

Wiggins, G. (1993): *Assessment, authenticity, context, and validity*. Phi Delta Kappan, November, 200–214.

Wu, Y. (2012). *A review on the reform of high school certificate examination in China*. Paper presented at the 8th conference of International Test Commission, Amsterdam, 3–5 July.

Xu, M. M., & Yan, M. G. (2008). Designing performance assessment for enquiring experiment in chemistry (化学探究性实验活动表现评价的设计探讨). *Journal of Chemical Education, 10*, 12–13 (In Chinese).

Zhang, J. P. (2005). *Instruction and assessment of senior physics* (物理教学与学业评价). Guangzhou: Guangdong Education Press. (In Chinese).

Zhao, B. G. (2005a). The characteristics of the assessment on scientific inquiry ability in GCE Physics in England (英国GCE物理课程科学探究能力评价的特点). *Curriculum, Teaching Material and Methods, 25*(9), 93–96 (In Chinese).

Zhao, B. G. (2005b). Experimental and inquiry ability—implication from GCE Physics test (实验与探究能力的评价—来自GCE物理考试的启示). *Journal of Physics Teaching, 23*(6), 1–2 (In Chinese).

Zhao, D. C. (2012). Review and reflection on the application of portfolio (成长记录袋应用的回顾与反思). *Curriculum, Teaching Material and Method, Beijing China, 32*(5), 21–26 (In Chinese).

Zhao, M. L., & Pan, S. D. (2010). Worksheet: An instrument for valid assessment of scientific process skills (工作单:一种有效评价科学过程技能的工具). *Education Research of Jiangsu, 1A*, 46–48 (In Chinese).

Chapter 10
Alignment Between the National Science Curriculum Standards and Standardized Exams at Secondary School Gateways

Xian Chen, Minyan Jiang, Li Cai, Ling L. Liang, Jing Du, and Yan Zhou

10.1 Introduction

China is well known for its exam-oriented education system, especially at the secondary school level. Although the most recent curriculum reform was launched a decade ago, progress in transforming secondary school science instruction has been slow and unproductive nationwide. In reality, the goal of "quality education" is generally acknowledged, but test preparation often overrides national curriculum standards (Zhao 2009). Many teachers continue to adopt traditional frameworks of science disciplines and teacher-centered instructional approaches in their classrooms. To ensure more successful implementation of standards-based science education, the following two aspects appear to be most critical: (1) reform the national college entrance exam and college admission system and (2) ensure consistency between the existing standardized exams at gateways and the national curriculum standards (Yuan et al. 2002). If certain topics and/or cognitive skills are

X. Chen (✉) • L. Cai
College of Teacher Education, Nanjing Normal University, 122 Ninghai Road, Nanjing, Jiangsu 210097, China
e-mail: cxwzx2006@163.com

M. Jiang
School of Physics and Technology, Nanjing Normal University, 1 Wenyuan Road, Nanjing, Jiangsu 210023, China

L.L. Liang
La Salle University, Philadelphia, PA, USA

J. Du
Nanjing Foreign Language School, 30 Beijing East Road, Nanjing, Jiangsu 210008, China

Y. Zhou
Jiangyin First Middle School, 88 Changjiang Road, Jiangyin, Jiangsu 214431, China

L.L. Liang et al. (eds.), *Chinese Science Education in the 21st Century: Policy, Practice, and Research*, Contemporary Trends and Issues in Science Education 45, DOI 10.1007/978-94-017-9864-8_10

consistently neglected in the exams at these gateways, the teachers are more likely to ignore those topics or cognitive skills in classroom teaching despite the curriculum standards requirements.

In this chapter, we focus our research on the second aspect mentioned above through examining the degree of alignment between the national science curriculum standards documents and the standardized exit exams at junior and senior high schools. We first briefly review the existing alignment research literature within and outside of China and then report the findings based on our own research studies. At the end of the chapter, we discuss implications and make recommendations for further research.

10.2 Alignment Research and Its Significance

In an effective educational system, important components such as content standards, curriculum, instruction, and assessment should be well aligned to send a consistent message about what is valued in the educational process (Webb 1999). Alignment research presents one way to evaluate the level of agreement or consistency between those abovementioned educational aspects or components. Alignment is also getting more attention as an aspect of the validity and reliability of an assessment and its uses (Beck 2007). Results of an alignment study may help policymakers, assessment developers, curriculum developers, and educators make informed decisions on further refinements of educational systems.

According to Bhola and colleagues (2003), alignment can be defined as the extent of agreement between an academic standards document and the assessment (s) used to measure student learning of these standards. Alignment can be measured by multiple models that vary in levels of complexity. At the simplest level, an alignment study might only identify the content match between the assessment and a specific set of curriculum standards. At highly complex levels, multiple aspects of alignment of standards and assessments would be examined. For instance, one of the most widely used alignment procedures used by individual states within the United States, Webb's model (1997, 1999) analyzes alignment using four criteria: (1) categorical concurrence, indicating the extent to which the test content is consistent with corresponding content standards; (2) depth-of-knowledge consistency, indicating whether the test content matches the specified level of cognitive challenge in the standards; (3) range-of-knowledge correspondence, indicating the extent to which the content in the standards is covered in an assessment; and (4) balance of representation, indicating the degree to which the test's content distribution is balanced across objectives and consistent with the standards. In Webb's alignment model, a criterion for each of the four measures is given. If all four criteria are met, then the alignment between the test and the corresponding standards is claimed to be acceptable (Webb 2007) (Table 10.1).

One may argue that the more dimensions involved in an alignment model, the more the findings drawn from the model would more accurately represent the

Table 10.1 Characteristics of the physics exit exams at the senior high school level by province

	Guangdong	Jiangsu	Shandong	Hainan	Ningxia
Feature	Economically well-developed region -------------------------→ economically underdeveloped region				
Time (min)	90	75	90	60	100
Total points	100	100	100	100	100 +20 (bonus)
Format	Multiple choice questions only	Multiple choice, fill-in-the-blank, word problem involving mathematical calculations or discussion/argumentation involving real-life situation	Multiple choice, fill-in-the-blank, word problem involving mathematical calculations	Multiple choice, fill-in-the-blank, experimentation-related word problem involving mathematical calculations	Multiple choice, fill-in-the-blank, graphing word problem involving mathematical calculations, discussion/argumentation involving real-life situation
Bonus question	No	No	No	No	Yes (20 points)

reality. However, it is also true that the more complex an alignment model is, the more difficult a high alignment between the assessment items and the standards can be achieved (Bhola et al. 2003; Fulmer 2011). As an example of a moderately complex model of alignment, Porter (2002) focuses the alignment analysis on both content and cognitive domain's dimensions using the data collected through the Surveys of Enacted Curriculum (SEC) (Council of Chief State School Officers 2004). Unlike the Webb model, the SEC approach does not rely on direct comparison of assessment items with content standards. Instead, a content taxonomy is first developed as the common framework by subject matter experts, and then trained content analysts map the standards and assessment items onto the taxonomies using two-dimensional matrices—one dimension for content and the other for cognitive demands, which allows differentiation of levels of content difficulties. The categories of Porter's cognitive demands are consistent with the revised Bloom's taxonomy for describing learning outcomes (Anderson and Krathwohl 2001). To evaluate the level of alignment, the matrices for standards and assessments are compared by cell, and an alignment index, P, is calculated to indicate the proportion of content in common (see Tables 10.2 and 10.3). The use of a common language and the quantitative alignment index produced in Porter's model allows the level of alignment to be calculated and compared across different standards documents, assessments, textbooks, classroom instruction, and many other components of the educational system. In addition, based on a simulation study, Fulmer (2011) further established critical values for Porter's alignment index, suitable for hypothesis testing at alpha levels of 0.05 and 0.10.

In a recent study, Polikoff et al. (2011) investigated the coherence of standards-based reform in the United States by analyzing 138 standards-assessment pairs spread across grades and the three tested subjects required by law. With the SEC approach, it was found that roughly half of standards content was covered on the corresponding test and roughly half of test content corresponds to the standards. Misalignment also occurred due to a mismatch on cognitive demands between the assessment items and the standards content. About 17–27 % of content on a typical test covers topics not mentioned in the corresponding standards. The authors again argued that in order to ensure that all students experience equal learning opportunities and demonstrate their achievements, the standards, assessment, and instruction must work together to deliver a consistent message.

10.2.1 Alignment Studies in China

Alignment research in China is relatively new. The first alignment study about standardized exams at high schools in China was a case study published in 2008, based on an analysis of the 2002 Chinese National Physics Syllabus (Grades 10–12) and the alignment between the guidelines and two recent twelfth-grade exit tests in Jiangsu Province (Liang and Yuan 2008). The data were also later used in an international comparison study of the physics curriculum standards and the physics

Table 10.2 Physics content standards (grades 10–12) by topic and by cognitive level

Topic	Sub-topic	Remember	Understand	Apply	Analyze	Evaluate	Create	Subtotal
Describing motion (DM)	Significance of experiment and mathematics method, physics model, particle model, displacement, speed, acceleration, law of uniform rectilinear motion	2(0.03)	2(0.03)	1(0.01)	0(0.00)	0(0.00)	0(0.00)	5(0.07)
Interaction forces and laws (IFL)	Static and kinetic friction forces, Hooke's law, forces in two dimensions, vector and scale, Newton's laws, overweight, weightlessness, the standard SI unit	3(0.04)	4(0.06)	1(0.01)	2(0.03)	0(0.00)	0(0.00)	10(0.14)
Mechanical energy and sources of energy (MESE)	Work, mechanical power, kinetic energy theorem, gravitational potential energy, the conservation of mechanical energy, energy in various forms and conservation	5(0.07)	8(0.11)	0(0.00)	1(0.01)	0(0.00)	0(0.00)	14(0.19)
Projectile and circular motion (PCM)	Projectile motion, circular motion, centripetal acceleration and force	1(0.01)	1(0.01)	1(0.01)	3(0.04)	0(0.00)	0(0.00)	6(0.08)
Classical mechanics achievement and limitation (CMAL)	Law of universal gravitation, artificial satellite, the second and third universe speed, Newton's and Einstein's views about gravitation, scientific methods	7(0.10)	2(0.03)	1(0.01)	0(0.00)	0(0.00)	0(0.00)	10(0.14)
Electromagnetic phenomenon and law (EPL)	Static electricity, electric fields, magnetic fields and flux, right-hand rule, electromagnetic induction, electromotive force, Maxwell electromagnetic fields	9(0.13)	3(0.04)	2(0.03)	0(0.00)	0(0.00)	0(0.00)	14(0.19)
Electromagnetic technology and society development (ETSD)	Generator, motor; telecommunication; electromagnetic wave; temperature sensor; relationship between science, technology, and society	4(0.06)	1(0.01)	0(0.00)	0(0.00)	2(0.03)	0(0.00)	7(0.10)

(continued)

Table 10.2 (continued)

Topic	Sub-topic	Remember	Understand	Apply	Analyze	Evaluate	Create	Subtotal
Electrical apparatus at home and application (EAHA)	Types of domestic electrical apparatus, applications of circuits at home, electrical safety at home, electrical safety knowledge, saving electricity	2(0.03)	1(0.01)	2(0.03)	1(0.01)	0(0.00)	0(0.00)	6(0.08)
Subtotal		33(0.46)	22(0.31)	8(0.11)	7(0.10)	2(0.03)	0(0.00)	72(1.00)

Note. The frequency of content statements is presented for each cell. The numbers in parentheses are the proportion of statements in the respective cell to the total number of statements coded

Table 10.3 Physics content standards (grades 7–9) by topic and by cognitive level

Topic	Sub-topic	Remember	Understand	Apply	Analyze	Evaluate	Create	Subtotal
Properties of matter (POM)	Physical properties and change of matter; structure of matter, scale, and application of new materials	17(0.13)	11(0.09)	3(0.02)	1(0.01)	1(0.01)	0(0.00)	33(0.26)
Motion and forces (M&F)	Variety forms of motion, force and mechanical movement	12(0.09)	8(0.06)	9(0.07)	0(0.00)	0(0.00)	0(0.00)	29(0.22)
Electricity (ELE)	Static electricity, electric circuits, magnetism, and effects of interacting fields	3(0.02)	5(0.04)	5(0.04)	0(0.00)	0(0.00)	0(0.00)	13(0.10)
Waves	Light, sound, wavelength, frequency, the speed of waves	12(0.09)	6(0.05)	0(0.00)	0(0.00)	0(0.00)	0(0.00)	18(0.14)
Energy	Types of energy, work, the conversion and transfer of energy, use of energy sources and sustainable development	17(0.13)	16(0.13)	1(0.01)	1(0.01)	0(0.00)	0(0.00)	35(0.28)
Subtotal		61(0.46)	46(0.37)	18(0.14)	2(0.02)	1(0.01)	0(0.00)	128(1.00)

Note. The frequency of content statements is presented for each cell. The numbers in parentheses are the proportion of statements in the respective cell to the total number of statements coded

tests at secondary school gateways among New York, United States; Singapore; and Jiangsu, China (Liu et al. 2009). In the 2008 study, Liang and Yuan adopted a method similar to the SEC model (Porter 2002) and analyzed the topics in the Syllabus and the tests. It was found that both the Syllabus and the standardized exams mostly emphasized student learning outcomes at the "understand" cognitive level. Furthermore, the two exams consistently overrepresented the Syllabus at both application and analysis cognitive levels. The study also indicated that neither the organization of the Syllabus nor the exit assessments encouraged creativity, critical thinking, or the development of students' abilities to conduct scientific inquiry.

Since 2008, more studies, normally conducted by college professors and their graduate students, have been conducted to examine alignment between the new secondary school curriculum standards and the gateway exams at the ninth- or the twelfth-grade levels, by applying either Porter's method as used by the SEC or Webb's model (Chen et al. 2010; Guo et al. 2010; Jiao and Chen 2010; Wang et al. 2010). In Chemistry, for instance, one of the most relevant alignment studies was conducted by Wang et al. (2010). Wang and colleagues (2010) examined alignment of the ninth grade exit tests in nine selected provinces and/or cities, covering provincial- and/or municipal-level exams in economically well-developed and underdeveloped regions. In some places, the chemistry content was tested in a single exam, while in other places, the chemistry content was combined with physics and biology in one comprehensive test. With a modified SEC type of approach, the content of chemistry was classified into seven main topics: Solution, States of Matter, Composition of Matter, Structure of Matter, Chemical Changes, Energy, and Chemistry Terminology. The concepts and subtopics within each main topic were then categorized according to the cognitive demands: (1) memorizing, (2) understanding, (3) simple or direct application, and (4) complex application and transfer. For the nine tests sampled, the Porter alignment indices range from 0.59 to 0.74. It appeared that better alignment was associated with the regions with higher economic development levels. However, the researchers did not report the statistical significance of their results. Compared to the municipal-level tests, the provincial-level exams seemed to be of higher quality and better aligned with the curriculum standards. Furthermore, the alignment was higher in a single chemistry exam than that in the combined science test. The identified problems with certain exams included an overly heavy emphasis on "remember" or complex computation in certain tests. Finally, the alignment findings seem to be independent of the types of textbooks in place (Wang et al. 2010).

10.3 Purpose of the Current Study

Over the years, physics courses and exams in China have been perceived as the most difficult of all by many students. We believe that alignment research on standardized physics exams will facilitate the conversation about implementation of standards-based curriculum reform and improvement of learning, instruction, and assessment in physics.

Building on the existing literature and our initial findings in an analysis of physics curriculum standards and the 2009 ninth grade exit physics test in one province (Chen et al. 2010), we launched the current research project by examining the physics exit exams at the secondary schools in several purposely selected geographic locations and by tracking the exams in one place for multiple years (2007–2011).

The research questions are as follows: (1) To what extent are the physics exit exams at the senior high school level aligned with the corresponding curriculum standards (Grades 10–12) in the selected provinces? (2) To what extent are the ninth grade physics exit exams aligned with the corresponding physics curriculum standards (Grades 7–9) in the selected cities?

10.4 Methods

In this section, we first provide some background information about the exit exams at secondary schools in China. We then describe the procedure of data analysis and calculation of the alignment indices. Given that the Porter's method could potentially make the results comparable across different standards documents, assessments, and other components of the educational system, we decided to adopt the Porter's model with a minor modification. We replaced Porter's five categories of cognitive demand with the revised Bloom's taxonomy, i.e., remember, understand, apply, analyze, evaluate, and create (Anderson and Krathwohl 2001) because of the similarity between the two classification schemes. Furthermore, the cognitive domain terminology described in Bloom's taxonomy or the revised Bloom's taxonomy has been widely adopted by Chinese education communities.

10.4.1 The Exit Exams at the Secondary School Gateways in Selected Regions

10.4.1.1 Physics Exit Exams at the Senior High School Level

In most provinces, students must pass a set of exit exams near the end of their senior high school year to be qualified to register for the National College Entrance Examinations in June. Each year, the senior high school exit exams are created by a group of expert high school teachers led by the educational testing agency within each province. Specific testing requirements and the number of subjects tested vary from one province to another. For instance, in provinces such as Guangdong, Jiangsu, and Hainan, physics exit exams are required for liberal arts students only, whereas in other provinces such as Shandong and Ningxia, physics exit exams are required for all students.

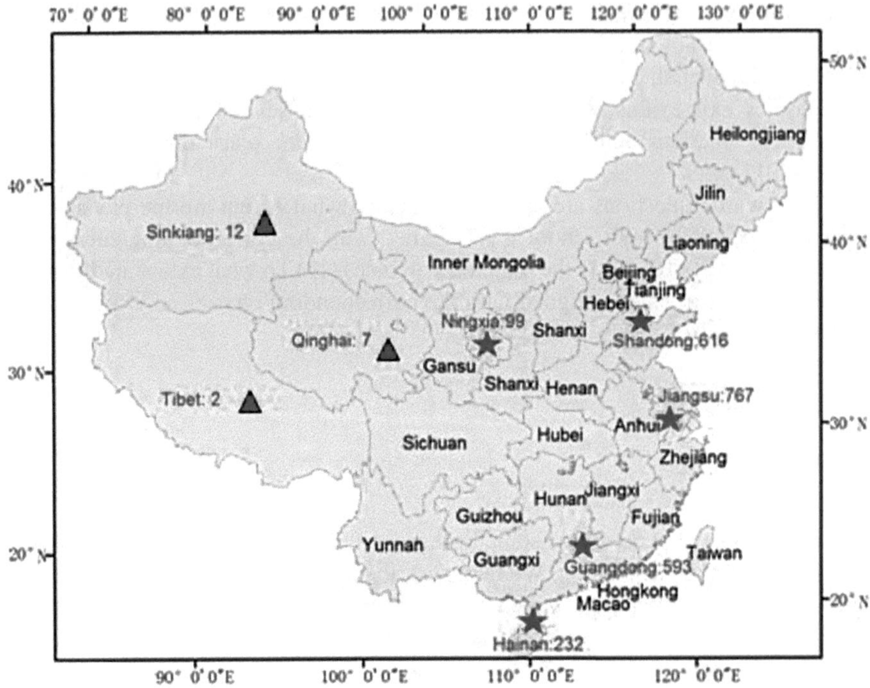

Fig. 10.1 Geographic locations of the sample provinces. Note: Locations labeled with stars represent the five samples. The number next to the name of each individual province represents its population density (number of people per square kilometer). The western China regions labeled with triangles are not included in this study. The population densities in these regions are much lower than those of the selected samples

In this study, we examined physics exit exams from five provinces, ranging from the economically well-developed to the underdeveloped areas in the country: Guangdong (well developed), Jiangsu, Shandong, Hainan, and Ningxia (underdeveloped). These provinces are selected to represent areas of different levels of economic and educational development in China (marked with stars on the map in Fig. 10.1). They are also among the most populated regions in China, representing the first and the second curriculum reform pilot cohorts in the nation. The evaluation of the new school curriculum in the selected provinces would provide evidence and/or lessons for the nationwide implementation of the curriculum reforms.

For the Jiangsu sample, we also examined the exit tests for 5 years (2007–2011) to identify potential trends and patterns. Table 10.2 presents some major characteristics of the physics exit exams in the selected regions.

10.4.1.2 Physics Exit Exams at Junior High School Level

As part of compulsory education in China, junior secondary school students (Grades 7–9) are required to take science courses either as separate or integrated subjects. In most provinces, students take biology in seventh grade, physics in eighth and ninth grades, and chemistry in ninth grade. All students must take the exit examinations before they graduate from the junior secondary school. In Nanjing city, for instance, the ninth grade exit exams for junior high schools are offered in June each year and have seven subjects, including Chinese (120 pts.), mathematics (120 pts.), English (120 pts.), physics (100 pts.), and chemistry (80 pts.). The biology exam is given at the end of eighth grade. Physics exam test items include multiple choice questions, fill-in-the-blank questions, problem-solving questions involving mathematical calculations, and graphing questions or questions related to inquiry-oriented experimental design. Additional required performance assessments (i.e., experiments) are conducted separately by individual teachers on a pass/fail basis. The exit exams serve two functions: first, the test scores gauge the level of student learning against the new curriculum standards and, second, students' total test scores determine the degree of prestige of the senior secondary school to which they can be admitted. Obviously, the ninth grade exit exams are the most important examinations for the students at the compulsory education stage. Parents, teachers, and school principals, as well as the students themselves, all take them very seriously.

According to Wang et al. (2010), the exams tend to have higher quality and better alignment with the corresponding curriculum standards in more economically developed regions. In this study, we purposely selected the Jiangsu Province ninth grade physics exit exams in four cities in 2010, including Nanjing, Changzhou, Suzhou, and Wuxi. Taking Nanjing as an example, we also tracked the exit tests for 5 years (2007–2011) to identify possible trends and patterns. The cities selected in Jiangsu represent economically well-developed areas with a history of regional/national educational leadership. Therefore, the findings of the study could provide valuable lessons for the rest of the country.

10.4.2 Analysis of the Curriculum Content Standards and Exams

The fundamental goal of the physics curriculum standards at the secondary school levels is to develop and improve scientific literacy for all students. Physics curriculum standards at both the lower and upper secondary levels share a common framework defined by three dimensions: knowledge and skills; processes and methods; and emotions, attitudes, and values (MOE 2001, 2003, 2011). In this study, we focus on the analysis of the clearly defined "knowledge and skills" dimension only. The statements on the other two dimensions are not assessed in standardized examinations and are too general for any meaningful productive analysis.

The physics curriculum content standards consist of main topics and subtopics (see Tables 10.2 and 10.3). A number of benchmark statements are also listed under each of the subtopics. To further classify the contents based on the level of thinking required, or the cognitive demands, we used the revised Bloom's taxonomy (Anderson and Krathwohl 2001) to define the cognitive levels with the following sample keywords:

- Remember: know, recognize, identify, recall
- Understand: interpret, translate, explain, illustrate
- Apply: use, implement, calculate
- Analyze: differentiate, distinguish
- Evaluate: critique, judge, reflect
- Create: generate, hypothesize, plan, design

For example, the benchmark statement in the physics standards (Grades 7–9), "students will *know* the relationship of wavelength, frequency, and the speed of wave," was mapped to "waves" and "remember" (Table 10.3). Each benchmark statement is weighted equally. Tables 10.2 and 10.3 present the physics curriculum standards (Grades 10–12 and Grades 7–9) in two 2-D matrices, respectively.

For the analysis of the exams, each test item was classified into a cell of the matrix identical to the curriculum standards analysis grid. After the classification of all test items was completed, the total points of the items in each cell were used as the cell value. The cell value was 0 if no corresponding test items were identified.

One physics education professor and four graduate students with physics teaching experience conducted the data analysis. The average inter-coder reliability was 0.90 for content topics and 0.85 for the cognitive demands. The final results were produced by resolving the disagreements through face-to-face discussions among the research team members.

10.4.3 Measurement of Alignment Between the Tests and Corresponding Standards

To determine alignment between curriculum standards and a test, we first created two tables (one for representing the curriculum standards and the other for the test), each using a two-dimensional matrix in which the rows represent topics/themes and columns represent levels of cognitive demand (such as Table 10.2 for the Grades 7–9 curriculum standards). The values in each cell were converted into proportions out of the grand total, indicating the proportion of total content in the standards document (or exam items) that emphasizes that particular combination of topic and cognitive demand. We then calculated an alignment index using the following equation:

$$P = 1 - \frac{\sum_{i=1}^{n} |(Xi - Yi)|}{2}$$

Here, *P* is the Porter alignment index, ranging from 0 (indicating no alignment) to 1 (indicating perfect alignment), X denotes cell proportions in Table X (e.g., the standardized exam matrix), Y denotes cell proportions in Table Y (e.g., the curriculum standards matrix), n represents the total number of cells, and i refers to a specific cell in each matrix. For instance, for the 8 × 6 matrix in our study (Table 10.1), $n = 48$. After calculating the alignment indices, we further tested the results for statistical significance by examining the estimated critical values derived from a simulation study (for a more detailed description of this method, see Fulmer 2011). It was found that the critical value was 0.67 for the 8 × 6 matrix and 0.78 for the matrix of five rows and six columns, with the statistical significance at the 0.05 level.

10.5 Results

10.5.1 Alignment of the Physics Content Standards and the Exit Exams at the Senior High School Level

Table 10.4 presents the alignment indices of the exit exams of five provinces in 2010 and the indices of the Jiangsu province exams over 5 years at the senior high school level. Given that the critical value is 0.67, none of the indices are statistically significant at the 0.05 level. However, the numbers are quite close or consistent across the regions and/or across time. The alignment index of the 2010 exam in Jiangsu (representing an economically well-developed region) is almost the same as the one in Ningxia (representing an economically underdeveloped region). This suggests that the test alignment results in a region are not necessarily related to the region's level of economic development. In addition, the test alignment indices in Jiangsu province range from 0.42 to 0.47 during 2007–2011, which indicates that

Table 10.4 The Porter alignment indices (P) of Jiangsu in 2007–2011 and the other four provinces in 2010 (senior high school exit physics exams)

Year	Guangdong	Jiangsu	Shandong	Hainan	Ningxia
2007	n.a.	0.43	n.a.	n.a.	n.a.
2008	n.a.	0.43	n.a.	n.a.	n.a.
2009	n.a.	0.44	n.a.	n.a.	n.a.
2010	0.54	0.42	0.55	0.55	0.41
2011	n.a.	0.47	n.a.	n.a.	n.a.

Note: The critical value is 0.67 with the statistical significance at the 0.05 level

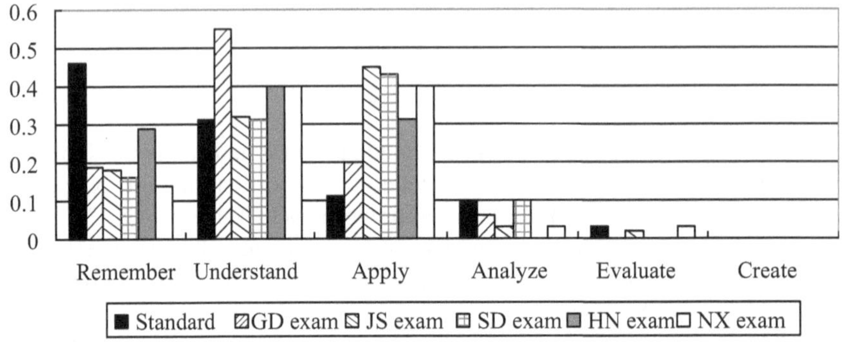

Fig. 10.2 Comparison between content standards (Grades 10–12) and exams in five provinces by cognitive level in 2010

the 5-year implementation of the national curriculum standards did not impact the stability or quality of the exit exams.

For further analysis, we examined the cognitive demands and content coverage separately. Figure 10.2 presents the comparison of alignment analysis across the five provinces by cognitive demands. Compared to physics content standards, the exam items overrepresented the "understand and apply" levels while underrepresented the "remember" level. In the curriculum standards (the black bars in Fig. 10.2), about 46 % of the content is required at the "remember" level and about 11 % at the "apply" level. However, in the five selected exit exams, only about 14–29 % of the test items are located at the "remember" level, while 31–55 % of the content is located at "understand" and 20–45 % at "apply" levels. Across the regions, the proportions of test items classified as "analyze" or "evaluate" seem to be consistently lower than what is required in the standards. Some provinces had excluded items at those two higher cognitive levels in testing. For instance, no items at the levels of analysis and evaluation were found in Hainan's test, while the tests from Guangdong, Shandong, and Hainan did not include any items at "evaluate" level.

In terms of the content coverage of the exams across the five provinces, it was found that most topics required in the standards were covered in the exams. Two exams (from Guangdong and Shandong) covered all main topics in the standards. However, the content distribution in the test did not seem to correlate with the level of emphasis of individual topics as described in the standards. The two topics consistently overrepresented in all five exams are "describing matter" and "interaction forces and laws." The consistently underrepresented topics in the exams include "classical mechanics achievement and limitation" (CMAL), "electromagnetic technology and society development" (ETSD), and "electrical apparatus at home and application" (EAHA). The last two topics (ETSD and EAHA) were not addressed in the exams from Ningxia and Jiangsu (Fig. 10.3).

Taking the Jiangsu exams in five consecutive years as examples, the distribution patterns of the cognitive demands and content are illustrated in Figs. 10.4 and 10.5,

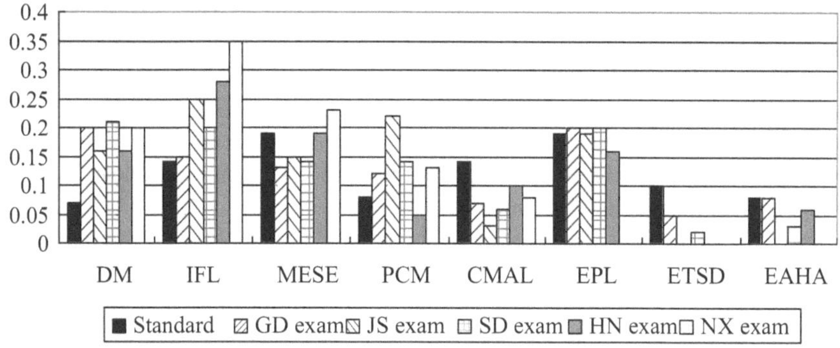

Fig. 10.3 Comparison between content standards (Grades 10–12) and exams in five provinces by topics in 2010

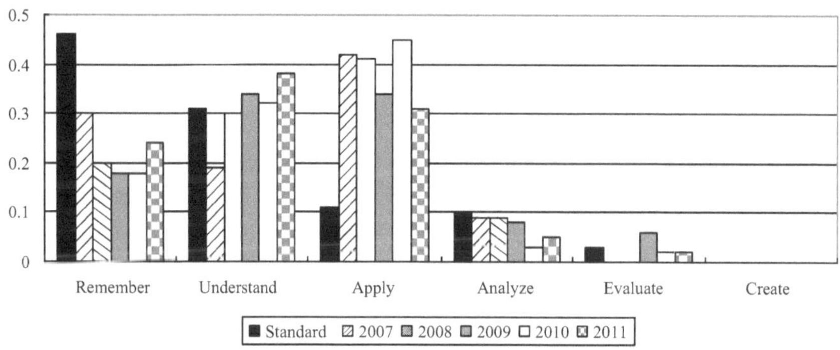

Fig. 10.4 Comparison between content standards (Grades 10–12) and exams for 5 years by cognitive level

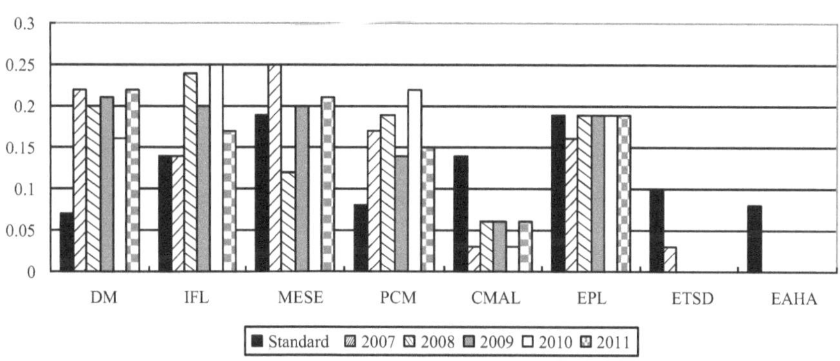

Fig. 10.5 Comparison between content standards (Grades 10–12) and exams for 5 years by topic

respectively. For the distribution of cognitive levels, on one hand, the percentage of the test items at the memory level over the 5 years ranged from 18 % to 30 %, compared to 46 % in the standards. On the other hand, about 31–45 % of the test items were classified as the application level, while the requirement in the standards is 11 %. Apparently, the test items in the Jiangsu provincial exit exams substantially overrepresent the standards in higher cognitive levels (such as application) while underrepresenting standards at memory levels.

In terms of content coverage, compared to the standards, the motion topics related to "describing motion" (DM) and "projectile and circular motion" (PCM) are consistently overrepresented in all exams, whereas the topic "classical mechanics achievement and limitation" (CMAL) is consistently underrepresented. From 2008 to 2011, both "electromagnetic technology and society development" (ETSD) and "electrical apparatus at home and application" (EAHA) topics were not tested.

10.5.2 Alignment of the Physics Content Standards and Exams at the Junior High School Level

Table 10.5 presents the alignment indices for the exams from the four selected cities in 2010 and for the Nanjing exams from 2007 to 2011. All indices are lower than the critical value of 0.78, indicating that none of the alignment results is statistically significant at 0.05. However, the alignment indices (ranging from 0.52 to 0.60) are quite stable across the time and the cities within Jiangsu Province.

Figure 10.6 shows misalignment of emphases between the physics curriculum standards and the content distribution in the exams at cognitive demands. Compared to the content standards, the test analysis results demonstrated an apparent shift toward higher cognitive skills by de-emphasizing "remember" and overemphasizing "apply" and "analyze." For instance, in the physics content standards, about 46 % of the content is required at the "remember" level and 14 % at the "apply" level. By contrast, about 30–40 % of points were coded as "apply" and 11–24 % of points as "remember" in the exams. At the "analyze" level, the test items representation was 6–9 %, while the standards requirement is 2 %.

Table 10.5 The Porter alignment indices (P) of Nanjing in 2007–2011 and the other three cities in Jiangsu Province in 2010, ninth grade exit physics exams

Year	Nanjing (NJ)	Wuxi (WX)	Changzhou (CZ)	Suzhou (SZ)
2007	0.55	n.a.	n.a.	n.a.
2008	0.53	n.a.	n.a.	n.a.
2009	0.52	n.a.	n.a.	n.a.
2010	0.54	0.58	0.60	0.55
2011	0.54	n.a.	n.a.	n.a.

Note: The critical value is 0.78

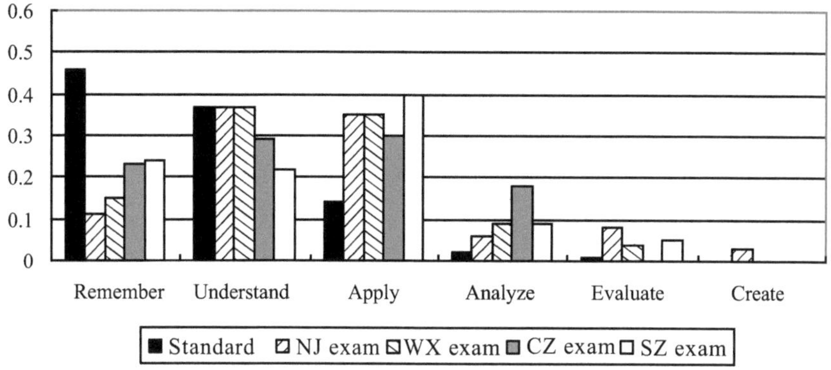

Fig. 10.6 Comparison between content standards (Grades 7–9) and exams in four cities by cognitive demands in 2010

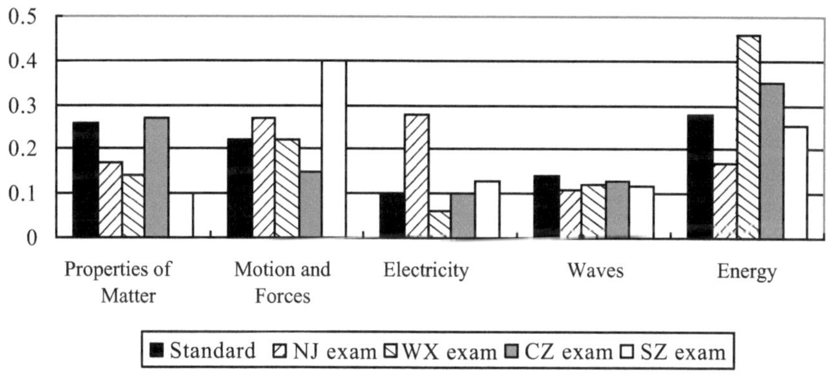

Fig. 10.7 Comparison between content standards (Grades 7–9) and exams in four cities by topics in 2010

The content distribution patterns among the four selected exams are presented in Fig. 10.7. Apparently, the content coverage varies from one place to another. For instance, the most emphasized topics in the standards include "energy" (28 %), "properties of matter" (26 %), and "motion and forces" (22 %), while the least emphasized topic is "electricity" (10 %). However, in the exams, the most emphasized area was "electricity" (28 %) for Nanjing, "energy" (46 %) for Wuxi, and "motion and forces" (40 %) for Suzhou in 2010. Interestingly, the topic on waves was consistently represented at a similar percentage across all four cities.

Following the Nanjing exams as examples, Figs. 10.8 and 10.9 present the overall distribution of cognitive reasoning skills and content coverage as measured by the test items over a 5-year period. Compared to the curriculum standards, the exams consistently de-emphasized memorization of contents and overemphasized higher reasoning skills such as "apply" and "analyze." As for the content coverage, it appears that the topic "electricity" has been consistently overemphasized in the

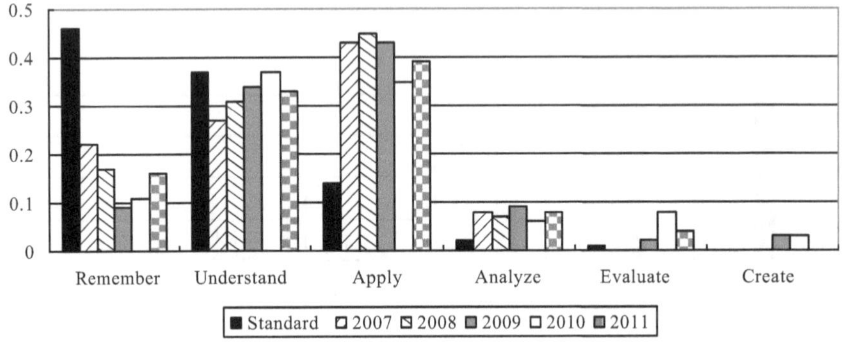

Fig. 10.8 Comparison between content standards (Grades 7–9) and exams over 5 years by cognitive demands

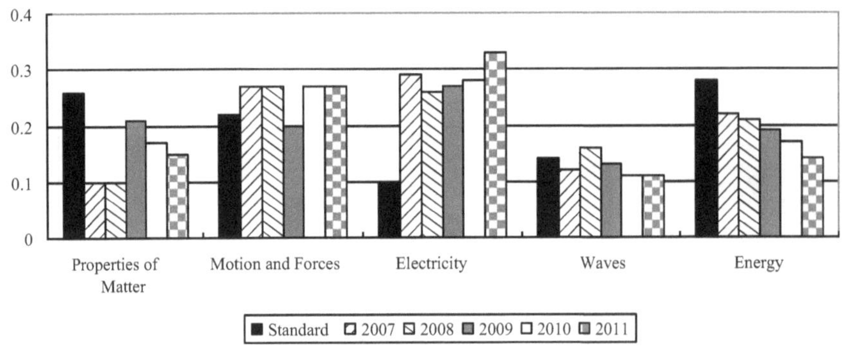

Fig. 10.9 Comparison between content standards (Grades 7–9) and exams over 5 years by topic

exams while other topics such as "properties of matter" and "energy" have been consistently underrepresented.

10.6 Discussion, Implications, and Recommendations

The findings of the current study can be summarized as follows:

1. The alignment indices between the standards and the exit exams at the secondary school level are quite stable for the 5-year window studied in the selected cities or provinces. For the Jiangsu provincial-level high school physics exit exams, the Porter indices range from 0.42 to 0.47, and for the Nanjing city-wide ninth grade exit exams, the Porter indices range from 0.52 to 0.55.
2. The alignment indices across regions and time are all below the corresponding critical values. In other words, none of the alignment results are statistically

significant at 0.05, indicating a general misalignment between the standards and existing physics exit exams

3. Unlike what was found in Wang's and her colleagues' (2010) report, the alignment indices in our study do not seem to be related to the level of economic development of varying geographic locations. Economically well-developed provinces do not necessarily produce physics assessments that are better aligned with the corresponding national curriculum standards
4. Compared to the requirements in the standards, the exams consistently underemphasized "remember" while overemphasizing "apply" and "analyze." Misalignment also occurred when the exams over- or underemphasized content relative to its proportion in the corresponding standards to a varying degree from one year to another.

What can we conclude based on the findings presented above? To what extent should the standardized exit exams at gateways be aligned with the respective curriculum standards? Such questions have not been completely answered here. For instance, by definition, the ninth grade exit exams are criterion-referenced benchmark tests. Students are required to meet the minimum benchmark standards for graduation. Meanwhile, the test scores are also used to "select" students into senior high schools with varying level of prestige. Therefore, the standardized exams are also of a norm-referenced nature. It is reasonable for an exam with a "selective" or "competitive" nature to place emphasis on the higher cognitive demands than those required in the standards. But where do we draw the line between "reasonable" and "unreasonable" levels? The Porter model does not specify the criteria for the alignment index to represent acceptable alignment. In the current study, we used the estimated critical values to determine the level of statistical significance of the results. However, we do not believe that such assessment issues can be solved by statistics alone. In China, the exit exam developers (normally expert teachers) meet and design tests each year based on their own understanding of the curriculum standards as well as the feedback received from teachers, students, and parents. Their knowledge, hard work, and dedication have been reflected in the consistency among the multiple-year alignment indices. We believe that alignment research can be used as a tool to help identify potential or existing problems and provide insights toward the establishment of a more valid and consistent assessment system. The misalignment between the standards and the exams may be addressed in two ways: by reexamining the appropriateness of content coverage and distribution in the curriculum standards and by improving the quality of test items. The test developers may consider constructing a test framework or test blueprint aligned with the respective standards. In addition, there have been concerns that standardized tests are not accurate measures of student achievement in the science education community. Given that a paper-and-pencil test may not be an effective tool for assessing student learning associated with higher levels of cognitive demands (e.g., "evaluate" and "create"), performance assessment or authentic assessment tools (e.g., students' written reports of inquiry projects, experimentation, etc.) should be considered in addition to the traditional paper-and-pencil tests.

The lack of alignment between the curriculum standards and high-stakes tests as identified in this study may influence teaching in both positive and negative directions. The positive aspect might include an emphasis on developing higher cognitive skills in instruction by teachers. During our analysis of the exams, we have noticed some encouraging trends. For instance, more test items involve real-life situations. Some items were designed to assess students' problem-solving and creative thinking skills related to the real world. This will certainly lead teachers to emphasize those higher-order thinking skills in teaching. On the other hand, the overrepresentation of "understand" and "apply" cognitive demands in the exit exams has led to a cognitive overload for many students caused by over-drill or practicing exam-type problems both inside and outside of the school. Students may have achieved high marks in the narrowly defined core subjects at the expense of student learning in other non-tested subject areas. Such misalignment may contribute to the failure in the development of positive attitudes toward school science, student creativity, and their abilities to conduct scientific inquiry in many schools. A substantial number of students have developed high levels of anxiety about the external exams that directly impact both their mental and physical health.

In the following sections, we discuss some limitations of the current study and provide several suggestions or recommendations for future research. First, our alignment research is based on the assumption that curriculum standards and the textbooks are valid and aligned with each other. In our analyses, the main topics and subtopics described in the curriculum standards are classified as general and key concepts or contents. In teaching, the amount of instructional coverage time devoted to each topic is different depending on the classification of the content (e.g., "know," "apply," etc.). We did not assign weight based on the instructional coverage time devoted to different topics, which may have contributed to the "misalignment" results. For future analyses, researchers may want to consider incorporating the suggested instructional time on various topics provided in the curriculum guides and textbooks companions (e.g., Teachers' Guides).

Second, policy researchers have argued that alignment of standards, curriculum, and assessment is the key to supporting implementation of standards-based reform efforts. In this chapter, we focused on the investigation of alignment between high-stakes exams and the corresponding curriculum standards in physics. Such alignment is necessary but not sufficient for successful standards-based curriculum implementation. In reality, the alignment between the standards and classroom instruction can never be assumed. Therefore, we suggest that future alignment studies examine the level of implementation of instructional strategies promoted in the standards. Whereas various approaches such as classroom observations and interviews can be used as fidelity measures, we think that using student/teacher surveys of instructional activities (Fulmer and Liang 2013) might be a more efficient way to get a bigger picture of what's happening at classroom levels in a region. Such surveys can provide information on the content of instruction, level of challenge in instruction, and/or teachers' use of inquiry-oriented, standards-based instructional strategies.

Third, one of the key features of the SEC tools is the use of a common language framework in describing the content areas. In application, however, for various reasons, Chinese researchers tended to modify the SEC framework or adopt a different content classification scheme for data analysis. This creates difficulty in comparing the results across the research studies. We suggest that science education researchers develop a common framework for content analysis based on the Chinese national curriculum standards documents, as a collaborative effort to improve the alignment of learning, teaching, and assessment. In addition, as suggested by other researchers (Martone and Sireci 2009; Polikoff et al. 2011), analyzing the same data with different alignment models (e.g., Webb's model) may provide us with additional insights into agreement and disagreement between standards and assessments.

Finally, based on our review of the alignment studies in China, the data collection and analyses were all done by college professors and their graduate students. There is little evidence that policy makers, test developers, educational administrators at city or provincial levels, and teachers are aware of any of the abovementioned research literature. We therefore suggest that alignment studies be integrated into professional development involving teachers as well as curriculum supervisors in the region.

For future research, we also suggest that different types of experts such as test developers be invited for participation. Different types of experts may view the content and rate exam items differently in terms of the depth of knowledge and the dimensionality of the items (Buckendahl et al. 2000). Involving multiple stakeholders in the alignment research will enhance the credibility of the findings and promote its dissemination to the broader education community more effectively.

It is our hope that our alignment research may prompt the Chinese science educators to reexamine the validity and quality issues related to curriculum standards, textbooks, classroom instruction, and standardized assessment. In addition, it is critical to engage teachers in regular professional development activities related to the above issues. We believe that it is the concerted efforts of stakeholders (e.g., science educator, teachers, educational policymakers, test makers, etc.) that will determine the level of success of the new curriculum reform in China.

Acknowledgment The authors thank Dr. Guo Yuying for her help provided during the preparation of this chapter. The authors also thank Drs. Xiufeng Liu and Gavin Fulmer for their insightful comments and many graduate students in Nanjing Normal University including Jia Cancan, Chen Nin, She Yafang, and Chen Hongli for their assistance in data collection and analysis. The population density and the map of China were provided by Drs. Liyun Lu and Changchun Huang. The authors appreciate it greatly.

References

Anderson, L. W., & Krathwohl, D. R. (2001). *A taxonomy for learning, teaching, and assessing: A revision of Bloom's taxonomy of educational objectives*. New York: Longman.

Beck, M. D. (2007). Review and other views: "Alignment" as a psychometric issue. *Applied Measurement in Education, 20*(1), 127–135. doi:10.1207/s15324818ame2001_7.

Bhola, D. S., Impara, J.C., & Buckendahl, C. W. (2003). Aligning tests with states' content standards: Methods and issues. *Educational Measurement: Issues and Practice* (fall), *22*, 21–29.

Buckendahl, C. W., Plake, B. S., Impara, J. C., & Irwin, P. M. (2000). *Alignment of standardized achievement tests to state content standards: A comparison of publishers' and teachers' perspectives*. Paper presented at the Annual Meeting of the National Council on Measurement in Education, New Orleans.

Chen, X., Jia, C. C., & Chen, N. (2010). Alignment between the physics content standard and the exam. *Curriculum, Teaching Material and Method* (课程·教材·教法), *30*(7), 67–71 (In Chinese).

Council of Chief State School Officers. (2004). *Surveys of enacted curriculum*. Madison: Wisconsin Center for Education Research.

Fulmer, G. W. (2011). Estimating critical values for strength of alignment among curriculum, assessments, and instruction. *Journal of Educational and Behavioral Statistics, 36*(3), 381–402.

Fulmer, G. W., & Liang, L. L. (2013). Measuring model-based high school science instruction: Development and application of a student survey. *Journal of Science Education and Technology, 22*(1), 37–46. doi:10.1007/s10956-012-9374-z.

Guo, Y. Y., Xing, T., Xu, G. Q., & Zheng, C. Y. *Alignment between physics curriculum standard and physics achievement qualifying examination of senior high school in China*. 2010 NARST annual international conference.

Jiao, & Chen (2010). Alignment between the biology curriculum standards and the 2009 college entrance biology exams in selected curriculum reform experimental sites. *Measurement and Evaluation in Education* (教育测量与评价), *3*(5), 54–58 (In Chinese).

Liang, L. L., & Yuan, H. Q. (2008). Examining the alignment of Chinese national physics curriculum guidelines and 12th-grade exit examinations: A case study. *International Journal of Science Education, 30*, 1823–1835.

Liu, X., Zhang, B. H., Liang, L. L., Fulmer, G., Kim, B., & Yuan, H. Q. (2009). Alignment between the physics content standard and standardized test: A comparison among US-NY, Singapore, and China-Jiangsu. *Science Education, 93*, 777–797.

Martone, A., & Sireci, S. G. (2009). Evaluating alignment between curriculum, assessment, and instruction. *Review of Educational Research, 79*(4), 1332–1361.

Ministry of Education (MOE). (2001). *National physics curriculum standards for the full-time compulsory education (grades 7–9). (trial version)*. (初中物理课程标准(实验稿)). Beijing: Beijing Normal University Press (In Chinese).

Ministry of Education (MOE). (2003). *National physics curriculum standards for senior secondary school* (Grades 10-12). (trial version). (高中物理课程标准(实验稿)). Beijing: People's Education Press (In Chinese).

Ministry of Education (MOE). (2011). *National physics curriculum standards for the full-time compulsory education* (Grades 7-9). (义务教育物理课程标准). Beijing: Beijing Normal University Press (In Chinese).

Polikoff, M. S., Porter, A. C., & Smithson, J. (2011). How well aligned are state assessments of student achievement with state content standards? *American Educational Research Journal, 48* (4), 965–995. doi:10.3102/0002831211410684.

Porter, A. C. (2002). Measuring the content of instruction: Uses in research and practice. *Educational Researcher, 31*(7), 3–14.

Wang, L., Huang, Q., & Liu, D. F. (2010) Alignment between the chemical curriculum standard (Gr.7-9) and the exams: A case study on chemical concepts. *Basic Education Curriculum* (基础教育课程), *3*, 75–80 (In Chinese).

Webb, N. (1997). *Criteria for alignment of expectations and assessments in mathematics and science education* (Council of Chief State School Officers and National Institute for Science Education Research Monograph No.6). Madison: University of Wisconsin-Madison, Wisconsin Center for Education Research.

Webb, N. L. (1999). *Alignment of science and mathematics standards and assessments in four states* (Research Monograph No. 18). Washington, DC: Council of Chief State School Officers.

Webb, N. L. (2007). Issues related to judging the alignment of curriculum standards and assessments. *Applied Measurement in Education, 20*(1), 7–25.

Yuan, Y. K., Zhao Z., & Yu, Z. Q. (2002). *Commentate on science curriculum standard for compulsory education* (7th–9th grade.) (p. 159) (全日制义务教育科学(7–9年级)课程标准(实验稿)解读). Wuhan: Hubei Education Press (In Chinese).

Zhao, Y. (2009). *Catching up or leading the way: American education in the age of globalization*. Alexandria: ASCD.

Chapter 11
Relationship Between Science Teachers' Conceptions of Assessment of Students' Academic Performance and Their Instructional Approaches

Weining Wu

11.1 Introduction

Since 2001, China has pursued a new basic education curriculum reform (Ministry of Education (MOE) 2001). Important questions about such a reform that may be raised include: is it successful or not? What achievements have been made through the reform? These issues are big concerns to policy makers as well as to practitioners involved in the curriculum reform.

Effectiveness of curriculum reform can be examined from a variety of perspectives. One important indicator of curriculum reform is change in teachers' conceptions of teaching, which describe the way they view teaching. Teaching conceptions are important because they influence teachers' decisions on their teaching approach that, in turn, may have a major impact on students' learning approaches and learning outcomes (Trigwell and Prosser 1999). Through many years of studies in Asia, John Biggs found that there were three major learning approaches among Asian students (e.g., Biggs 1987). They were surface approach, deep approach, and achievement approach. Later on, in 1996, by a combination of qualitative and quantitative research, Trigwell and his colleagues (1999) found that, in classes where teachers describe their teaching approach as having a focus on knowledge dissemination, students are more likely to report that they adopt surface approach to the learning of that subject. Conversely, students who adopt significantly deep approaches to learning are found to be taught by staff who adopt approaches to teaching that are significantly more oriented toward students (Trigwell and Prosser 1999). As assessment is an important aspect of teaching, this research investigates

W. Wu (✉)
Institute of Physics and Electronic Science, Hubei University, 368 Youyi Ave., Wuchang, Wuhan, Hubei 430062, China
e-mail: palmer8888@163.com

L.L. Liang et al. (eds.), *Chinese Science Education in the 21st Century: Policy, Practice, and Research*, Contemporary Trends and Issues in Science Education 45, DOI 10.1007/978-94-017-9864-8_11

science teachers' general view on assessment issues, using the combination strategy of qualitative and quantitative methods. And on the basis of conceptual investigation, it explores the relationship between teachers' conceptions of assessment and their teaching approaches.

Investigating teachers' conceptions of assessment, among other teaching conceptions, is especially significant. As Brown (2004) pointed out:

> the implementation of new standards from professional bodies or state authorities, while well intentioned, may be reduced in effectiveness, if teachers' conceptions of assessment remain unchanged or unchallenged, or if teachers remain unaware of their own conceptions. Simply introducing an assessment innovation ... even if it is accompanied by appropriate teacher professional development, will not necessarily achieve policy objectives unless the differing, interlocked conceptions of teachers are exposed and addressed. Otherwise, quite possibly few teachers will adopt and utilize the innovation in a manner consistent with the intentions of developers of the innovation. (p. 314)

This prior research creates a precedent for secondary school teachers' assessment conception research in China, providing an important basis for course evaluation and an important reference for teachers' professional development as well. There is some research on teachers' conceptions of assessment in other countries or regions (Brown 2004; Watkins et al. 2005), but the participants in these studies are either primary school teachers or university lecturers. Whether there are any commonalities or differences between the results of this and other research also needs to be examined.

11.2 Methodology

In this section, I will introduce the design of the entire research, which includes three parts: research questions, research samples, and research process. Given the complementary nature of qualitative method and quantitative method, a growing number of researchers have begun to consider combining the use of the two methods into a third paradigm: mixed research methods (Johnson and Onwuegbuzie 2004; Johnson et al. 2007). This study will use a sequential design with a qualitative method first, mainly for in-depth interview of teachers' assessment conceptions, and then a quantitative method, mainly for assessing the teachers' assessment conceptions and their impact on their instructional approaches.

11.2.1 Research Questions

The main questions for the research are as follows: (a) What kinds of conceptions of assessment do secondary school science teachers have? (b) What are the characteristics of these conceptions? (c) What is the relationship between science teachers' conceptions of assessment and their instructional approaches?

11.2.2 Research Sample

In this study, two stages of sampling, an in-depth interview (qualitative) and a survey (quantitative), were adopted. In the qualitative research phase, four secondary schools were chosen that represent two different levels and two different types of schools. Two belong to the district authority, one belongs to the town authority, and one is an art high school. Finally, a sample group of 16 high school physics teachers was chosen for in-depth interviews.

In order to measure science teachers' conceptions of assessment and teaching approach in a quantitative way and to examine the relationship between the two variables, the researcher took a sample of science teachers from ten local high schools (with the help and assistance from the Foshan educational authority) and provided each of them with a "science teachers' conceptions of assessment and teaching approach questionnaire" to complete. The total number of distributed questionnaires was 220 (which is the same as the total number of science teachers within the ten sampled schools), and 196 valid completed questionnaires were returned (a response rate of 89 %).

11.2.3 Research Procedure

The three research questions mentioned above correspond to the three phases of this research:

Phase 1: in-depth interview survey on high school physics teachers' assessment conceptions.
Phase 2: questionnaire survey on secondary school science teachers' conceptions of assessment, describing their basic characteristics in a quantitative way. This survey was taken in the Nanhai and Chancheng Districts of Foshan City, Guangdong Province.
Phase 3: a questionnaire survey to assess the teaching approaches of science teachers and examine the relationship between teachers' conceptions and their teaching approaches.

The following is the detail of the procedure.

11.2.3.1 Phase 1

In this phase, in-depth interviews with 16 physics teachers were carried out at four secondary schools in the Nanhai District. This stage can be divided into the following steps:

Step 1: Class Observation

Table 11.1 Sample teachers who accepted interview

Teacher	Gender	Years of teaching	Grade to teach	Type of school
Teacher A	Male	3	12	District-supervised HS (type A)
Teacher B	Male	23	12	District-supervised HS (type A)
Teacher C	Male	26	10	District-supervised HS (type A)
Teacher D	Female	34	11	District-supervised HS (type A)
Teacher E	Female	17	10	Town-supervised HS
Teacher F	Male	5	11	Town-supervised HS
Teacher G	Male	11	10	Town-supervised HS
Teacher H	Female	12	10	Town-supervised HS
Teacher I	Male	14	10	District-supervised art HS
Teacher J	Male	26	10	District-supervised art HS
Teacher K	Male	19	11	District-supervised art HS
Teacher L	Male	8	12	District-supervised art HS
Teacher M	Male	3	11	District-supervised HS (type B)
Teacher N	Male	9	10	District-supervised HS (type B)
Teacher O	Male	10	11	District-supervised HS (type B)
Teacher P	Male	12	10	District-supervised HS (type B)

Note: Basically, Type A schools obtain more financial aid from the government than Type B schools. So, usually, Type A schools can enroll much more academically talented students than Type B schools. "HS" stands for high school

Before the interview, each interviewee arranged a class to be observed by the interviewer. One-to-one individual interviews were carried out immediately after each class. The researcher's main object for observation was instructional behaviors which related to assessments of students' performance, such as classroom questioning, assessment among students, teacher's comments on students' classroom activities and/or comment on their test paper, and homework. The aim of class observation before the interview is to build up a platform of communication between the researcher and respondent. It also facilitates starting the interview from a specific event which might have happened in the observed class. Starting an interview with a specific question rather than an abstract one is also usually more acceptable to interviewees and functions as a "warm-up" for the whole process of the interview.

Step 2: Individual Interviews

As mentioned above, the purpose of individual interviews is to understand the teachers' specific views about student assessment. The researcher, as the interviewer in this case, has taken a semi-constructive interview format. The implementation period was mid-November through early December 2006. The interviewees were the 16 high school physics teachers listed in Tables 11.1 and 11.2. The time length for each interview was 40–60 min. Each interview was recorded using a mini-recorder. Immediately after each interview, the interviewer transcribed the record of the interview into a text version. The text versions of the recordings were then sent via email to the teachers interviewed. The teachers were asked to check the text version of the recording to see if there

Table 11.2 Sample teachers who completed the questionnaires

Category		Number	Total
Gender	Male	104	196
	Female	92	
Subject taught	Physics	68	196
	Chemistry	73	
	Biology	55	
Level of school	Middle school	8	196
	High school	188	
Type of school	District supervised	108	196
	Town supervised	88	

was anything wrong or not appropriate. The teachers were also asked to give the feedback to the interviewer as soon as possible. The researcher sent 15 emails to 15 of the interviewees and then received their feedback about the text (one of the teachers did not leave an email address). Finally, based upon the feedback of the respondents, the researcher revised the text versions of recordings based on these responses.

Step 3: Identification of Teachers' Conceptions of Assessment

After revising the interview texts, the researcher immediately analyzed the interviews. Through systematic analysis of the interviews, five types of conceptions on students' assessment were obtained (for more detail, see Sect. 11.2.4.1).

11.2.3.2 Phase 2

The second phase of the study uses quantitative methods to investigate assessment conceptions of science teachers and contains the following two steps:

Step 1: developing the questionnaire

In this paper, a self-designed structure-format questionnaire entitled "Science Teachers' Assessment Conception Questionnaire (STACQ)" was used as the research instrument to investigate science teachers' conceptions of assessment in Nanhai and Chancheng Districts, Foshan City. Development of the questionnaire included four steps: identifying items, wording, compiling, and pilot testing. The pilot test was carried out at a middle school in Guangzhou, and the subjects were all science teachers at that school. Some items were deleted after reliability analysis (those items which may lower the whole reliability coefficient of the test were identified for deletion). Finally, 36 items were retained. Then the questionnaire was ready for the survey on science teachers' conceptions of assessment (for more details, see Sect. 11.2.4.2).

Step 2: surveying science teachers' conceptions of assessment

With the assistance of the local educational authority, the researcher handed out 200 questionnaires to science teachers in Nanhai and Chancheng. There were two main purposes of the survey: first, to investigate science teachers' conceptions of assessment in the districts and describe their characteristics and, second, to lay the basis for the final examination of the relationship between teachers' conceptions of assessment and their teaching approaches.

11.2.3.3 Phase 3

The third phase of the study used the quantitative method to examine the correlation between teachers' conceptions of assessment and their teaching approaches. In this study, the instrument for measuring teachers' teaching approach is the "Approaches to Teaching Inventory" (ATI; for more detail about ATI, see Sect. 11.2.4.2). Since the ATI had been repeatedly tested and used in many countries, this study adopted the modified version of the ATI as the instrument to assess teachers' teaching approaches. This phase included the following two steps:

Step 1: revision, pilot tests, and use of ATI
First, the ATI was translated into Chinese by the researcher. Next, based on considerations of cultural convention and habits of expression in practice, its wording was modified slightly. Since the ATI was first developed in the West and used for university faculty, the researcher did some slight revisions and adjustments to the ATI based on consultation with experts[1] in order to make it applicable to middle school teachers in China before it was officially used. In this study, the ATI went through two rounds of pilot testing: the first was carried out in Xuancheng City, Anhui Province, with a sample of 30 science teachers at an ordinary high school in this city. The results of the reliability coefficients of the test were unacceptable, so the ATI was modified slightly and then taken for the second round of pilot testing in the Nanhai District, Foshan City, Guangdong Province, with a sample of 40 high school science teachers. The aim of the pilot tests was to examine the internal consistency coefficient of the ATI and to see whether or not its expressions and wordings were readable for Chinese middle school teachers. At the end of December 2006, the revised ATI was administered officially to 196 science teachers from 10 middle schools in the Nanhai District to measure their teaching approaches (for more detail about the process, see Sect. 11.2.4.2).

Step 2: examining the correlation between teachers' conceptions of assessment and their approaches to teaching
Finally, with the results of the Science Teachers' Assessment Conception Questionnaire and ATI survey, using correlation analysis, the researcher analyzed the

[1] The three experts who did validity inspection of the questionnaire were a university professor who specializes in curriculum and instruction, an instructional researcher from a local educational authority, and an experienced high school science teacher. The main feedback from the experts was that using the phrases "college entrance exam" and "high school entrance exam" instead of "official exam" would be better because it was much more specific and readable for the teachers completing the questionnaire. As the result, three items were revised.

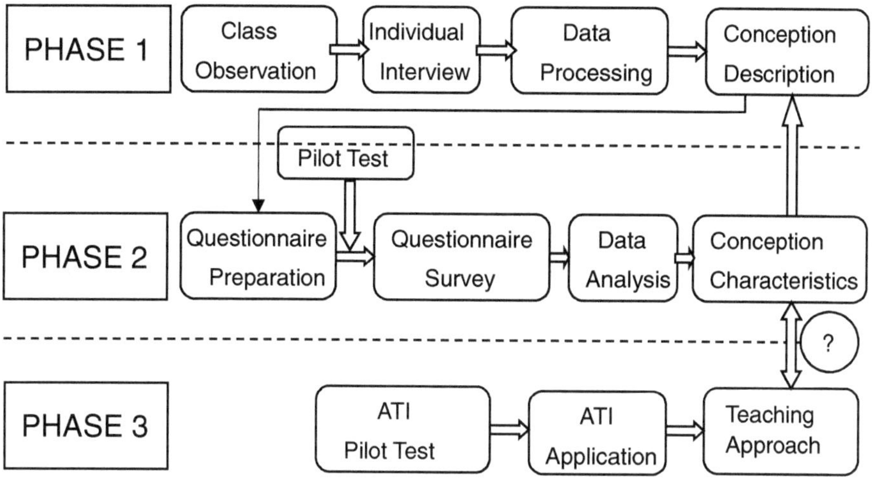

Fig. 11.1 Research procedure

data in order to assess the relationship between teachers' conceptions of assessment and their approaches to teaching.

The whole research procedure can be described in the following flowchart (Fig. 11.1).

11.2.4 Research Methods

In this part, I will introduce the qualitative method used in Phase 1 and the quantitative methods used in Phases 2 and 3.

11.2.4.1 Qualitative Method

In the 1980s, Ference Marton, a Swedish scholar, began to use phenomenography to identify students' conceptions (Trigwell 1994). Since then, phenomenography has been a widely used qualitative method to investigate variations in how a group of people experience various educational phenomena (Watkins et al. 2005). It is not a method in itself but a framework for identifying the experiencers' conceptions of a phenomenon in some fields (Gao 2004, p. 70). It focuses on human conception. Using one-on-one in-depth interviews, it collects and analyzes data in accordance with a certain framework. Its interview strategy starts from specific issues preceded by a class observation. Finally, it comes up with certain categories of the conceptions of the interviewees. In mainland China, the first person who used phenomenography to analyze the teaching conception of teachers is Prof. Gao

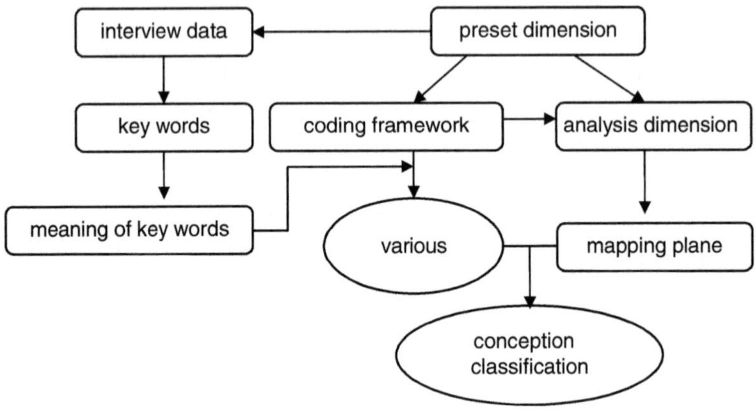

Fig. 11.2 Process of data analysis

Lingbiao. He made an investigation using phenomenography on physics teachers' teaching conceptions with a large sample in Guangdong, China, in the 1990s (Gao and Watkins 2001). Afterward, his graduate students conducted a series of surveys on teachers' teaching conceptions in different subjects and on different teaching conceptions using the same research method (Jin 2003; Zhang 2004, 2005; Liang 2005, 2006).

This section will describe the qualitative method used in the first stage of my investigation. It includes the framework for data collection, the process and strategy of the interviews, and strategies for data analysis (Fig. 11.2).

Data Collection Framework

The data collection framework for this study includes the choice of interview structure and preset dimensions. A semi-structured interview type was adopted based on previous theoretical and empirical studies' results. Four dimensions were adopted as preset dimensions. They are purpose of assessment, content of assessment, subject of assessment, and method of assessment.

Interview Process

All interviews followed a similar protocol. They started with 10 min of free talk as a "warm-up" in order to create a relaxed and natural conversation atmosphere. The interviews started with the specific issues, for example, asking the interviewee to give a brief description of the design of the observed class and to explain the assessment behavior that occurred in the class. After briefly talking about specific issues related to the content of classroom teaching, the researcher then changed the topic of the interview to more abstract issues.

Table 11.3 Sample questions in the interviews

Preset dimension	Sample question
Purpose of assessment	Why do you encourage students to make peer assessments?
Content of assessment	Generally, what kind of questions will you ask your students in class?
Method of assessment	Have you ever tried any method of assessing the students' innovation ability?

(1) Because there are not enough data collected to support the subject of assessment dimension, this dimension was finally deleted. (2) For the entire interview protocol, see the appendix to this chapter

Generally, the questions were raised in preset dimensions.

Most interviews lasted for 45–60 min (Table 11.3).

Data Analysis Strategy

Data analysis for this research went through three stages. The first stage was extracting the key words and key phrases from the data. The second stage was encoding the teachers' conceptions of assessment. Finally, in the third stage, the researcher classified the conceptions of assessment of the science teachers with the following mapping plane.[2]

The mapping plane is used for encoding the science teachers' conceptions of assessment and is basically a process of arrangement of teachers' conceptions (indicated by the extracted key words), in the order from the most teacher centered to the most student centered, in different analytic dimensions. Following the encoding task is identification of teachers' assessment conceptions. Identification is actually a classification of teachers' assessment conceptions, based on qualitative data collected, in accordance with certain rules. The mapping plane is the carrier of these rules. In Table 11.4, each row represents a dimension, and each column represents a conception. As mentioned above, this research has identified several analytic dimensions and there are several conceptions in each dimension. The mapping plane is arranged such that each dimension occupies one row, arranged in order from the extreme teacher centered on the left to the extreme student centered on the right. Thus, the most teacher-centered conception will be located in the leftmost column on the mapping plane, and the most student-centered conception will be in the rightmost column. Columns between the leftmost and the rightmost are conceptions that range between the most teacher centered and the

[2] The mapping plane is used to identify teachers' conceptions of assessment. Its structure is determined partly by preset dimensions and partly by data collected. For this case, the original four preset dimensions were reduced to three because the data collected did not support one of the four. The five Cs representing the five conceptions were also determined by collected key words and phrases.

Table 11.4 Mapping plane for identifying teachers' conceptions

	←			→	
	Teacher-centered			Student-centered	
	Conception				
Dimension	C1	C2	C3	C4	C5
Purpose of assessment	Keep order	Prepare for exam	Inform teaching	Motivate interest	Inspire thinking
Content of assessment	Classroom performance	Exam skill	Homework	Physics ability	Comprehensive ability
Method of assessment	Classroom exercise	Problem-solving with paper and pen	Classroom quiz	Lab work	Multiple assessment

most student centered. The number of columns on the mapping plane represents the number of categories of teachers' conceptions of academic assessment. The mapping plane is used to identify patterns in the teachers' conceptions of assessment overall and possible similarities or differences that may exist in specific categories such as the teachers' school type or subject area.

11.2.4.2 Quantitative Method

As mentioned above, in addition to investigating teachers' concepts of academic assessment of students using a qualitative method, this research has also used quantitative methods to assess teachers' concepts. Specifically, quantitative research methods are used to complete the following three tasks: (a) measure the teachers' assessment concepts; (b) measure teachers' teaching approaches; and (c) do correlation analysis between teachers' assessment concepts and their teaching approaches.

As mentioned above, in this research, the quantitative tool used to measure teachers' assessment concept is called the "Science Teachers' Assessment Conception Questionnaire" (STACQ). STACQ was entirely compiled by the researcher on the basis of the previously carried out interviews. Specifically, STACQ's development and application included the following four steps:

(a) Design of the items. STACQ has five subscales that correspond to the five categories of assessment conception obtained in the interviews. Taking the teachers' typical statements in the interviews as references, I designed STACQ to make full use of the key words that appeared in the interviews as basic materials.
(b) Revision of the items. After the item pool (containing 77 items) was built, it was sent to relevant experts for a quality check. At the same time, by email, four interviewed teachers were invited to give suggestions about item revision after reading the item pool. Then, based on the feedback from the experts and teachers, several changes were made to the draft items. Five items were deleted and some were modified. Seventy-two items remained for the pilot test.

(c) Pilot test of draft items. The pilot test was conducted at a high school in Guangzhou. Thirty-four science teachers (teaching physics, chemistry, and biology) took part in the test. Thirty-six items were deleted according to the reliability analysis and content analysis. The remaining 36 items became the final version of STACQ for formal survey (see Appendix II).
(d) Quality inspection. This should be done after any formal survey. For this research, analysis of reliability and construct validity of STACQ was conducted after the survey.

In the formal survey using STACQ, the reliability analysis specifically calculated the subscales' coefficient of internal consistency, Cronbach's coefficient α. The results of the reliability analysis are shown in Sect. 11.3.2.1. The construct validity study of STACQ used confirmatory factor analysis (CFA). In the process of CFA, the researcher found that nine items did not fit the model. Then these nine items were deleted from STACQ. Ultimately, 27 items remained. These items constituted the final version of STACQ. When conducting the correlation analysis between assessment conception and teaching approach, this final version of STACQ was used to provide data for teachers' conceptions of assessment. The final results of STACQ's construct validity analysis and loading of each factor are also shown in Sect. 11.3.2.1.

Another quantitative measurement tool used was the Approaches to Teaching Inventory (ATI; Trigwell and Prosser 2004). It consists of four subscales: teacher-focused information transmission (ITTF), teacher-focused conceptual acquisition, student-focused conceptual development (CDSF), and student-focused conceptual change (CCSF). Each subscale contains four items. Trigwell and Prosser (2004) reported two of the four subscales' Cronbach's α: Alpha ITTF = 0.73 and Alpha CCSF = 0.75. In this study, a teaching approach inventory (new ATI) was developed on the basis of Trigwell and Prosser's ATI (old ATI).

The new ATI's preparation and revision process include the following steps: pilot test, revision, and re-pilot test.

(a) Pilot test. After the old ATI was translated into Chinese, it was sent to a secondary school in Xuancheng, Anhui Province. All science teachers in this school were asked to complete the ATI. Of the 30 questionnaires distributed, 25 valid responses were returned. The results of the reliability analysis (Cronbach's coefficient α) for the new ATI's four subscales were 0.52, 0.54, 0.56, and 0.05. The scales' reliability values were unacceptably low, so the researcher started to revise the questionnaire.
(b) Revision. This revision had three tasks. The first task was to adjust the framework by changing the four original subscales into the following four new subscales: student-centered tendency (A), student-centered strategy (B), teacher-centered tendency (C), and teacher-centered strategy (D). The second task was to modify the original items of old ATI. Since the old ATI was developed in Australia and written in English, using the ATI in China inevitably

led to some problems due to language and teaching context differences between the two countries. Thus, the researcher modified the items where such problems occurred, using the typical Chinese language and making the statement of items more consistent with an ordinary teaching environment in China. The third task was to write new items. Considering that the subsequent pilot test would delete some items, the researcher wrote 16 new items for the ATI in the new framework. Thus, together with the originally revised 16 items, the total number of items of new ATI reached 32.

(c) Re-pilot test. The second pilot test was conducted at a high school in Nanhai, Guangdong. The subjects were all of the school's science teachers. Forty copies of questionnaires were handed out, and 38 valid copies were returned. In the process of reliability analysis, 14 items that led to a lower value of Cronbach's α were deleted. Finally, 18 items remained, forming the final version of ATI (see Appendix III). At this point, reliability values of four subscales of the new ATI are as follows: Alpha A = 0.71, Alpha B = 0.74, Alpha C = 0.75, and Alpha D = 0.76. These observed values of reliability are more than acceptable compared with those of previous version of ATI.

It is important to point out that the above two measurement instruments are Likert-type scales,[3] in which each item score can be summed up to form a total score. And these total scores can be used to do correlational analysis.

11.3 Results

In this part, I will introduce qualitative research results from Phase 1 and quantitative research results from Phases 2 and 3.

11.3.1 Results of Qualitative Research

As a report on the first phase of research, this section will describe in detail the five categories of secondary school science teachers' conceptions of assessment extracted from interview recordings. Specifically, it will include key phrases and key words and science teachers' five kinds of conceptions of assessment. Finally, two kinds of higher-level assessment orientations will be introduced based on the abovementioned research findings.

[3] For more details about Likert scale, see https://en.wikipedia.org/wiki/Likert_scale.

11.3.1.1 Key Phrases Extracted from Different Analysis Dimensions

The analysis dimensions adopted in this section are the same as the previously used preset dimensions forming a framework for interview information collection (see Sect. 11.2.4.1). From the data collected, I found that interview data (key words and phrases) are sufficient to support the three dimensions, and different conceptions of assessment can obviously be seen from this framework. These three dimensions are purpose of assessment, content of assessment, and method of assessment. Key phrases extracted from different dimensions are as follows:

Sample key phrases in dimension of assessment purpose are:

Bring back the distracted students.
Develop standardized problem-solving habits.
Check learning outcomes.
Increase their interest in learning; develop interactive skills.

Sample key phrases in dimension of assessment content are:

Classroom discipline
College entrance examination objectives
Knowledge mastering
Lab work ability
Communication skills

Sample key phrases in dimension of assessment method are:

Classroom questioning
Learning competition
Classroom test
Aircraft-model making competition
Portfolio assessment

11.3.1.2 Categories of Science Teachers' Conceptions of Assessment

From the results of the above analysis, it can be seen that obviously different conceptions of assessment in the above three dimensions do exist. These conceptions, in different dimensions, are arranged from teacher-centered orientation to student-centered orientation. They form the five columns in the mapping plane that is used to differentiate science teachers' conceptions of assessment. These columns, in turn, represent five different conceptions of assessment. In this way, there are five categories of science teachers' conceptions of assessment.

The first type of conception was defined as "management-oriented conception." It focuses on the student's classroom performance by classroom reviewing, aiming at maintaining classroom discipline. The second type is defined as "examination-oriented conception." It focuses on critical knowledge that may be on the college entrance examination (CEE) by assessing students in a way that is significantly similar to that of the CEE, aiming at being well prepared for the CEE. The third type of conception was defined as "teaching-oriented conception." This conception stresses teaching effectiveness and improvement of student learning. Specifically, it focuses on the quality of students' assignments and students' mastery of teaching content. Usually, it takes the form of classroom questioning and paper-and-pen testing to assess the students. The fourth type of conception was defined as "ability-oriented conception." This conception aims at stimulating the students' interests of all kinds and cultivating their abilities. It focuses on examining students' overall quality and all kinds of ability in the process of students' scientific and technology exploration activities. The fifth type of conception was defined as "subject-oriented conception." It induces students to think actively and encourages the interaction between the teacher and students. In addition, it focuses on students' emotions and attitudes, advocating process and multiple-subject assessment.

Classroom Management-Oriented Conception

This conception takes student assessment as a tool of classroom teaching control. Usually, it takes the measure of classroom questioning as student assessment, either asking the whole class to get a collective answer or asking an individual student who is absent-minded during the class. Both ways of questioning students are aimed at "waking them up" in the instructional process. The following are typical statements from teachers in the interviews (Fig. 11.3):

> Questioning can be used as a means of teaching organization. In today's class, something occurred concerning this issue. A student was looking around during the class. I guessed that he would like to see the reaction of the teachers sitting in the back of the classroom. Just at this moment, I asked him a question. If a students' attention is distracted in the class, you can ask him a question. In this way, you are telling him: this is in the class!
>
> Teacher E
>
> Sometimes asking a question to a student during the class is simply a reminder for him. In other words, it is to organize teaching or maintain the discipline. For example, if there is a student who fails to listen to me and pursues his own affairs during the class, I will give him a reminder by asking him a question.
>
> Teacher F

Examination-Oriented Conception

"Examination" generally refers to out-of-school and educational-authority-organized official tests used to select candidates to receive higher education. The interviewees in this research were all high school teachers, so "examination" here refers to the college entrance examination (CEE). The examination-oriented conception takes assessment as the pre-training for the CEE. It requires the assessment in the learning process to be implemented in accordance with the CEE. Specifically,

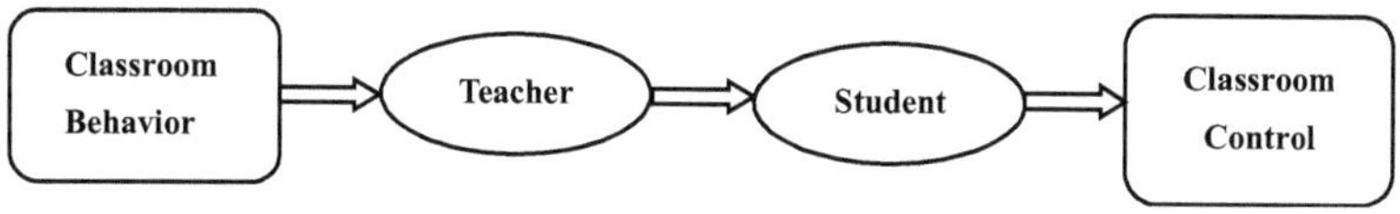

Fig. 11.3 A model of management-oriented assessment conception

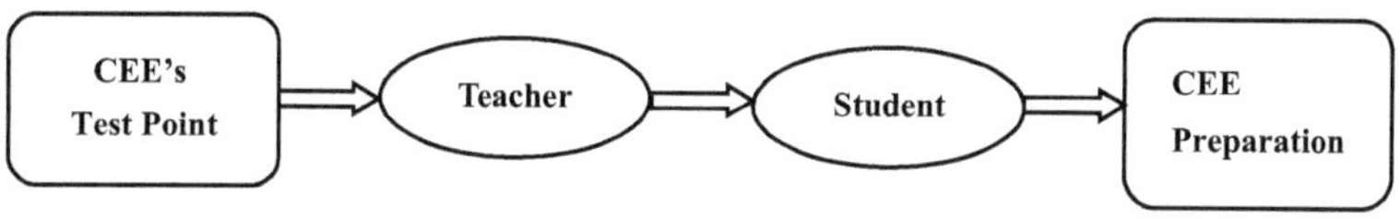

Fig. 11.4 A model of exam-oriented assessment conception

it is the same as CEE in assessment content, assessment standard, and assessment method. For example, it emphasizes the paper-and-pen test and focuses on content knowledge that might be tested in CEE. Furthermore, it takes problem-solving ability and problem-solving speed as assessment standards, so students' scores in various simulated CEE tests are highlighted. The following are typical statements of the interviewees (Fig. 11.4):

> As a teacher of a grade 12 class, I am very concerned about whether they are able to pass the CEE in order to go to the college. I will adopt all useful methods to assess the students.......so-called "useful methods" refers to any methods that can make it possible for them to enter college.
>
> Teacher A
>
> A few years ago, in order to change the dull atmosphere before the CEE, I put forward a proposal to our principal to carry out a learning competition which aimed at stimulating the teachers and students alike. If a student made great progress, he received RMB500 as a bonus. Now the bonus has been canceled, but the success of the competition was recognized. Since then, we have had this contest every year.
>
> Teacher B

Teaching-Oriented Conception

This conception views the assessment as an important means of diagnosis of teaching and good methods for improving student learning. Here, the assessment methods used include classroom questioning, assignment correction, and classroom checking. By doing this, teachers can collect necessary information about how well the students learned in order to decide on their own teaching pace. In short, assessment here is an integral part of the teaching process and has become an important tool for teachers to achieve teaching objectives. The following are typical statements of the teachers interviewed for the related issues (Fig. 11.5):

> I personally believe that [classroom questioning] stimulates students' thinking and guides its direction. Through the process of Q&A, the teacher can induce the students to gradually deepen their understanding of the content knowledge. In fact, the process of Q&A happening in the classroom is really a process of deepening the understanding of knowledge taught in the class and is an integral part of teaching.
>
> Teacher B

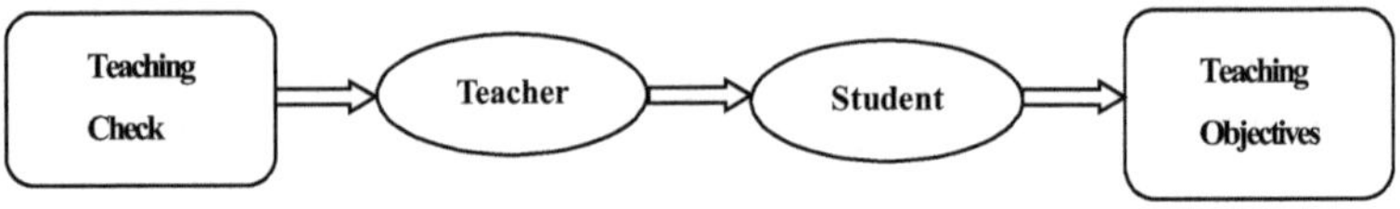

Fig. 11.5 A model of teaching-oriented assessment conception

> Through questioning, students can be guided from one knowledge point to the next. So it's actually playing a connecting role. Another purpose of questioning is to examine the effect of student learning.
>
> Teacher C

Ability-Oriented Conception

This conception takes assessment as a significant means of cultivating students' ability and stimulating their interest in learning. Here, the ability refers to capacity of learning, including hands-on ability, creativity, communication skills, interpersonal skills, organizational skills, and sense of responsibility. Interest here refers to not only interest in textbook knowledge but also interest in relevant subject knowledge and its application, understanding of society, and participating in social practice. The following are typical statements of the teachers interviewed about related issues (Fig. 11.6 and Table 11.5):

> We do have some initiatives for cultivating and assessing students' abilities. For example: we have carried out "a second class" (extra-curriculum activities), innovative experiments, had students participate in some social activities, etc. Furthermore, we have a "sports and arts festival" annually which is entirely organized by students. This is to help students practice their organizational abilities. With respect to management, we have had students participate and run their own dormitory management committee. This is a way of practicing "student-self-management." In fact, these activities are not only an effective way of cultivating students' abilities, but also an effective means of ability assessment.
>
> Teacher C
>
> Our teaching is not merely for the college entrance examination. Physics students (students who will take the physics exam in CEE) need to know some things about physics in their daily lives, such as the principles behind the operation of electrical appliances and other knowledge of physics applications. So I have been paying much attention to cultivating and assessing students' abilities. We intend to organize "creative experiment" activities for students, having them design the experiment and write the lab reports by themselves. Obviously, we can see students' abilities in this process.
>
> Teacher D

Subject-Oriented Conception

This conception takes assessment as an important way of inducing students' critical thinking, cultivating their personality, and encouraging their independent innovation. It emphasizes that assessment should be based on students' need, focusing on their emotion and attitude development and cultivation of their personality. Aiming

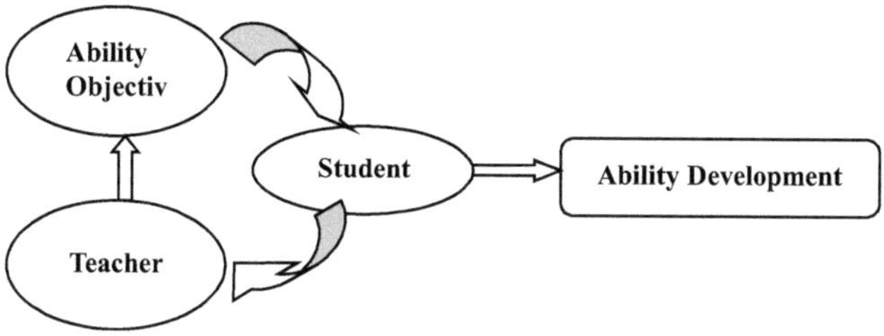

Fig. 11.6 A model of ability-oriented assessment conception

Table 11.5 Science teachers' conceptions of assessment

Dimensions	Conceptions				
	Management oriented	Examination oriented	Teaching oriented	Ability oriented	Subject oriented
Assessment purpose	Maintain discipline	CEE preparation	Encourage learning	Cultivate ability	Individual development
Assessment content	Discipline performance	CEE's test points	Master knowledge	Comprehensive ability	Process and method
Assessment method	Classroom comments	Practice prior to CEE	Paper-and-pen test	Scientific inquiry	Alternative assessment

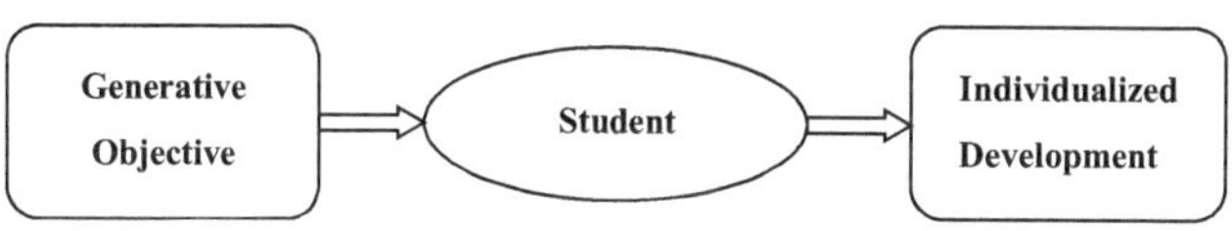

Fig. 11.7 A model of subject-oriented assessment conception

at cultivating students' lifelong ability and their creative consciousness and innovation ability, it emphasizes process assessment and multi-subject assessment as methods. The following are typical statements of the teachers interviewed about related issues (Fig. 11.7):

> We have been doing this [process assessment] every semester since Grade 10. As a matter of fact, we have all these archived materials such as process assessments and growth portfolios. If these are expected to be made into electronic archives in the future, it will not be a hard work to do. Personally, I think it is beneficial to measure student's performance as a college admission basis. At least it can benefit those who failed in the exam incidentally. And that growth record will reflect their strong points.
>
> Teacher A
>
> One of the problems in our evaluation system is that we use only one mode to evaluate all the students. Just like wearing clothes, you wear this suit, I wear that suit, each of us will feel free and comfortable. Standards and modes should be diversified. And that might be the goal of our evaluation reform.
>
> Teacher C

11.3.2 Results of Quantitative Research

In this section, the overall results of measurement of science teachers' conceptions of assessment, their teaching approaches, and correlation between the two variables, as well as the reliability and construct validity of the measurement tools, will be reported.

11.3.2.1 The Survey Result of Science Teachers' Assessment Conception

From Table 11.6, we can see that among STACQ's five subscales' Cronbach's α values, four exceed 0.7 and only one is between 0.6 and 0.7, meeting the basic indicator requirement for subscales (Wu 2003).

In general, if the RMSEA is smaller than 0.08 (the smaller, the better) and NNFI and CFI are bigger than 0.9 (the bigger, the better), then the model is a "good" one (Hau et al 2004). From Table 11.7, we can see that the three indices of goodness of fit, RMSEA, NNFI, and CFI, have met the good model requirements. Thus, we can say that the 27-5 model is a good one. In other words, the STACQ has good construct validity. Here, "27" denotes that STACQ has 27 questions, and "5" represents that STACQ has five subscales which were used to measure five kinds of conceptions of assessment. The following chart is the parameter estimation of the 27-5 model, which also shows a good result of construct validity examination (Fig. 11.8).

From Table 11.8 we can see that the fractions of agrees and disagrees are almost the same in management dimension. (The sum of column 1 and column 2 is 34.7 %,

Table 11.6 Subscale reliability of STACQ

Subscale	Mean	S.D.	Cronbach's α
Management oriented	15.06	3.67	0.72
Examination oriented	18.29	3.43	0.71
Teaching oriented	14.98	2.59	0.61
Ability oriented	24.78	5.10	0.83
Subject oriented	21.56	3.82	0.73

Table 11.7 Goodness of fit statistics

Sample size	196
Model to be examined	27-5
Degree of freedom	314
Chi-square	571.59
Root mean square error of approximation (RMSEA)	0.06
Non-normed fit index (NNFI)	0.92
Comparative fit index (CFI)	0.93

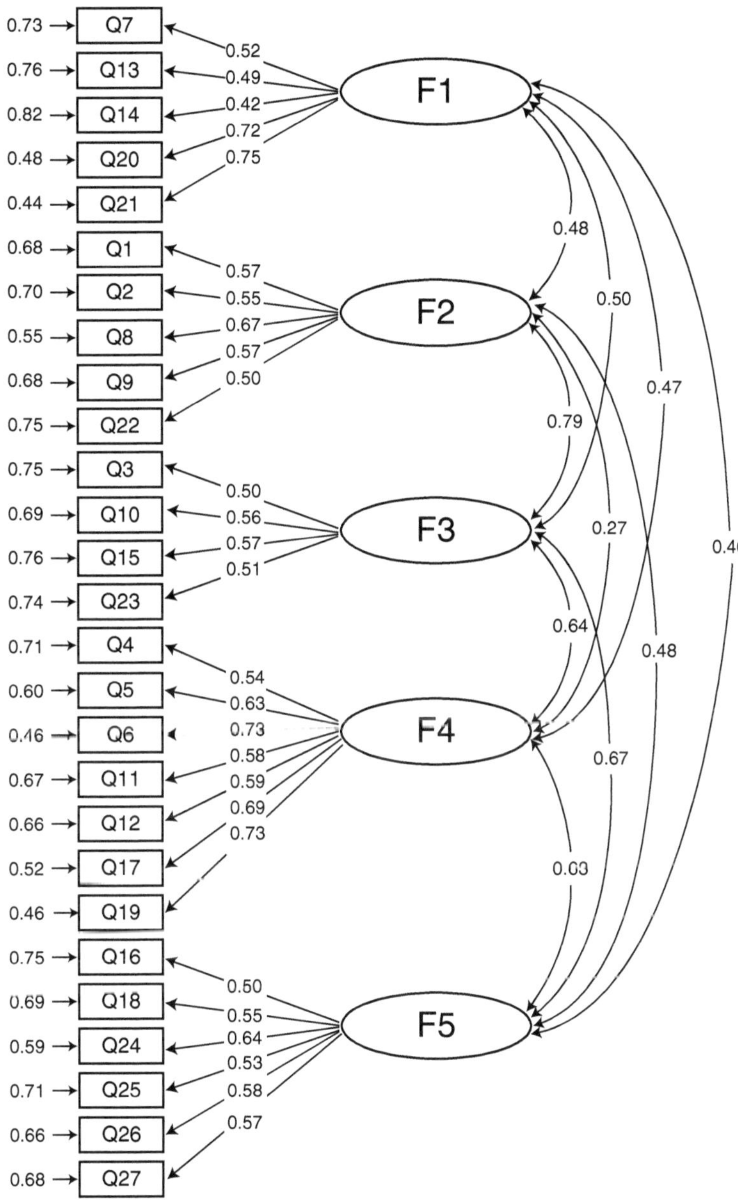

Fig. 11.8 Parameter estimation of 27-5 model. Note. In the figure, F1–F5 represent the five subscales (factors) of the STACQ, and Q1–Q27 denote the 27 items (questions) of the STACQ

while the sum of column 4 and column 5 is 35.6 %. The difference between the two is 0.9 %.) But in the other four dimensions, the fraction differences between agrees and disagrees are obviously significant. They are 46.2 % for examination oriented, 51.8 % for teaching oriented, 36.0 % for ability oriented, and 42.8 % for subject

Table 11.8 The statistical result of science teachers' response to the STACQ

	Disagree		Neutral		Agree	
Assessment conception	N	%	N	%	N	%
Management oriented	340	34.7	291	29.7	349	35.6
Examination oriented	148	15.1	231	23.6	601	61.3
Teaching oriented	94	11.9	190	24.2	500	63.7
Ability oriented	246	18.4	371	27.0	748	54.4
Subject oriented	175	14.9	323	27.5	678	57.7

Note: N in the table represents the total number of attendees who chose that option for an item

Table 11.9 Subscale reliability of ATI

Subscale	Mean	S.D.	Cronbach's α
Student-centered orientation	15.35	2.68	0.65
Student-centered strategy	15.26	2.68	0.66
Teacher-centered orientation	16.51	3.51	0.69
Teacher-centered strategy	16.50	3.03	0.58

oriented. Among them, the differences between agrees and disagrees in the examination- and teaching-oriented dimensions are the most significant. This implies that among this group of subject teachers, most of them take the assessment as a tool to facilitate their teaching and help students get high scores on the CEE. Simultaneously, they also emphasize that assessment should be used for cultivating students' ability and developing their individualized personalities. On the one hand, this result reflects the real state of basic education in China: teaching and CEE are the eternal themes that cannot be shaken. On the other hand, it reflects the influence of the new curriculum on the teachers, which emphasizes that capacity building and self-development of the students will benefit them for their whole life. In addition, from the data analysis, it was found that the mean value for each scale of STACQ was more than 3. So, it appears that any individual teacher's conceptions of assessment are not single, but multiple and potentially conflicting.

11.3.2.2 The Survey Result of Science Teachers' Teaching Approach

From Table 11.9 we can see that among the four subscales of the ATI, three subscales' reliability indexes meet the minimum requirement (bigger than 0.60), but one value is comparatively lower (smaller than 0.60).

The data in Table 11.10 indicate that in the four subscales, the numbers of agrees are all larger than the numbers of disagrees (the sum of columns 4 and 5 is larger than that of columns 1 and 2). Closer inspection shows that the ratios of difference between agrees and disagrees among four subscales are significantly different. The numbers of the specific differences are student-centered orientation, 55.73 %; student-centered strategy, 56.00 %; teacher-centered orientation, 23.38 %; and teacher-centered strategy, 23.36 %. So we can say that the majority of the science teachers are for or have carried out student-centered teaching.

Table 11.10 The statistical result of science teachers' responses to the ATI

	Disagree		Neutral		Agree	
Teaching approach	N	%	N	%	N	%
Student-centered orientation	69	8.8	209	26.7	506	64.5
Student-centered strategy	72	10.5	181	23.1	521	66.5
Teacher-centered orientation	209	21.3	333	33.9	438	44.7
Teacher-centered strategy	194	19.8	363	37.0	423	43.2

Table 11.11 Correlation between teaching approach and assessment conception

	Teaching approach			
	Student-centered orientation	Student-centered strategy	Teacher-centered orientation	Teacher-centered strategy
Assessment conception	Corr. coef.	Corr. coef.	Corr. coef.	Corr. coef.
Management oriented	0.13	0.12	0.52**	0.47**
Examination oriented	0.27**	0.17*	0.49**	0.48**
Teaching oriented	0.48**	0.50**	0.37**	0.39**
Ability oriented	0.35**	0.36**	0.21**	0.19**
Subject oriented	0.58**	0.40**	0.34**	0.39**

*$p < 0.05$; **$p < 0.01$

11.3.2.3 Correlation Analysis of Teaching Approaches and Conceptions of Assessment

The data in Table 11.11 shows that except for a nonsignificant correlation between the management-oriented conception and the student-centered teaching approach, the other four conceptions of assessment are all correlated with the two different teaching approaches (teacher centered and student centered). It indicates that the teachers who have management-oriented assessment conceptions mainly adopt a teacher-centered teaching approach. Other four conception holders may take a student-centered approach and teacher-centered approach simultaneously. Its implication is analyzed in detail in Sect. 11.4.3.

11.4 Conclusion and Discussion

This section will address two parts. One is the main conclusion of the research, including the qualitative and quantitative conclusion that teachers can hold multiple conceptions of assessment that have less influence on teachers' teaching approaches than the influence of the CEE. The other is a limitation analysis of the research and a discussion of what further research needs to be done.

11.4.1 Interview Analysis of Science Teachers' Assessment Conception

As mentioned above, the research result of the first phase is the identification of science teachers' five conceptions of assessment: management-oriented, examination-oriented, teaching-oriented, ability-oriented, and subject-oriented conceptions. Through further analysis of this result, we can see that there are two main intentions among these conceptions, i.e., utility-oriented and development-oriented intentions. The first three conceptions of assessment can be classified as utility-oriented intentions, while the latter two conceptions of assessment can be classified as development-oriented intentions. The specific differences between the two intentions are as follows:

(1) The objectives concerned are different. From a time-dimension perspective, utility-oriented intentions focus on short-term objectives, such as whether or not classroom discipline is effectively controlled, whether or not the teaching goal has been reached, and whether or not the students' test scores are satisfactory. Development-oriented intentions pay more attention to long-term objectives, such as students' literacy and ability for lifelong development. From a space-dimension perspective, a utility-oriented intention focuses on local targets such as students' book-knowledge learning and their ability to cope with examinations. Development-oriented intentions emphasize the whole target, such as students' comprehensive literacy and practical and innovation ability.
(2) Standards for judgment are different. Utility-oriented intentions emphasize external standards of assessment, such as whether or not it is beneficial to or conform to teaching and curriculum standards and the authorities' examination explanations, while development-oriented intentions focus on internal standards such as students' personality traits and ability characteristics.
(3) The ways of conducting activity are different. Here, activity mainly refers to assessment methods. Utility-oriented intentions usually involve the use of paper-and-pen tests to assess students, while development-oriented intentions will mainly use activity-based and context-rich assessment methods.

Generally, the assessment conceptions of high school physics teachers have the following three characteristics: First of all, teachers' conceptions of assessment are multiple. In the process of the interviews, the researcher found that the interviewees tend to have more than one conception of assessment. For example, there are some teachers who have argued that assessment should be able to cultivate students' comprehensive abilities and serve to develop their personalities. On the other hand, teachers put great emphasis on assessment's teaching and management functions. Second, conceptions of assessment are situation dependent. We found that teachers' conceptions of assessment have a close relationship with their teaching environment. Specifically, a difference in school and the student quality determined teachers' conceptions of assessment to some extent. For example, teachers from town-supervised schools usually put more emphasis on the management function of assessment. District-owned school teachers gave more emphasis to the examination function of assessment. Ordinary high school teachers focus on the teaching function of assessment, while art school teachers put more emphasis on the assessment function of promoting students' personalities. Third, conceptions of assessment are subject to external environment. Here, external environment refers to the impact of the external examination, i.e., the National College Entrance Examination. In the interviews, almost all the interviewees talked about the impact of the CEE.

From my findings on teachers' conceptions, it can be seen that in the basic education stage school teachers are always in a dilemma. On the one hand, they have to implement exam-oriented education due to the tremendous impact of the CEE. This is the bottom line for schools and teachers to survive. On the other hand, schools and teachers are overwhelmingly criticized by people from all walks of life in society. This is unfair for schools and teachers. In fact, the crux of the problem does not lie in the schools and teachers, but rather the social and external examination (e.g., CEE) system. The results of the study show that to solve the deep-seated problems in school education, it is necessary to relieve the pressure of external examinations. Exam pressure, however, usually stems from the pressure of employment. From this it can be seen that CEE is not merely an educational issue but a social problem as well. Therefore, as a bottleneck in the development of basic education, the college entrance examination must be reformed. But how should it be reformed? We cannot just work with the technical issues such as test content and test method, and we cannot be concerned only with the internal problems of education, but the entire enrollment system of higher education and the corresponding social support system must be reformed from the overall perspective of society.

11.4.2 Measurement Analysis of Science Teachers' Assessment Conception

Based on measurement, the following two characteristics of the sample teachers' conceptions of assessment can be seen.

First, their conceptions are multiple (this is consistent with the previous interview result). In other words, a teacher may have more than one conception of assessment. By examining the average score of each subscale of the questionnaire, we found that the mean value of each subscale is greater than 3 (the median value of the Likert five-point subscale). This shows that most teachers may recognize more than one assessment conception. These conceptions may be similar, contradictory, or even totally opposite. It may indicate that teachers' conceptions are complicated. Faced with different situations, teachers may have different ways of conceptual expression. This recognition of multiple conceptions is also related to the issue of the distinction between "ideal conceptions" and "practical conceptions." This research does not distinguish between the two types of conceptions. In the questionnaire, there are some ideas that are consistent with those of the mainstream society. These are ideal conceptions. Other ideas may be taken into practice by teachers, and these are so-called practical conceptions. Therefore, on the one hand, the subjects tend to choose practical conceptions, because they are thought to be beneficial to their teaching. On the other hand, they may agree with ideal conceptions, which coincide with the mainstream values of society.

Second, science teachers put great emphasis on the teaching and examination functions of assessment. The measurement results of teachers' assessment conceptions showed that the proportional differences between agreement and disagreement in teaching-oriented and examination-oriented subscales are the biggest, 51.8 % and 46.2 %, respectively (for more detail, see Table 11.8). This shows that the assessment purpose of the vast majority of teachers is for teaching inspection and college entrance examination preparation. At the same time, data on the other two subscales show a proportional difference of 36.0 % for capacity-oriented and 42.8 % for subject-oriented. On the one hand, this may be the external manifestation of the multiplicity of science teachers' conceptions of assessment. On the other hand, it may be the positive impact of the curriculum reform being carried out in China.

11.4.3 The Relationship Between Teaching Approach and Assessment Conception

During the preparation and revision of items for the science teachers' teaching approach questionnaire (ATI), the researcher's assumption for the two teaching approaches is as follows. The teacher-centered approach focuses on one-way transmission of knowledge and emphasizes academic knowledge, while the student-centered approach focuses on interaction and communication between teachers and students and it cares about students' feelings, attitudes, and preconceptions.

The test results of the ATI in general show that science teachers' teaching approach is student oriented. In other words, most of them emphasize the interaction and exchange between teachers and students and concern about the students' attitude, motivation, and original life experience. This result seems good, because existing research shows that a teacher's teaching approach will affect students' learning, and student-centered teaching will lead students to adopt a deep learning approach which in turn will lead to higher-quality learning outcomes. These self-reported results should not be interpreted overly optimistically; they need to be reverified by classroom observations. However, these results show that at least teachers know what the ideal approach of teaching should be. This must be the result of teacher training programs sponsored by governments at all levels prior to and during the process of the new round of curriculum reform.

Furthermore, the correlation analysis between conceptions of assessment and teaching approach shows that the management-oriented conception is closely related to the teacher-centered approach. Even though the number of teachers who hold a management-oriented conception is not so large, this conception may lead to a teacher-centered approach, resulting in a surface-level learning approach by the students and a lower-level learning outcome. Therefore, this result should raise the attention of teacher education and teaching management personnel. The overall results of the correlation analysis show that, except for the situation where management-oriented conception is correlated only with a teacher-centered approach, the other four conceptions have consistently and highly correlated with both approaches of teaching. This indicates that teachers' conceptions of assessment as a whole actually have nothing to do with their teaching approach. In other words, the impact of science teachers' conceptions of assessment on their teaching approaches is very weak. This is something connected with our teaching environment and external examination environment.

Compared with teachers in Western countries, Chinese teachers have less teaching autonomy and a greater pressure from external examination. They do not have much freedom on issues such as textbook selection or what and how to teach. Teachers' teaching arrangements must be strictly subordinated to the relevant curriculum documents, such as curriculum standards, syllabuses, and exam descriptions. Therefore, although the teachers may have their own ideas and opinions about assessment and teaching issues, it is difficult to reflect these ideas or beliefs in their teaching practice. The fundamental route to changing teachers' teaching approaches and improving their teaching quality should be to give them greater teaching autonomy and reduce the pressure of external high-stakes examinations.

There are two major studies in the relevant literature to date that focus on assessment conceptions of teachers, Brown (2004) and Watkins et al. (2005). The similarities and differences between this study and these two studies in research paradigms, objects, frameworks, and findings are as follows:

(a) Research paradigm. Brown's study is a survey using a questionnaire and statistical analysis, i.e., a quantitative method. Watkins uses in-depth interviews, i.e., a qualitative method. This study combines the two methods, using the mixed approach. In general, the mixed approach is more difficult, because it is usually more time-consuming and requires more funding and staff. But its advantage is also obvious: the two methods can support and verify to each other. In this study, the interview data laid the foundation for the development of questionnaire, and the results of the survey also supported the results of the interviews.

(b) Subjects. The subjects of Brown's research are primary school teachers in New Zealand, while Watkins's are university teachers in Hong Kong and Sweden. This study takes secondary school teachers in China as subjects.

(c) Research framework. Brown asks the questions mainly around the role of assessment. His outcome space is one-dimensional, while Watkins's outcome space is two-dimensional. One dimension focuses on the backwash effect of assessment, and the other is about the relation between teaching and assessment. The final classification of assessment conception is based on the second dimension, that is, teachers' understanding of relation between teaching and assessment. The outcome space of this study is also two-dimensional, but the content is different. The two dimensions are the assessment element dimension (assessment purpose, assessment content, and assessment method) and the orientation dimension (management oriented, teaching oriented, examination oriented, ability oriented, and subject oriented). The final classification of assessment conception is based on the latter dimension.

(d) Conclusions. Because of different research frameworks of the relevant studies, the conclusions of these studies are also in different formats. Brown's study concluded that the teachers' conceptions of assessment included the following four categories: (1) assessment is related to improvement of the students' learning and the teachers' instruction. (2) Assessment makes students accountable for learning. (3) Assessment evaluates the quality of schools and teachers. (4) Assessment is irrelevant to the work of teachers. Watkins's study results are shown in Fig. 11.9.

In Fig. 11.9, there exist eight categories of assessment conception distributed in a two-dimensional outcome space. The researchers summarized eight categories of assessment conception into three basic types of conception according to the teachers' understanding about the relation between teaching and assessment (T/A relation). The first type of conception consists of Categories 1–3. The common

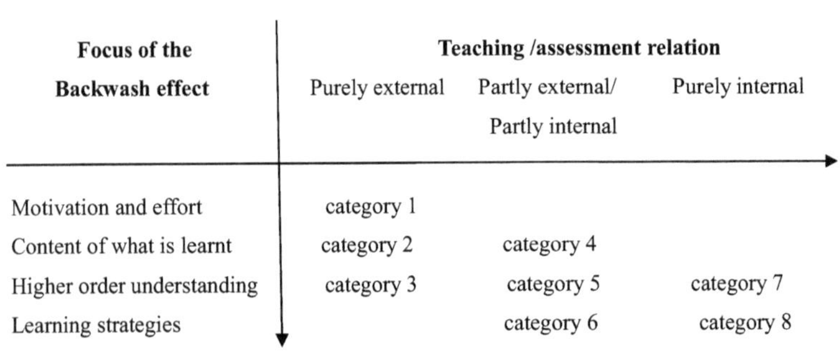

Fig. 11.9 Watkins' outcome space. Note. This is a two-dimensional outcome space for conceptions of the actual role of assessment in student learning. *Arrows* here indicate progressions toward more inclusive conceptions

feature of these conceptions is that assessment is viewed as something very separate from the teaching and learning process. The T/A relation is therefore thought of as purely external. The second type of conception consists of Categories 4–6. Here, teachers seem to be aware of the process of teaching and learning and sometimes even the strategies students used in order to learn. But they believe that students need to learn some "basic knowledge" before using more sophisticated learning strategies. So they view the T/A relation as partly external and partly internal. The third type of conception includes the remaining Categories 7–8. In this situation, teachers believe that deeper strategies, such as understanding, reflecting, interpreting, and analyzing, play a much more significant role in student learning than "basic knowledge." They view assessment as an integral part of the teaching process. In other words, they think that the T/A relation is purely internal.

Comparing the conclusions of these studies, it is not difficult to find that there are some common features among them even though their research frameworks and outcome spaces are quite different. For example, teachers from different regions and different levels all considered that assessment can improve teachers' teaching and students' learning. On the other hand, differences in conceptions do exist among teachers in what aspects of teaching need to be improved and how they can be improved through assessment. These are what our teacher educators and educational administrators need to consider seriously.

11.5 Implications

Implications of the results of this research are as follows: (1) The results provide evidence of science teachers' conceptions of assessment in mainland China. Teachers' assessment conceptions are their general views on issues of assessment. They are not only different from the assessment conceptions of experts and scholars, but may also differ from those which are advocated by the government or society. They are the result of interaction between teachers and their environment. The research findings about teachers' conceptions of assessment are teachers' assessment conceptions in reality. As an empirical study of science teachers' conceptions of assessment, these study results can be used to improve the relevance and effectiveness of teacher education or teacher training, which will provide the benefit of improvement of teachers' teaching approaches and thus enhance the quality of teaching. (2) A reliable and effective measurement tool has been developed. The questionnaire used in this research for testing teachers' assessment conceptions (STACQ) is based on the previous qualitative study. It was developed through a rigorous process of item preparation, expert examination, pilot testing, revision, and formal testing. When put into use, it was tested with a large sample to help demonstrate its reliability and construct validity. The test results show that the questionnaire has good reliability and construct validity. It not only effectively guaranteed the quality of this research, but also can be used as a qualified test tool for follow-up studies. (3) The results provide a reference for further study. The results of this study, including findings of qualitative and quantitative research, provide a reference or norm of teachers' conceptions of assessment in Guangdong Province or even in China. For example, the findings from the qualitative phase of five conceptions of assessment and the findings on the subscales' dimensions in the development process of the questionnaire and the correlation between assessment conception and teaching approach can all be used as a reference for later studies in other contexts.

11.6 Limitations

This is an empirical study using a sequential mixed-methods design. It is a combination study of personality and commonality. The two methods, qualitative and quantitative, complement and confirm each other. The research process is normative and the result is credible; however, for subjective and objective reasons, this research still has some shortcomings and limitations. One limitation is the measurement tools used in this research, which were two questionnaires: the STACQ and the ATI. The author-developed STACQ is geared toward science teachers in secondary schools. It is not necessarily suitable for teachers of humanities or social sciences. The teaching approach questionnaire (ATI) was developed to be a questionnaire for testing foreign university faculty. Although its reliability indicators were good enough in the pilot test, these indicators are not ideal in the formal test (reliability for each subscale was less than 0.7, with one less than 0.6). Although these reliability indices can be accepted statistically, there is still great room for improving the questionnaires. The other limitation is of research content. As mentioned earlier, science teachers' conceptions of assessment can be divided into "ideal conception" and "practical conception." This research has been struggling to differentiate between teachers' practical conceptions and ideal conceptions, but the two conceptions are not distinguished in the present research. Of course, probing ideal conceptions is also necessary. Because ideal conceptions are the product of teachers' interactions with their environment, especially with the conceptions of mainstream society, they partly reflect the outcome of teacher education in recent years. Practical conceptions, however, have more direct impact on science teachers' teaching than do ideal conceptions. Some important questions therefore remain unanswered: how are "ideal conception" and "practical conception" differentiated? What causes their creation? And in what ways do they influence science teachers' teaching? All these questions need to be answered in further studies.

Acknowledgments I really thank the reviewers and editors who provided insightful comments and valuable suggestions. I would also like to express my appreciation to the local educational authorities and secondary schools that provided crucial support to my investigation and to the teachers who accepted my interviews or/and completed my questionnaires.

Appendices

Appendix I: Interview Protocol

Core Questions

(1) About the nature of academic assessment

   (A) Have you ever heard of the concept of academic assessment? Are you accustomed to using it?
   (B) How do you understand academic assessment? What do you think of the essence of it?
   (C) Is academic assessment just testing? If so, why? If not, what is the relationship between them?

(2) About the purpose of academic assessment

   (A) Why do you assess students' learning? Does it help you realize your expected goal of instruction?
   (B) Does academic assessment affect your instruction and/or the learning of your students? If so, what are the effects?

(3) About the method of academic assessment

   (A) What methods do you know that can be used to assess students? Among these, what are suitable for assessing students in the physics classroom?
   (B) What methods have you used to assess students? What are the most frequently used ones?

(4) About the standard of academic assessment

   (A) What standard do you think should be used to assess students' learning?
   (B) From an academic view, what kind of student do you like?

(5) About the subject of academic assessment

   (A) Who do you think should implement the academic assessment?
   (B) In the process of implementation of academic assessment, what roles should be played by teachers, students, parents, and community?

(6) About the object of academic assessment

   (A) In general, what should be the object(s) when assessing students?
   (B) In physics teaching, what aspect(s) will you pay special attention to when you assess students?

(7) About the occasion of academic assessment

   (A) When do you assess your students? For example: during the process of instruction, when students are doing lab work, after a unit test, and after a term exam.

(B) Generally, in what circumstances do you assess (praise) your students? For example: when students get good marks on an exam, when they perform well in class, and when they show originality in the lab.

(8) About the interpretation and use of assessment results

(A) How do you interpret the result of assessment on your students? For example: explain to the parents about the result of student's unit test and of united exams conducted by local educational authorities.
(B) How do you use the results of academic assessment? For example: take students' classroom performance as their daily grade, take students' classroom test results as the basis to diagnose instruction, and take students' test scores as the basis for classifying students.

Relevant Questions

(1) What do you think about the general circumstances of academic assessment at schools in our country? What are the favorable and unfavorable factors?
(2) Do you believe that the college entrance exam in our country has an impact on your assessment of students? If so, in what way?
(3) Have you ever used portfolio assessment? If not, why? If so, how do you feel about it?
(4) Have you ever assessed students' lab work ability? If so, what means of assessment have you used? If not, why?
(5) Have you ever used open-ended physics problems to assess students? If so, what kinds of problems have you used? What has been the effect of their use? If not, why?

Note. This protocol was used for teacher interviews. It contains two sections. Section 1 contains core questions which must be used for each interview, while Sect. 2 has relevant questions which may not necessarily be used for each interview; instead, it can be used as a complementary pool of questions.

Appendix II: Science Teachers' Assessment Conception Questionnaire (STACQ)

A 36-item version of the questionnaire in which instructional and demographic parts are omitted

(1) My aim of conducting paper-and-pen test is to prepare students for the college entrance examination (CEE).
(2) It's important to examine how well students have mastered the key points of the knowledge tested by CEE.

(3) Classroom questioning can help me to be informed about the extent in which students understand my lesson.
(4) I get myself informed of the mastery of students' understanding of teaching content by classroom questioning.
(5) I question students in order to improve their speaking ability.
(6) It is important to examine the hands-on ability of the student.
(7) It is an effective way of assessing students by doing creative experiment.
(8) I make comment on the performance of students in order to maintain discipline in the classroom.
(9) It is an important content with which a student should be assessed whether his performance is consistent with the teacher's criterion.
(10) I make comment on students' homework in order to make them aware of the problem-solving rules requested by the CEE.
(11) I use paper-and-pen test to examine students' experiment ability because they examine the same ability in the same way in CEE.
(12) It is an important content with which a student should be assessed to inspect if he/she finishes his/her homework in time.
(13) I use classroom test to examine whether students have mastered teaching content.
(14) It is important to examine students' ability of oral expression.
(15) Extracurriculum is an effective way to assess students' comprehensive ability.
(16) It is important to assess students according to each one's ability characteristic.
(17) Natural observation is an effective way to assess a student.
(18) My aim of questioning students is to stop their unexpected behavior in the classroom.
(19) It is important to assess a student by examining the extent in which he/she focuses on the teacher's lecture.
(20) It is an effective way to arouse students' attention by questioning them.
(21) Because the speed of problem-solving is required in CEE, usually I ask my students to fulfill each specific test within the given time.
(22) Classroom questioning and quiz can help me know whether students have mastered the key and difficult points of my instruction.
(23) It is important to check if a student dare to challenge academic authorities.
(24) Exploratory experiment is an effective way to assess student.
(25) I question students in order to arouse their thinking.
(26) It is important to assess students' ability of creativity.
(27) It is important to assess a student by knowing if he/she make notes seriously in the classroom.
(28) It is necessary to check students' classroom notes.
(29) It is necessary to use problems in CEE as everyday test problems.
(30) I adopt the method of "learning competition" to guide the students to a state of examination preparation.
(31) Because lag-behind students stand for the lowest level of mastering teaching content, I pay special attention to checking their homework.
(32) I question students in order to arouse their learning interest.

(33) It is important for students to be assessed by the society.
(34) Classroom questioning can guide the students to think of problems in a right direction.
(35) It is important for the students to be assessed by their parents.
(36) It is important to assess the way of thinking of students.

Subscales of STACQ

Management-oriented scale: items 8, 9, 18, 19, 20, 27, and 28
Examination-oriented scale: items 1, 2, 10, 11, 21, 29, and 30
Teaching-oriented scale: items 3, 4, 12, 13, 22, 31, and 34
Ability-oriented scale: items 5, 6, 7, 14, 15, 24, and 26
Subject-oriented scale: items 16, 17, 23, 25, 32, 33, 35, and 36
Items dropped in data analysis: items 4, 9, 12, 16, 17, 19, 21, 30, and 34

Answer Sheet of STACQ

There are five numbers ranging from 1 to 5 beside each item number, and these numbers stand for specific responses:

- Strongly disagree or is never true (SD)
- Disagree with reservations (D)
- Not specified or is true about half the time (NS)
- Agree with reservations (A)
- Strongly agree or is always true (SA)

Show your choice by ticking the number which is closest to the way you want to respond.

Item No.	SD	D	NS	A	SA	Item No.	SD	D	NS	A	SA
1	☐	☐	☐	☐	☐	19	☐	☐	☐	☐	☐
2	☐	☐	☐	☐	☐	20	☐	☐	☐	☐	☐
3	☐	☐	☐	☐	☐	21	☐	☐	☐	☐	☐
4	☐	☐	☐	☐	☐	22	☐	☐	☐	☐	☐
5	☐	☐	☐	☐	☐	23	☐	☐	☐	☐	☐
6	☐	☐	☐	☐	☐	24	☐	☐	☐	☐	☐
7	☐	☐	☐	☐	☐	25	☐	☐	☐	☐	☐
8	☐	☐	☐	☐	☐	26	☐	☐	☐	☐	☐
9	☐	☐	☐	☐	☐	27	☐	☐	☐	☐	☐
10	☐	☐	☐	☐	☐	28	☐	☐	☐	☐	☐
11	☐	☐	☐	☐	☐	29	☐	☐	☐	☐	☐
12	☐	☐	☐	☐	☐	30	☐	☐	☐	☐	☐
13	☐	☐	☐	☐	☐	31	☐	☐	☐	☐	☐
14	☐	☐	☐	☐	☐	32	☐	☐	☐	☐	☐
15	☐	☐	☐	☐	☐	33	☐	☐	☐	☐	☐
16	☐	☐	☐	☐	☐	34	☐	☐	☐	☐	☐
17	☐	☐	☐	☐	☐	35	☐	☐	☐	☐	☐
18	☐	☐	☐	☐	☐	36	☐	☐	☐	☐	☐

Appendix III: Approaches to Teaching Inventory

An 18-item and simplified version of ATI

(1) It should be encouraged that a student ask questions to the teacher in the classroom.
(2) In the classroom, I encourage my students to discuss the problems they met in their learning process.
(3) Showing what are the key points of learning to the students is an important way to make them aware of the core knowledge of the subject.
(4) I am used to giving a lecture to the class, from the very beginning to the end.
(5) It is necessary for a teacher to be aware of the ideas of his/her students.
(6) I try my best to let the students express their ideas in the class.
(7) Problem-solving ability is a key indicator when assessing a teacher.
(8) When preparing a lesson, what I pay special attention to is how to ensure that students attend the class in a way I wanted.
(9) It is an essential requirement that the teacher be aware of the students' attitude to learning.
(10) In the class, I never miss any chance to encourage my students to work harder.
(11) To give a good lesson, the key point is to provide students with a great deal and sufficient materials.

(12) My main task is to be aware of the teaching content when I prepare a lesson.
(13) Assessing a student is for the student to understand his/her own learning state.
(14) I usually encourage my students to sum up the knowledge they learned.
(15) It is appropriate to see a teacher as a treasury of knowledge.
(16) In the classroom, student's activities should be reduced to the least in order to make sure that the teachers have enough time to lecture deeply.
(17) When teaching the class, I pay special attention to training the students according to my preset target.
(18) I lay emphasis on lecturing content that is relevant to CEE in my class.

Subscales of ATI

Student-centered intention: items 1, 5, 9, and 13
Student-centered strategy: items 2, 6, 10, and 14
Teacher-centered intention: items 3, 7, 11, 15, and 16
Teacher-centered strategy: items 4, 8, 12, 17, and 18

Answer sheet of ATI is omitted here because it is very similar to that of STACQ.

References

Biggs, J. B. (1987). *Student approaches to learning and studying*. Melbourne: Australian Council for Educational Research.

Brown, G. (2004). Teachers' conceptions of assessment: Implications for policy and professional development. *Assessment in Education, 11*(3), 301–318.

Gao, L. (2004). *A study of Chinese teachers' conceptions of teaching (English version)*. Wuhan: Hubei Education Press.

Gao, L., & Watkins, D. (2001). Identifying and assessing the conceptions of teaching of secondary school physics teachers in China. *British Journal of Educational Psychology, 71*(3), 443–469.

Hau, K.-T., Wen, Z., & Cheng, Z. (2004). *Structural equation model and its applications (结构方程模型及其应用)* (p. 45). Beijing: Educational Science Publishing House (In Chinese).

Jin, S. (2003). *A study on scientific literacy conceptions of primary school science teachers* (小学科学教师科学素养观研究). Unpublished Master's thesis. South China Normal University. Guangzhou (In Chinese).

Johnson, R. B., & Onwuegbuzie, A. J. (2004). Mixed methods research: A research paradigm whose time has come. *Educational Researcher, 33*(7), 14–26.

Johnson, R. B., Onwuegbuzie, A. J., & Turner, L. A. (2007). Toward a definition of mixed methods research. *Journal of Mixed Methods Research, 1*(2), 112–133.

Liang, S. (2005). *A study on environment education conceptions of primary school teachers* (小学教师环境教育观研究). Unpublished Master's thesis. South China Normal University. Guangzhou (In Chinese).

Liang, J. (2006). *A study on learning assessment conceptions of middle school english teachers* (中学英语教师学习评价观研究). Unpublished Master's thesis. South China Normal University. Guangzhou (In Chinese).

Ministry of Education (MOE). (2001). Outline for basic education curriculum reform (trial version) (基础教育课程改革方案(试行)). Retrieved from http://www.edu.cn/20010926/3002911_2.shtml (In Chinese).

Trigwell, K., Prosser, M., & Talor, P. (1994). Qualitative differences in approaches to teaching first year university science. *Higher Education, 27*, 75–84.

Trigwell, K., & Prosser, M. (1999). Relations between teachers' approaches to teaching and students' approaches to learning. *Higher Education, 37*, 57–70.

Trigwell, K., & Prosser, M. (2004). Development and use of the approaches to teaching inventory. *Educational Psychology Review, 16*(4), 409–424.

Watkins, D., Dahlin, B., & Ekholm, M. (2005). Awareness of the backwash effect of assessment: A phenomenographic study of the views of Hong Kong and Swedish lecturers. *Instructional Science, 33*, 283–309.

Wu, M. (2003). *SPSS statistical application in practice* (p. 109). Beijing: Science Press.

Zhang, C. (2004). *A study on information literacy conceptions of information technology teachers* (信息技术教师信息素养观研究). A Master thesis of South China Normal University. Guangzhou (In Chinese).

Zhang, L. (2005). *A study on teaching conceptions of middle school English teachers* (中学英语教师教学观研究). Unpublished Master's thesis. South China Normal University. Guangzhou (In Chinese).

Part V
Science Learning in Informal Settings

Xiufeng Liu

Editor's Introduction: Part V

In China, informal science education is a completely separate system than school education. Informal science education is the responsibility of Chinese Association of Science and Technology (CAST), while school science education is the responsibility of Ministry of Education. Although CAST is a nongovernmental organization, equivalent to AAAS in the USA, CAST is directly affiliated with and fully funded by the Chinese Ministry of Science and Technology. In fact, officials of CAST enjoy equivalent senior governmental official status and the associated privileges. Different from many developed countries, China has a national hierarchical system to carry out science popularization. With CAST at the central government level, there are branch and subbranch organizations of CAST at the provincial, county, and even township levels. Therefore, CAST is a complete and hierarchical national system with one main task: science popularization. This feature of Chinese science popularization is clearly unique compared to that in many developed countries, where informal science education activities are often uncoordinated, limited in scope, and lack sustainability.

Because of its separation from school science education, informal science education research is almost exclusively conducted outside normal universities or colleges that are under the jurisdiction of Ministry of Education. Specifically, research on informal science education is primarily conducted by organizations within the system of CAST and a few limited non-normal universities. China Research Institute for Science Popularization (CRISP) was created by the State Council in 1980 dedicated to research on all activities related to informal science education; it is directly affiliated with CAST. Because of its unique and authoritative status of CRISP, all the five chapters in this section are authored by researchers from CRISP.

In China, the major government policy document guiding all informal science education activities in the country is the *Outline of the National Scheme for*

Scientific Literacy (2006–2010–2020) (State Council of China 2006). Chapter 12 by Zhang and Shi reviews the background, major content areas, and implementation measures of the *Outline*. As can be seen from this chapter, the outline is a further articulation of two previous government policy documents, the *Law of the People's Republic of China on Popularization of Science and Technology* and the *Outline of the National Program for Long- and Mid-term Scientific and Technological Development (2006–2020)*.

The development of the outline was influenced by similar science education initiatives in developed countries, such as Project 2061 in the USA and the EU's recent science education reform policies. It is interesting to note that the *Outline* identifies four target groups for science popularization efforts: youth, farmers, urban workforce, and leading cadres and public servants, and specific activities of science popularization for each of the groups are proposed. The chapter also gives specific examples of activities targeting youth.

The *Outline* is ambitious given its national scope and diverse target groups. The overall goal of the *Outline* is to increase Chinese citizens' scientific literacy, i.e., "knowing some necessary knowledge of science and technology, mastering basic methods of science, building up scientific thoughts, advocating a scientific ethos, and having the ability to apply them to solve practical problems and participate in public affairs" (Chap. 12, pp. xxx). This notion of scientific literacy is consistent with what Roberts (2007) refers to as Vision I scientific literacy, which emphasizes science as being authoritative in knowledge, knowing, and problem-solving. Scholars have questioned this notion of scientific literacy and called it a "deficit model" (Bauer et al. 2007; Layton et al. 1993; Liu 2009). Research has also called to expand Vision I scientific literacy to incorporate science technology as applied in contexts (i.e., the Vision II scientific literacy) (Roberts 2007) and to include engagement with social, cultural, political, and environmental issues involving science and technology (Liu 2014).

The recent National Research Council Report (NRC 2009) identifies the following six strands of learning outcomes for informal science learning:

1. Experience excitement, interest, and motivation to learn about phenomena in the natural and physical world
2. Come to generate, understand, remember, and use concepts, explanations, arguments, models, and facts related to science
3. Manipulate, test, explore, predict, question, observe, and make sense of the natural and physical world
4. Reflect on science as a way of knowing; on processes, concepts, and institutions of science; and on their own process of learning about phenomena
5. Participate in scientific activities and learning practices with others, using scientific language and tools
6. Think about themselves as science learners and develop an identity as someone who knows about, uses, and sometimes contributes to science

The above NRC outcomes are more comprehensive than those in the *Outline*; they overlap with outcomes of school science in addition to some unique learning outcomes for informal science education (i.e., stands #1 and #6).

Chapter 13 by Ren and Wang reviews the history of Chinese informal science education and the scope and current status of informal science education research and practices in China. From this chapter, we see that while informal science education activities in China may date back to more than 100 years ago, when China first introduced science from the West, and students are engaged in a variety of informal learning activities, research in informal science education is a very recent phenomenon. According to Ren and Wang in Chap. 13, research papers and dissertations on informal science education began to increase only after 2007, which is likely due to the 2006 Chinese Central Government's major initiative for science popularization, the *Outline* reviewed in Chap. 12. Based on the number of publications shown in Figs. 13.5 and 13.6, in recent years, theoretical inquiries and empirical studies of informal science education are still few, suggesting that informal science education research in China is still in its beginning stage. In fact, the only research journal devoted to informal science education, *Science Popularization*, was only established in 2005.

Even in developed countries, informal science education as a distinct discipline has a short history. Although research activities on informal science education, particularly in museum learning, have been ongoing since 1970s, significant advancement and systemic synthesis of research in informal science education theories and research methods have taken place only recently (e.g., Falk 2001; NRC 2009). It must also be pointed out that the integration of informal science with school science education is becoming a trend in recent years in developed countries, but this integration remains in a very preliminary stage in China.

Chapter 14 by Chen and Yan reviews empirical studies on the effects of media on the science learning of Chinese youth. Chen and Yan define media to include print, broadcast and TV, and the Internet. The study of the effect of media on science learning is based on a science communication framework that considers effects to be in any of the following forms: awareness, enjoyment, interest, opinions, and understanding. Overall, based on the studies reviewed, the authors conclude that all media can potentially produce one or more of the above intended outcomes of science communication. Given the limited number of empirical studies reviewed, and that the chapter does not give details of the research designs of the studies, the conclusions made in this chapter should be accepted with great caution.

Research on the effect of media on students' and adults' science learning has been a research area in the West for a long time. The National Research Council report (NRC 2009) devotes one whole chapter to reviewing studies on media for science learning. Although a large number of studies have been conducted over the past few decades, an unequivocal conclusion on the effect of any type of media (e.g., print, TV, or internet) for science learning remains unavailable; mixed results and anecdotal findings are common. This reflects the current state of research on effects of media on science learning. More rigorous and systematic studies on the effects of media for science learning are needed. In fact, the NRC report (NRC

2009) recommends the following research areas related to media for science learning: (a) who uses media to learn science in informal environments, (b) how does audience identity interact with media, (c) does format matter and how, (d) how does media promote science as a process, and (e) how do people learn through media longitudinally and cross-media.

Chapter 15 by Li reports a case study of a science teacher guiding students to prepare for a contest in the Intel International Science and Engineering Fair in China. Using a cognitive apprenticeship theoretical framework, Li details how the teacher mentored his students in an out-of-school science club. Research in out-of-school science programs has a long history in the West, and a general conclusion has been that out-of-school science programs can have positive effects on participants' attitudes toward science, grades, test scores, graduation rates, and specific science knowledge and skills (NRC 2009). Research related to out-of-school science programs has been primarily in the evaluation type; studying the specific mechanisms or strategies of an out-of-school program such as a science club is rare. The case study reported in Chap. 15 makes a unique contribution to research in this area. As recommended by the National Research Council committee, more research is needed to identify a set of best practices that can be applied across programs, such as the interaction among curriculum choices, staff training, management issues, and physical, social, and cultural contexts (NRC 2009).

Chapter 16 by Wang reports a case study of visitors to science museums in Beijing. Specifically, the author reports findings related to school children's perceptions of the features of museum learning, patterns of interacting with museum exhibits, and perceived relationship between museum visits and school learning. The findings support the contextual model of learning proposed by Falk and Dierking (2000). The findings are also consistent with the established conclusion in research in Western countries that learning in designed spaces such as museums can contribute to learning in all the six strands identified by the NRC committee (NRC 2009) and are reviewed early in this commentary.

Overall, the set of five chapters in this section provides a good overview of informal science education policies, practices, and sample research studies in China. Research in informal science education in China is still at a very preliminary stage. However, given the Chinese Central Government's complete and hierarchical national organization structure for science popularization and its emphasis on improving its citizens' scientific literacy, we can anticipate much growth and achievements in informal science education research in China in the coming years.

References

Bauer, M. W., Allum, N., & Miller, S. (2007). What can we learn from 25 years of PUS survey research? Liberating and expanding the agenda. *Public Understanding of Science, 16*, 79–95.

Falk, J. H. (2001). *Free-choice science education: How we learn science outside school.* New York: Teachers College Press.

Falk, J. H., & Dierking, L. D. (2000). *Learning from museums: Visited experienced and the making of meaning*. New York: Alta Mira Press.
Layton, D., Jenkins, E., Macgill, S., & Davey, A. (1993). *Inarticulate science? Perspectives on the public understanding of science and some implications for science education*. East Yorkshire: Studies in Education Ltd., University of Leeds.
Liu, X. (2009). Beyond science literacy: Science and the public. *International Journal of Environmental & Science Education., 4*(3), 301–311.
Liu, X. (2014). Expanding notions of scientific literacy: A re-conceptualization of aims science education in the knowledge society. In N. Mansour & R. Wegerif (Eds.), *Science education for diversity in knowledge society* (pp. 23–39). New York: Springer.
National Research Council. (2009). *Learning science in informal environments: People, place and pursuits*. Washington, DC: National Academies Press.
Roberts, D. A. (2007). Scientific literacy/scientific literacy. In S. K. Abell & N. G. Lederman (Eds.), *Handbook of research on science education*. Mahwah: Lawrence Erlbaum Associates Publishers.
State Council of China. (2006). *Outlines of the national scheme for scientific literacy (2006–2010–2020)*. Beijing: State Council of China [全民科学素质行动计划纲要 (2006–2010–2020 年)]. (in Chinese)

Chapter 12
An Examination of National Policy on Youth Science Learning in Informal Education Settings and Its Implementation in China

Huiliang Zhang and Shunke Shi

12.1 Introduction

The Chinese government has always devoted much attention to youths' science learning in the context of informal education. In 1949, just a few days before the founding of the new China, the *Common Program*, which was the precursor to the constitution of the country, was laid out by Chinese People's Political Consultative Conference (CPPCC). It stated an intent in Article 43 "to strive to advance natural sciences to serve the country in its industrial, agricultural, and national defense construction, to reward science discoveries and inventions, and to popularize scientific knowledge" (CPPCC 1949). In 1981, the Central Committee of the Communist Party of China (CCPC) and the State Council (SC) issued the *Outline of the Chinese Science and Technology Development Approach* drafted by the State Science and Technology Commission. In Article 6, Part 2, the report aimed at "strengthening youths' science activities, such as various science lectures, science competitions, extracurricular science inquiries" (CCPC-SC 1981). In 1991, the NPC formulated the *Law of the Juveniles' Protection of the People's Republic of China*, which stated in Article 30 that "the educational bases, libraries, and youth and children's centers should be free for visiting children and youth; science museums, exhibition halls, art galleries, gyms, cinemas, zoos, parks, and the like should be free or provide discount tickets for children and youth." Moreover, Article 32 states that "the state should encourage science research institutes and science and technology associations to carry out science popularization activities for youth" (NPC 1991).

H. Zhang (✉) • S. Shi
China Research Institute for Science Popularization, No. 86 Xueyuannanlu, Haidian District, Beijing, China
e-mail: leanne2005@sina.com; zhangleanne@gmail.com

L.L. Liang et al. (eds.), *Chinese Science Education in the 21st Century: Policy, Practice, and Research*, Contemporary Trends and Issues in Science Education 45, DOI 10.1007/978-94-017-9864-8_12

In late 1994, the *Instructions for Strengthening Engagement in Science and Technology Popularization* (the *Instructions*) was jointly issued by the CCPC and the SC, calling for "improving science popularization activities in three aspects of popularizing education: science knowledge, scientific methods, and scientific thought." The *Instructions* clearly stated the importance of youth in science popularization, namely, "providing a variety of ways for youth; cultivating their thinking ability, practicing, and creating; helping them shape appropriate attitudes towards science, philosophy, and a world outlook" (CCPC-SC 1994). The *Instructions* was the first-ever science popularization policy at the state level dealing comprehensively with all types of science popularization practices in China. At that time, China was at a critical moment in its ongoing economic and social reforms. The *Instructions* raised the importance of scientific culture to a strategic height in order to promote national prosperity and strength.

Entering the twenty-first century, Chinese informal science learning entered the diversified developmental stage and became increasingly important in improving youths' scientific literacy. In November 2000, the *Guideline for China's Youth Science and Technology Popularization Activities between 2001 and 2005* was jointly issued by the Communist Youth League of China, the China Association for Science and Technology (CAST), and three other governmental departments. Its purpose was to standardize and guide youths' science learning activities for government departments, schools, communities, public media, enterprises, rural areas, and families. In 2006, the SC issued the *Outline of the National Scheme for Scientific Literacy (2006-2010-2020)* (the *Outline*). It regarded youths as one of the key groups to carry out scientific literacy action plans and considered youths' informal science learning as an important way to improve youths' scientific literacy. The *Outline* has great significance in Chinese informal science education, creating a positive social environment for youths' science learning. This chapter takes the *Outline* as an example of a national policy, reviews the development of youths' science learning, and discusses the importance of youths' science learning in the context of Chinese informal education.

12.2 Background to the Outline

12.2.1 Domestic Background

During the development of the *Outline*, the *Law of the People's Republic of China on Popularization of Science and Technology* (the *SP Law*) and the *Outline of the National Program for Long- and Mid-Term Scientific and Technological Development (2006–2020)* were promulgated and implemented. These played an important impetus toward the development of and provided guidance for the *Outline*.

The NPC enacted the *SP Law* in June 2002, taking on scientific culture improvement from a legislative perspective, and sought to bring science and technology popularization into the realm of laws and institutions. For informal youths' science learning, its third chapter encouraged schools, research institutions, media, national organizations such as trade unions, women's federations and youth leagues, enterprises, rural grassroots organizations, urban communities and public places such as parks and shopping centers, and airports to perform science popularization within their resources. "[All] schools and other educational institutions should consider science popularization as an important content of literacy education; should organize students for various science outreaches, especially science and technology museums, science and technology activity centers, and other science popularization bases; and should organize youths' science and technology popularization activities outside schools" (NPC 2002). The *SP Law* clearly declares science and technology popularization to be a social task, creates social responsibility, and emphasizes how science and technology popularization plays an important role in mobilizing initiatives in all social sectors.

In January 2006, Beijing hosted the National Science and Technology Conference. It explicitly stated that its 2020 goal for China was to strengthen its innovative ability and become one of the most innovative countries worldwide. In February of the same year, the SC promulgated the *Outline of the National Program for Long- and Mid-Term Scientific and Technological Development*, which incorporated into the law the improvement of the quality of the national scientific culture and the establishment of important policies and measures for the social environment of science and technology innovation. It clearly stated its intent to carry out the national scientific culture action plan for science and technology popularization education for farmers, youths, and leading cadres and civil servants as its main tasks. It set a proper approach and provided fundamental references for formulating the *Outline*.

12.2.2 International Background

Since 1984, the European Union (EU) has been implementing research policy through successive "Framework Programs." In the EU's Framework Programs (FPs), dissemination of results to the public, especially to youth, is a contractual obligation for participation in research initiatives. Since FP6 (European Commission 2002), beneficiaries of EU funding are also required to develop public communication activities. With a view to enhancing the impact of research funded by the EU and to foster dialogue and debate, the FP7 grant agreement requires project participants to communicate and engage with stakeholders beyond the research community. Plans for these outreach activities for the public, especially for youth, should also be outlined at the proposal stage. These plans are taken into account during the evaluation process. The specific aims of this provision are to promote knowledge sharing, greater public awareness, transparency, and education. Consortia of researchers are required to provide tangible proof that collaborative research not only exists but also pays dividends in terms of academic excellence, industrial competitiveness, employment opportunities, environmental improvements, and enhanced quality of life for all. The EU also formulated and promulgated other series of important documents, such as *Science, Society and the Citizen in Europe*; *Strategic Objectives 2000–2005: Shaping the New Europe*; and *Plan of Science and Social Action*. These documents try to promote citizens' scientific literacy appropriate to the developmental strategies of all EU countries.

In 1985, the American Association for the Advancement of Science (AAAS) founded Project 2061, a long-term initiative to help all Americans become literate in science, mathematics, and technology. In 1989, *Science for all Americans* (Project 2061, 1989) was published by the AAAS, showing what a scientifically literate person should know about science, mathematics, and technology. In 1993, *Benchmarks for Science Literacy* (Project 2061, 1993) demonstrated how scientific literacy may develop from kindergarten through Grade 12. To ensure a STEM-skilled workforce for the future, the US Federal Government has considerable assets that can engage youth so that the pathway through their education leads to challenging STEM-related careers. Investments that focus on engagement are designed to increase learners' involvement and interest in STEM, inform their view of STEM's value in their lives, or positively influence the perception of their ability to participate in STEM. The scope of STEM engagement is vast and includes investments in a wide range of areas, such as development of learning materials; programs at museums, science centers, or parks; games, simulations, and virtual environments; public talks; and educational broadcast programming. Thus, there are many avenues available to the Federal Government for reaching learners and many possibilities for STEM-learning youth.

In the UK, the House of Lords issued the report *Science and Society* in 2000. The report defines public understanding of science as the understanding of scientific matters by non-experts, and it includes understanding of the nature of scientific methods, including the testing of hypotheses by experiment. It may also include awareness of current scientific advances and their implications. The issue of *Science and Society* report resulted in a wide development of public engagement with science, for example, *Café Scientifique* by Wellcome Trust from 2002 to 2005 and RCUK public engagement with research team in 2005.

These documents provided useful references for the *Outline*. "All of these documents inspired the formulation of the *Outline*, which is based upon China's situation and will create a great improvement in general national science and technology literacy by implementing a long-term plan" (Report on National Action Plan for Science Literacy 2011).

12.3 Tasks of and Approaches to Youths' Informal Science Learning

The *Outline* is the most ambitious science popularization scheme ever enacted with the aim of greatly advancing science popularization. The long-term objective of the *Outline* is that by 2020 it will see an enhanced development of science- and technology-related education, communication, and popularization that will bring its citizens' scientific literacy up to the early twenty-first century level of major developed countries. It points out that "scientific literacy makes an important part of the quality of a citizen. The basic scientific literacy of citizens generally refers to having key knowledge of science and technology, mastering basic methods of science, building up scientific thoughts, advocating a scientific ethos, and having the ability to apply these to solving practical problems and participating in public affairs." Today, the *Outline* is a dominant policy for running science communication across the country. As the document states, the principal strategy in carrying out the *Outline* is "Boosting the government, creating mass participation, raising scientific literacy, and promoting harmony" (State Council 2006).

Funding for science communication soared after the issuance of the *Outline*. According to statistics from the Ministry of Science and Technology, the total investment in science and technology popularization was about 2.4 billion yuan in 2004, 4.6 billion in 2006, 10 billion in 2010, and 13 billion in 2013 (DPR 2010, 2015). Table 12.1 sheds some light on the results.

The *Outline* lists youths as one of the four key groups (the other three being farmers, the urban workforce, leading cadres and public servants) and includes the *Youths' Scientific Literacy Action Plan*; it considers youths' informal science learning as an important way to improve youths' scientific literacy. The *Youths' Scientific Literacy Action Plan* outlines a set of missions, tasks, and measures for youths' informal science learning.

The missions stated in the *Youths' Scientific Literacy Action Plan* are to familiarize primary and middle school students with important basic scientific information and skills and to allow them to experience the process and method of scientific

Table 12.1 Output of science popularization investment in 2004–2013

Year	S&T museum Number	S&T museum Visitors (m)	Science book (title)	TV program (h)	Science website	Lecture (m)	Exhibition (m)
2004	185	N/A[a]	2,523	74,959	995	38.13	7.06
2006	239	17	3,162	113,758	1,465	72.33	10.31
2008	380	39	3,888	219,168	1,899	95.51	11.53
2009	505	54	6,787	243,094	1,978	84.94	13.02
2013	678	98	8,423	223,610	2,430	91.21	16.13

Source: Science popularization statistics of China 2010, Science popularization statistics of China 2014, Department of Policy and Regulations of the Ministry of Science and Technology

[a]There is no individual number of visitors for a comparison with 2004. The total number of visitors who visited science centers, science and technology museums, and youth centers or science stations in that year was less than 30 million

exploration activities. It states that efforts should be made to foster a scientific attitude, passion, and sense of value among youths; to develop their preliminary capability for scientific explorations; and to enhance their awareness of innovation and practice.

Specifically, the tasks that are needed to accomplish the missions are initiating science outreach activities and social practices in diverse forms; enhancing youths' enthusiasm and interests in science and technology; allowing them to gain knowledge of the nature of science and the relationship among science, technology, and society; enhancing youths' sense of social obligation and communication and cooperation abilities; and enhancing their ability to solve problems by integrating newly learned information.

The means of accomplishing these tasks are:

- Developing informal educational activities for youths outside the formal rural educational system and initiating science outreach activities for improving youths' general living and working skills.
- Organizing extracurricular science and technology activities to increase youths' innovation awareness and useful skills through an array of initiatives, including hand in hand between scientists and adolescents, teenager science outreach actions for youths, and middle school students visiting research institutes and labs.
- Institutions and organizations such as media and culture organizations should strengthen their efforts in the communication and popularization of science and technology by attracting youths using excellent, beneficial, and vivid public communication of science and technology works and creating a favorable public opinion environment for healthy youth development.
- Consolidating science education resources outside schools and establishing an effective mechanism linking outside school science and technology activities with school science courses and improving youths' scientific literacy by taking advantage of diverse education resources, including science and technology museums, research institutes, and other science and technology education bases.
- Strengthening the popularization of science and technology function of such comprehensive after-school activity sites for youths as the youth and pioneers' palaces and children's centers, establishing exclusive science and technology activity centers for children and youths where conditions permit, and taking advantage of the role played by community education in youths' science education outside the context of school.

12.4 Major Implementation Models

The following account of science activities, carried out by the Children and Youth Science Center (CYSC) of CAST, was developed from the ideas described in the *Outline*. CYSC is engaged in the spread of science knowledge for youths and the public, organizing pilot science activities as well as science contests. As a nongovernmental organization of scientists and engineers, CAST performs its obligations according to the *Law of the People's Republic of China on Popularization of Science and Technology* and the *Outline* and considers the promotion of public understanding of science and the enhancement of science literacy as its missions. The CYSC is a driving force of CAST for youths' science education outside the school, as well as the popularization of science and technology in the general public.

Since the 1970s, the CYSC has developed various science activities for youths to nurture their interest in science, build up their scientific mindset and spirit, as well as promote their scientific literacy as future citizens.

12.4.1 China Adolescents Science and Technology Innovation Contest

The China Adolescents Science and Technology Innovation Contest (CASTIC) is a national contest focusing on adolescents' science education and science research. It has a nearly 30-year history and is jointly organized by CAST, the Ministry of Education, the Ministry of Science and Technology, the National Environment Protection Administration, the National Sports Administration, the Chinese Natural Science Foundation, the Central Committee of the Chinese Communist Youth League, the All-China Women's Federation, and provincial and municipal governmental sponsors. Now CASTIC is one of the most popular science education activities for all students in China. Every year there are more than ten million students engaged in different levels of contest, and around 500 students and 200 teachers among them participated in national level of contest.

12.4.2 Youths' Scientific Investigations and Experimental Activities

In 2006, the Youths' Scientific Investigations and Experimental Activities program was jointly launched by the Ministry of Education; the Civilization Office of the CCPC; the State Administration of Radio, Film, and TV; the Communist Youth League of China; CAST; and other organizations, with the aim of improving youths' scientific literacy. Scientific investigation, scientific experience, and scientific exploration were the main content. Through these activities, youths master basic science and technology knowledge and skills, experience the process of the scientific method and scientific exploration, cultivate a positive scientific attitude, and improve basic scientific research ability. To organize youths to gain relevant knowledge, they develop their own scientific investigation activities at home and in schools (classes or groups), launch relevant scientific experiences and scientific explorations in the agricultural industry, and learn how to collect and arrange relevant material and data. Based on these activities, they perform data summaries and analyses and provide investigation reports with proposed reasonable suggestions. These activities have been held in 31 provinces nationwide, and more than three million students engaged in these activities.

12.4.3 Science Activities from Science and Technology Museums on Campus

By 2010, the number of museums in China had reached 900, of which about 300 were science and technology museums. The CYSC of CAST set up and developed the "Science activities from science and technology museums on campus" project in 2006. The activities include training, science show, hands-on experimentation, and science workshops and clubs which aim to integrate the resources of science and technology museums into school curricula, especially science courses, comprehensive practical activities, and research studies. It also improves the effective connection between science and science education technology activities inside and outside school.

In 2014, the second competition of museum-based science education projects was held. There were 145 projects from 130 science museums and centers nationwide. After two rounds of preliminary competing, there were 64 projects from 48 science museums and centers that participated in the final competition in Beijing with three parts, which were online voting, promotion of the project, and face-to-face debating. The judges were from the public, school teachers and professionals. Furthermore, two training programs were given by the experts at home and abroad to advance project leader's capacity in design and implement science activities from science and technology museums on campus.

12.4.4 Science and Technology Activity Popularization for Youths' Interest in Space

To stimulate youth interest in aerospace technology and promote more talent in science and technology innovation, CAST, the China Manned Space Program Office, and the Chinese General Company of Astronautics Science and Technology directed a series of science and technology popularization activities intended to involve youths in space flight. Running from 2010 to the end of 2013, it was named "Opening the Dream of Space" and was intended for all youths. It included four parts:

(1) Exploring space: science and technology experiment activities for youths.
(2) Seeking space: a space flight-related science and technology competition for youths. This competition was widely promoted by TV, the Internet, journals, magazines, and so on, to encourage more youths to learn about astronautics science and technology.
(3) Searching space: youths researching articles to increase their science and technology knowledge about space flight.
(4) Realizing space: an experiential activity for youths in astronautics science and technology. From 2011 and annually thereafter, various youth groups are organized to participate in experiential activities about astronautics science and technology. These include visiting an astronautic control center, a spacecraft launching base, and an astronaut training base and communicating with astronautics experts.

In 2014, 80 tour reports of popular science of space were given by scientists and experts to students in Jiangsu, Qinghai, Inner Mongolia, and Chongqing, with the audience of 100,000. It stimulated students' learning interest of space science.

12.4.5 Youths' Informal Education Programs in the Countryside

Since 1982, CAST has cooperated with the United Nations Children's Fund to advocate youths' education and development in poor areas of China. From 2011 to 2015, joint programs have been developed in 22 poor counties in 11 provinces and autonomous regions in the central and western regions of China, with the purpose of helping youths aged 14–17 years transition from school to work, improving their work and social lives, strengthening their sense of participation with social problems affecting them, and guiding them to become socially responsible citizens.

This program plans to fulfill this purpose by a series of activities, such as establishing "youths' development training centers," organizing expert groups, developing informal educational training courses and youths' science and technology popularization activities in a program named *Go to Society—Life Job Development*, holding youths' development forums, promoting participation for countryside youths, and appointing and retaining good role models.

In 2014, 188 training programs with 5,518 trainees from rural areas were held; the educational programs embedded in school curriculum were promoted in four counties; moreover, the relating training programs was given to 514 rural young people. The revision of *Guidebook of Girls' Living Capacity and Skills* started as well in 2014.

12.4.6 Youths' Science and Technology Camps in Colleges and Universities

In 2012, CAST and the Ministry of Education, with the support of the Chinese Academy of Science, organized youths' science and technology camps in colleges and universities. Several colleges and universities and many provincial science and technology associations and educational bureaus took part in the program.

This program aims to explore the essence of science and technology camps and gain experience in doing activities such as integrating scientific research with science popularization; promoting the development and sharing of science education resources; utilizing the abundant educational resources of colleges and universities to advance the function of colleges and universities in disseminating scientific knowledge, scientific ideas, and methods in order to improve youths' scientific literacy; inspiring youths' interest in science and encouraging youths to enter science-related careers in the future; encouraging youths' engagement in scientific research; cultivating youths' scientific spirit, innovative mindsets, and problem-solving ability; and laying a solid foundation for science and technology talent.

In July of 2014, 10,610 high school students from four regions across the straits (9,050 from mainland, 200 from Taiwan, 1,000 from Hong Kong, and 360 from Macau) from 1,500 schools spent a week of science and technology camp in 49 first-class universities and colleges in mainland. There were 176 seminars held, and 17 national key laboratories were open to these students, with 139 hands-on science and technology activities, 44 outreach activities, and 452 arts and sports activities carried out; 325 college campus and cultural sites were toured as well.

12.5 Discussion

12.5.1 Impact of Science Popularization Activities

In the context of informal science learning, the extent of students taking part in science activities plays an important role. A nationwide survey, sponsored by CAST, was done by the China Research Institute for Science Popularization in 2009. The survey was conducted in a total of nine provinces, four middle/high schools (two urban and two rural) and four primary schools (two urban and two rural) in each province. It aimed to study the progression of Chinese students' creative imagination from elementary through high school. With regard to extracurricular science activities, the study (Ren et al. 2012) shows that students who actively participated in science project competitions (including CASTIC) and Olympiads on various subjects and those who visited science venues (including science and technology museums, natural history museums, zoos, and botanical gardens) had a higher level of creative imagination. Students who "have no idea" about science project competitions and Olympiads got the lowest total creative imagination scores, those who "know about" got higher total scores than those who "have no idea," and those who "have participated" got higher total scores than those who "know about." Those who "have won prizes" got the highest total scores. Students who "always" go to science venues got the highest scores, students who "sometimes" go to those places got the second highest scores, and students who "seldom" go to those places got the lowest scores.

Since China is a huge developing country, however, the regional development of culture and economy is not balanced; therefore, the allocation of these educational resources such as these science venues is far from balanced. Most qualified science venues are located in big cities; few of these places are seen in towns and counties in China. Therefore, the form of science venues should be flexible, using mobile science museums, science wagons, online science museums, etc., which can do science exhibitions, tours, and virtual visiting for youth in small towns, counties, and villages.

12.5.2 Implementation of Chinese Youths' Informal Science Learning

Shi and Zhang (2012) argue that science popularization in China has developed in an idiosyncratic way, as part of an organized and mobilized effort. They state that after the *Outline* was implemented, a state-leading group was formed, including 23 members from cross-border governmental departments and national organizations. Tasks were divided among the 23 members, and 9 guidelines corresponding to each action and project were developed in detail. Local governments soon took up the *Outline* as a state assignment. Each province worked out similar working patterns and corresponding packages of programs. Furthermore, a series of national policy documents had been issued and enacted in accordance with the *Outline*, such as *Instructions on Opening of Institutes, Colleges and University to Society and Developing Science and Technology Popularization* (2006), *A Program of Science Activities from Science and Technology Museums on Campus*(2006), *Instructions on Strengthening the Ability of State Science and Technology Popularization Establishment* (2006), *Implementation on Youths' Science Literacy Action* (2007), and *Information on Promoting Provincial Youths' Outside School Activity and Developing Sharing a Place of Science and Technology Popularization Education* (2008). These documents dealt with specific aspects of informal education, such as place, institute, members, activity, support, and promotion of the development of informal education.

An example is the "Science activities from science and technology museums on campus" program, which was started as a way to implement the *Outline* in

June 2006. In August of that year, CAST expanded this program countrywide. During 2006–2009, more than 30 science and technology museums, youths' science and technology centers, and youths' science offices took part annually in this program. In total, 48 pilot units joined in the 3 years. In May 2010, the Ministry of Education and CAST jointly issued a program to extend the pilot unit for "Science activities from science and technology museums on campus" from 2010 to 2012. Through applications, selections, and reviews, they formally established 36 demonstration pilots spreading to 15 provinces, autonomous regions, and centrally administrated municipalities and deepened pilot units in 19 science and technology museums. This program had already become one of CAST's important assignments for youths' science popularization and has promoted some of the relevant youths' educational programs, further strengthened the science education function of places of youths' activities outside of the school context, and engaged in the public enterprise of science and technology popularization.

12.5.3 Flexible and Diverse Forms for Youths' Informal Science Learning

In informal science learning, youths learn science outside schools. Therefore, this kind of learning must be flexible and diverse. It could involve producing and broadcasting science programs for youths; organizing and developing science experiential activities, lectures, reports, and competitions; cultivating innovative ideas in youths; carrying out practical activities for youths' science and technology popularization; fostering youths' science ideas; connecting museums with schools for education; etc. These youths' science experiences support, strengthen, and play an important role in improving their science literacy.

Taking the experiential activities of youths' scientific investigation as an example, the main forms are "scientific investigation," "scientific experience," and "scientific exploration." One scientific investigation takes a low-carbon theme. Through visiting recycling stations, youths develop an understanding of ways to recycle and its global significance. For the scientific experience, there is a "farm tour" program with the theme of "saving food." This affords youths the experience of working on a farm, becoming familiar with the types and processes involved in growing food, and teaching them about the challenges in producing food. One scientific exploration is about water conservation, giving youths the opportunity to design 15 mini scientific experiments about water using the aid of the *Guideline of Fun Water Experiments*, which guides youths through water experiments to improve their understanding of water.

12.5.4 Attracting More Youth Participation in Informal Science Learning, Especially Those of Lower Social Economic Levels

Sixty-two percent of people aged 15–30 years in the Chinese countryside are estimated to leave their villages seeking jobs in cities and towns (International Labor Organization 2009). The program for countryside youths' informal education outside schools has launched informal educational course training and youth science and technology popularization activities, but some vulnerable groups may not participate. The most vulnerable group is teenage girls with the least education and skills who drop out earlier than teenage boys and then gradually become more vulnerable to violence. While some policies and practical programs may encourage participation of these groups, they are insufficient, thus missing the goal of reaching these vulnerable groups. Therefore, in order to comprehensively accomplish the goals of the *Outline*, more optimally practical programs in informal science learning need to be established that are geared toward vulnerable groups. To achieve this goal, the main effort is devoted to building a mechanism that emphasizes the best allocation of human and material resources and to constructing the required infrastructure.

References

American Association for the Advancement of Science. (1989). *Science for all Americans, Project 2061*. Retrieved from http://www.project2061.org/publications/sfaa/default.htm

American Association for the Advancement of Science. (1993). *Benchmarks for science literacy, Project 2061*. Retrieved from http://www.project2061.org/publications/bsl/default.htm

CCPC-SC (Central Committee of the CPC and State Council). (1981). *Outline of the Chinese Science and Technology Development Approach* (关于我国科学技术发展方针汇报提纲). CCPC-SC. Policy Document.

CCPC-SC (Central Committee of the CPC and State Council). (1994). The instructions on strengthening engagement in science and technology popularization (中共中央、国务院关于加强科学技术普及工作的若干意见).CCPC-SC. Policy Document.

CPPCC (Chinese People's Political Consultative Conference). (1949). Common program of CPPCC(中国人民政治协商会议共同纲领). Policy Document.

DPR (Department of Policy and Regulations of the Ministry of Science and Technology of the People's Republic of China). (2010). *Science popularization statistics of China2010* (中国科普统计2010). Beijing: Scientific and Technical Documents Publishing House.

DPR (Department of Policy and Regulations of the Ministry of Science and Technology of the People's Republic of China) (2015). *Science popularization statistics of China 2014* (中国科普统计2014). Beijing: Scientific and Technical Documents Publishing House.

European Commission. (2002). Framework program 6, Retrieved from http://ec.europa.eu/research/fp6/pdf/fp6-in-brief_en.pdf

International Labor Organization (ILO). (2009). *Protecting and promoting the rights of China's vulnerable young migrants*. Retrieved from http://www.ilo.org/beijing/what-we-do/projects/WCMS_143208/lang--en/index.htm

NPC (National People's Congress of the People's Republic of China). (1991). *Law of the juveniles protection of the People's Republic of China* (中华人民共和国未成年人保护法). Retrieved from http://www.mps.gov.cn/n16/n983040/n1928424/n1928499/1930204.html

NPC (National People's Congress of the People's Republic of China). (2002). *Law of the People's Republic of China on popularization of science and technology* (中华人民共和国科学技术普及法). Beijing: Popular Science Press.

Office of implementing the Outline of the National Scheme for Scientific Literacy (2006-2010-2020). (2011). *The report on national action plan for science literacy* (全民科学素质行动计划纲要年报). Beijing: Popular science press.

Ren, F., Li, X., Zhang, H., & Wang, L. (2012). Progression of Chinese students' creative imagination from elementary through high school. *International Journal of Science Education*, *34*(13), 2043–2059.

Shi, S., Zhang, H. (2012). Policy perspective on science popularization in China. In B. Schiele, M. Claessens, S. Shi (Eds.), *Science communication in the world: Practices, theories and trends* (pp. 81–95). Dordrecht/New York: Springer.

The State Council. (2006). Outline of the National Scheme for Scientific Literacy (2006-2010-2020) (全民科学素质行动计划纲要). Beijing: Popular science

Chapter 13
An Overview of Research on Informal Science Learning Among China's Youth

Fujun Ren and Jingying Wang

13.1 Historical Development of Informal Science Education

Since the time of the Opium Wars (1842) in the mid-eighteenth century, democracy and science have evolved into a collective pursuit for the Chinese people. Informal science education in modern China can be traced to the emergence of community organizations that were founded in the early twentieth century and prevailed in the 1920s and 1930s. Saving the nation through science and technology was the internal driving force of these organizations. This can be contextualized in view of the coercive inception of China's modernization, marked by the Opium Wars. After Western gunboats blew open the nation's door, Chinese intellectuals began to explore ways of safeguarding the country's survival, power, and resources. To these ends, they first drew on advanced Western science and technology, with a particular military emphasis. Local informal science education societies first appeared during this time. Scientific societies that emerged during the late Qing era and in the succeeding Republic of China proliferated rapidly, focusing on applying science and technology to everyday life. Libraries and educational and science museums became important modern public education assets that promoted the development of informal science education. Lecture halls, public schools, and

F. Ren (✉)
China Research Institute for Science Popularization, No. 86, Xueyuannanlu, Haidian District, Beijing 100081, China
e-mail: hljrenfujun@126.com

J. Wang
China Research Institute for Science Popularization, No. 86, Xueyuannanlu, Haidian District, Beijing 100081, China

Capital Normal University, Beijing 100048, China

L.L. Liang et al. (eds.), *Chinese Science Education in the 21st Century: Policy, Practice, and Research*, Contemporary Trends and Issues in Science Education 45, DOI 10.1007/978-94-017-9864-8_13

exhibition rooms were also created as venues for informal science education. From the late Qing period onward, informal science learning activities developed gradually. A system of informal science education was initially formed, led, and operated by nonprofit and nongovernmental societies and associations, with government advocacy, support, and promotion.

In the mid-1980s, rural and urban economic development, rising social production forces, and improvements in the industrial, agricultural, and service sectors created new cultural and technical demands. The sense of urgency emanating from communities led to the emergence of accounting, sewing, English language classes in the cities, and tractor driving and agricultural techniques in rural areas, as required by the times. The elimination of illiteracy was aided by mobile teaching, whereby a few teachers taught students in different districts and engaged in mastery teaching that focused on selected students, enabling them to learn well.

Short-term correspondence courses, broadcasts, television, and popular science lectures also began to emerge. As public's appreciation of science grew in China, and the government promoted scientific literacy as part of a harmonious modern society, informal science education expanded. This was achieved through the 2002 *Science Popularization Law* and the 2006 *National Scientific Literacy Action Plan*, which raised informal lifelong education to its peak as a necessary aspect of China's social and economic development.

13.2 The School Science Education Context of Informal Science Learning in China

The history of science and technology in China is both long and rich with many contributions to their development. Science education reforms in China initially concentrated on the knowledge and skills that students acquired from classrooms, gradually shifting to solving real-life problems through scientific inquiry. This process propelled informal science learning forward and highlighted the important role that informal science learning plays in science education.

13.2.1 The First Phase: Cultivating Basic Knowledge and Skills

Early Chinese education was reconstructed during this initial phase based on the education system and experiences of the former Soviet Union (USSR), which importantly influenced the direction of Chinese science education. China established the education goals of "basic knowledge and skills" and focused on systematic knowledge building.

13.2.2 The Second Phase: Cultivating Scientific Ability

This phase began in the 1960s with an explicit focus on cultivating scientific ability. Teaching outlines were established for all elementary and secondary school subjects that simultaneously emphasized basic knowledge and skills and cultivating students' analytical and problem-solving ability. During this period, scientific basics, ability, and education were emphasized for achieving well-rounded development. The teaching goal was to cultivate and develop scientific thinking, observation, experimenting, and self-studying abilities. Although knowledge-centered theory emphasizing systematic knowledge and its verification prevailed, STS (science, technology, and society) education and an integrated curriculum were also focal areas. Additionally, the mastery learning method, curriculum theory, and Bloom's taxonomy were influential during this phase of science education.

13.2.3 The Third Phase: Cultivating Scientific Literacy

This phase began during the 1990s and entailed three kinds of transformation that emphasized the functions of informal science education. The first was evident in the goal of science education that focused on all students and emphasized the development of scientific literacy, creativity, and practical ability (Wang and Yuri 2012). The second related to the science curriculum and emphasized the relationship between the curriculum and students' practical and experiential worlds, promoting the interdisciplinary study of technological development in the society. The last transformation entailed a practical view of science education that led to an understanding of the scientific process, the nature of science, and the value of science, cultivating the ability for scientific inquiry, sensibility, attitude, and valuation in students. Some of the critical literacy abilities associated with self-directed study of science such as critical thinking, information source select and critique, and languages of science began to be noticed as part of the scientific literacy framework only in China in recent years.

13.3 Theoretical Framework for Studying Informal Science Learning in China

Informal science learning is directly guided by core administrative theory on the division and mutual transformation of explicit and tacit knowledge. Commencing with the existing form of knowledge, knowledge management divides knowledge into explicit and tacit knowledge (Polanyi 1967). Explicit knowledge refers to words, images, and symbols recorded through print or electronic media. It is composed of programmed knowledge such as facts, natural principles, and scientific knowledge. Explicit knowledge has been publicized because of its effective and cost-efficient structure. People, therefore, have easy access to dominant knowledge in libraries, in the Internet, or through databases. Tacit knowledge is composed of cognition, emotion, beliefs, experience, and skills. These five elements are components of a person's mind and skills and are revealed through behavior. Because tacit knowledge is an inherent attribute, it is hard to express by means of language and written words. In informal science learning, the mutual transformation of explicit and tacit knowledge includes four steps: socialization, externalization, fusion, and internalization. Informal science learning is a process of implementing knowledge management (Tal 2012). This requires the following steps: (1) transforming knowledge from informal science learning into personal knowledge, (2) turning our individual experience and inspiration into understandable explicit knowledge, and (3) socializing personal knowledge and putting it into practice. Knowledge management of tacit knowledge is the key theoretical foundation of informal science learning in China.

Informal science learning occurs anytime and anywhere in natural settings. Learners can choose what to learn based on their interests. This choice is influenced by social construction and social constructionism. Both of these view society as an orientation and emphasize the social influences affecting individual development. However, social construction reaches further than social constructionism by putting society above individuals and studying its effect on individual learning within social settings. This approach is supportive of the basic concept that social interaction promotes human development.

13.4 Scope of Informal Science Learning Engaged by Chinese Youths

In addition to school time, two-thirds of students' time is spent in informal learning (Bell et al. 2009). Their activities during this time, therefore, have a far-reaching influence on their individual development (Cheng 1989). In order to understand the scope of informal science learning engaged in by Chinese youths, in 2010 we developed a questionnaire to survey Chinese youths. The questionnaire consisted of three sections pertaining to adolescents: ways of acquiring scientific knowledge and information, reasons for using out-of-school facilities for science education, and forms of adolescent representation and participation in out-of-school science education (Dillon 2012).

Based on stratified sampling, we chose 60 middle schools in Jilin, Hebei, Henan, Hubei, and Guizhou provinces. We sent out 4470 questionnaires and received 3670 completed questionnaires from students, including 1730 junior students and 1940 senior students.

13.4.1 *Ways for Adolescents to Acquire Scientific Knowledge and Information*

According to the status of informal science education, and of respondents as youths, the questionnaire listed 11 items relating to how youths could acquire scientific knowledge and information. This study was conducted in three districts of P. R. of China. From east to west, we chose Beijing, Henan, and Guizhou; we randomly selected 8th grade and 11th grade classes of four middle schools in each district with a total of 7126 students to participate in the study. The questionnaire contained six comprehensive questions, which contained two multiple-choice question and four description questions.

The survey indicated that class learning was the main and appropriate method for youths to acquire scientific information (73.6 %). However, a significant proportion of the youth widely used the Internet (54.5 %), television (41.4 %), and libraries (39.5 %) as their main method of knowledge acquisition. Fewer youths participated at scientific venues, and in scientific activities, or acquired this knowledge from newspapers and magazines. They rarely listened to broadcasts (see Fig. 13.1).

As shown in Fig. 13.2, urban youth made good use of the internet for communication. Given their location advantage, 5.1 % more urban youths used scientific venues than suburban youths. However, almost 16 % more suburban youths, compared with urban youths, made greater use of television for acquiring scientific information.

There were also gender differences, with 7.4 % more boys than girls browsing science websites and Internet videos and 7.6 % more boys discussing scientific topics. Girls were more inclined to acquire scientific knowledge and information from traditional media, such as libraries and magazines, with 9 % more girls than boys using magazines.

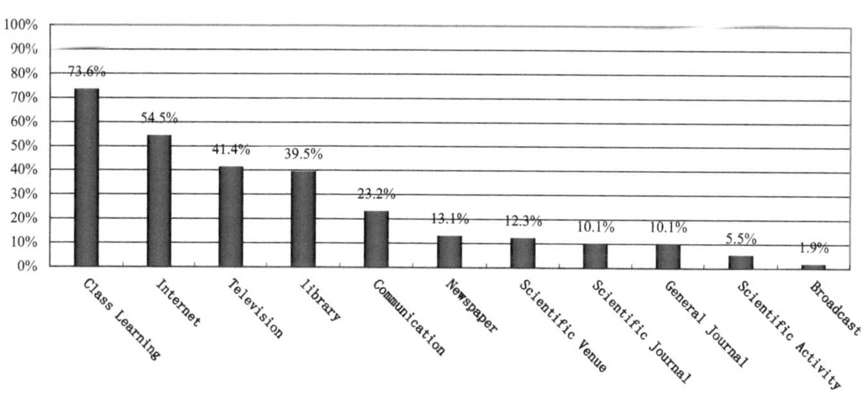

Fig. 13.1 Ways for adolescents to acquire scientific knowledge

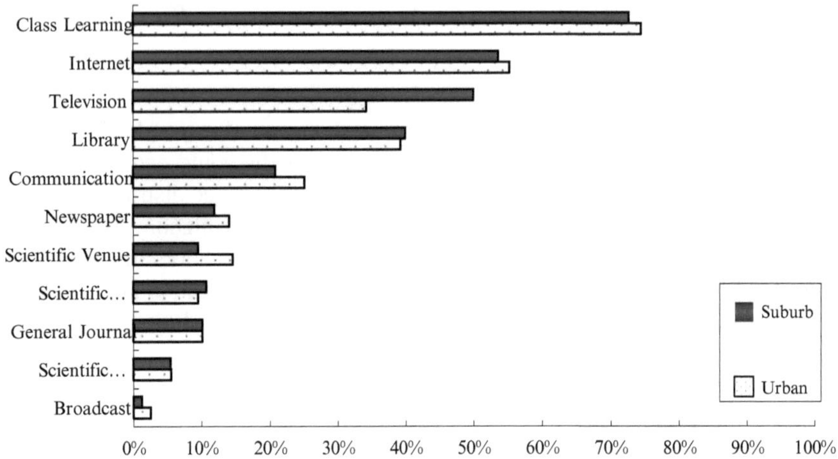

Fig. 13.2 Differences in acquisition of scientific knowledge and information between urban and suburban youth

13.4.2 Reasons for Using Out-of-School Facilities for Science Education

Adolescents widely used out-of-school facilities, such as zoos, aquariums, and botanical gardens (93.5 %), science and technology museums (87.5 %), and public libraries (83.9 %). At school, they also frequented library reading rooms (83.4 %). However, the use of some kinds of professional science and technology venues was evidently lower. In descending order, these were industrial and agriculture production zones (33.2 %), science and technology models or popular science activities and sites (28.6 %), and college or scientific research laboratories (25.1 %). Table 13.1 shows specific reasons for using and not using informal science education facilities. Visits by adolescents to the first four items listed as out-of-school scientific education facilities were based on their own interests. The main reason for their low usage of other professional science and technology sites was ignorance of their locations. Specifically, almost 45 % of adolescents did not know the locations of college or scientific research laboratories. A section of them expressed a lack of interest in these places.

The survey reveals that most adolescents acquired knowledge from popular science sites and facilities, with few youths perceiving that they had acquired nothing. Over half of the adolescents viewed public libraries (59.4 %) and science and technology museums (58.0 %) as useful facilities. Integrating these data with the data on less useful facilities, the highest percentages were found for science and technology museums (89.8 %), zoos (89.2 %), public libraries (86.5 %), and library reading rooms (81.3 %). Although professional science and technology sites were less commonly used, their percentages were also above 65 %. Even the least popular science galleries or public columns reached 61.6 %.

Table 13.1 Conditions of using out-of-school facilities for science education (%)

Place name	Been to				Haven't been to				
	Interest	With friends	School activity	By chance	None location	Ticket expensive	Less exhibits	Unknown location	No interest
Zoo, aquarium, botanical garden, aquarium	43.3	28.9	10.4	10.9	1.1	1.1	0.5	0.5	3.3
Science and technology museum, planetarium	36.0	18.8	24.5	8.2	1.6	1.4	0.3	2.7	6.5
Public library	55.9	13.6	3.5	10.9	0.3	0.8	0.5	4.9	9.8
Art gallery or exhibition	22.6	13.4	15.5	10.1	4.1	0.8	0.8	13.1	19.6
Popular science gallery or publicity column	13.1	6.5	7.9	19.6	4.4	0.8	1.4	19.1	27.2
Library reading room	51.2	9.0	7.1	16.1	0.3	0.3	0.5	7.4	8.2
Scientific and technology model or popularization of science activity place	7.1	5.2	6.8	9.5	7.6	0.5	1.4	34.6	27.2
Industrial and agriculture production zone	4.6	8.7	12.3	7.6	6.8	0.3	0.5	34.3	24.8
College or scientific research laboratory	5.2	1.9	10.1	7.9	9.3	0.8	0.3	45.0	19.6

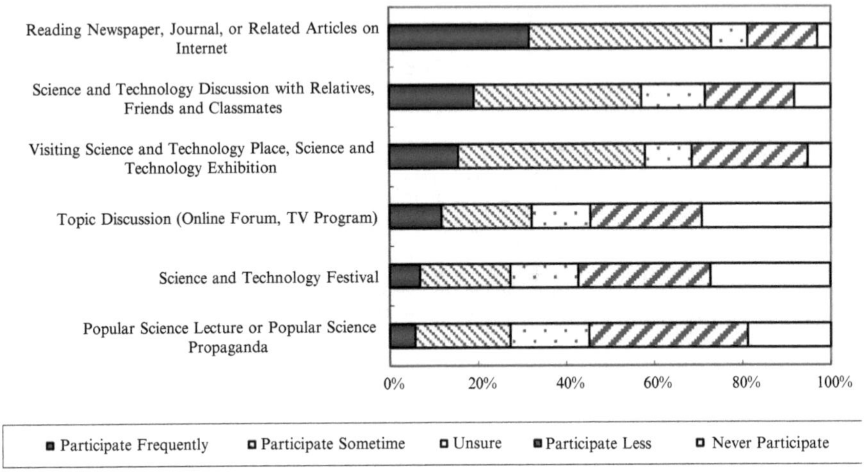

Fig. 13.3 Forms of adolescent participation in out-of-school scientific education

13.4.3 Forms of Adolescent Participation in Out-of-School Science Education

Adolescents were generally aware of many kinds of popular science activities, an exception being scientific broadcasts, of which 45.8 % of adolescents were completely unaware. However, in general, all adolescents who participated in this activity were enthusiastic. Most had not previously participated, but desired to do so after gaining some understanding. However, participation was generally low in these kinds of popular science activities, with science and technology exhibitions being the highest in this category at 44.4 %. As Fig. 13.3 shows, adolescents preferred to read newspapers, journals, or related articles on the Internet. While also communicating with others and visiting science and technology venues, they participated less in interactive discussions and in practical science and technology activities. Because most youths were the single child in their families, they also lacked communication and practical abilities. Moreover, these kinds of activities were not popular and attractive among adolescents. In particular, popular science lectures and other forms of popular science propaganda were unpopular.

13.5 Research on Informal Science Education in China

Although science education was first evident in China in the 1800s, the systematic study of informal science learning emerged rather late. Research on informal learning sprang up in Chinese academic circles in the late 1990s and gathered momentum by the early years of this century. Using journals retrieved from the

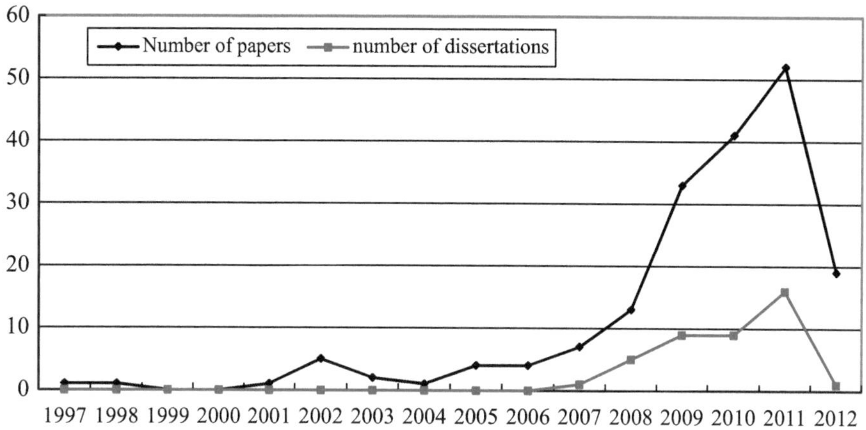

Fig. 13.4 Numbers of references related to informal learning (January 1997–May 2012)

Chinese National Knowledge Infrastructure (CNKI) database (www.cnki.com) as our major data source for analysis, we retrieved only a few articles by applying the keywords "informal science learning" and "informal science education." Considering research convenience and aiming to broaden the research scope, we then expanded the retrieval range and selected "informal learning" as our keyword sort and analyzed the current status and historical development of informal learning research.

By using the keyword "informal learning" in the CNKI search engine, we retrieved a total of 274 studies, including 26 dissertations and 248 academic papers published between 1997 and May 2012. After excluding irrelevant papers, 225 remained. Chronology statistics are shown in Fig. 13.4. It is evident that there was little domestic research done in this field before 2005. However, since 2008, "informal learning" has become a hot research topic. There is also a concentration of dissertations in this field during the last few years. From 2007 to the present, interest in informal learning research has shown an increasing trend. However, because our sample only extended to May 2012, this survey did not represent all informal learning studies published in 2012.

Our analysis of abstracts and keywords revealed six focal areas of informal learning research: theoretical studies, practical applications, virtual technology, current status, evaluation studies, and resource constructing. Numbers and percentages of articles on these subtopics are shown in Fig. 13.5.

To further understand the historical development of informal learning, we obtained chronological statistics in the six main research areas described above. The results are shown in Fig. 13.3. We divided the period from 1997 to 2012 into three phases: before 2005, 2005–2009, and 2009–2012. This was because the number of papers on informal learning began to quickly rise in 2009. Figure 13.6 shows the corresponding characteristics and focus areas during these three phases.

Fig. 13.5 Components of informal learning research

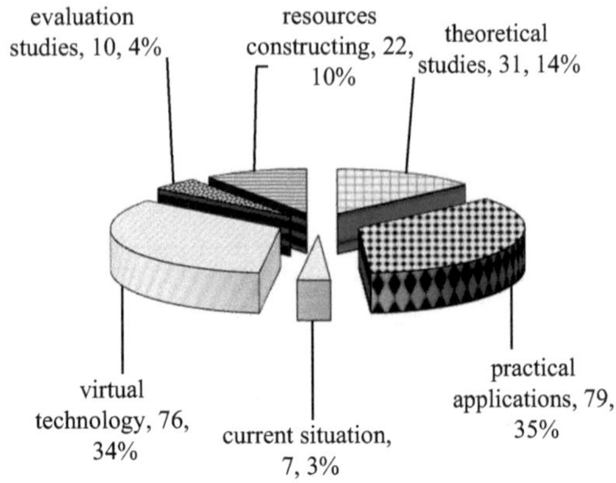

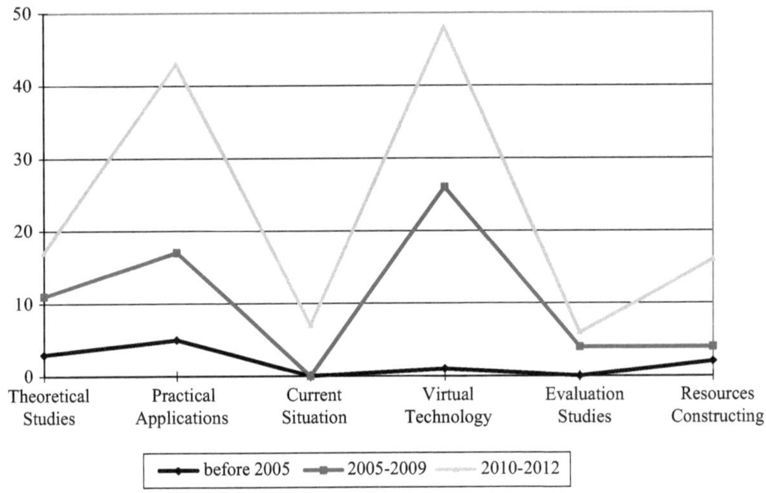

Fig. 13.6 Change of informal learning topics in different years

For a more detailed understanding of these trends, we analyzed the abovementioned research directions and their key aspects, as shown in Table 13.2.

During the first phase (up to 2005), "informal learning" investigations by Chinese scholars had only just started. Theoretical discussions on this subject mainly focused on the importance and significance of "lifelong learning." More practical studies were applied to adult and community education, but related research was relatively sparse.

During the second phase (2005–2009), studies in this area increased gradually. Theoretical research deepened, and related models, such as long-tail theory and tacit knowledge, were introduced by Chinese scholars. With the development of Internet networks and software, research on informal learning in virtual technology fields increased, as did the development of machine-based evaluation systems and resources.

Table 13.2 Research studies by different topics

Research topics	Contents	Number
Theoretical studies	Function and value of informal learning	5
	Conception and theoretical basis	15
	Review	11
Practical applications	Training	11
	Students' learning	14
	Classroom teaching	16
	Teachers professional development	24
	Informal learning of adults	14
Current situation	Undergraduates	4
	Teachers	3
Virtual technology	Digital media	4
	Micro-mobile terminal	21
	Micro-blog	9
	Society software	6
	Internet learning	35
	Game learning	1
	Distance education	1
Evaluation studies	Assessment of learning outcome	6
	Comparative research	4
Resource constructing	Platform construction	20
	Construction of learning resources	2

During the third phase (2010–2012), informal learning research conducted by Chinese scholars greatly accelerated. Chinese investigators began to theoretically focus on research results from the European Union, the United Kingdom, and the United States. During this period, more empirical investigations were carried out, which mainly addressed the status of informal learning among college students, graduate students, and teachers. This provided a reference for understanding the development of informal learning in China at a time when virtual technology was overtaking traditional technologies. Other mobile media such as microblogs, mobile phones, and network learning were evolving, and the study of new media also accounted for a large proportion of research. Research on resource development showed an upward trend, and the design and creation of learning communities and societies became the main research direction.

13.6 Discussion

In-class learning is the main method for Chinese adolescents to acquire scientific knowledge and information. At the same time, the Internet and television programs play an important role, whereas newspapers, journals, and broadcasts are less used.

Visits to science and technology venues and participation in popular science activities also have less appeal for adolescents. Regarding differences between urban and suburban youth, the former can acquire information by visiting science and technology venues because of their location advantage, whereas the latter acquire this through television programs (Paris 1997). Boys prefer to use the Internet, whereas girls prefer reading traditional print media to acquire information. Class learning should, therefore, be closely integrated with out-of-school science education. During classes, teachers should provide students with cutting-edge science and technology information that imparts knowledge connected to everyday life. This can motivate adolescents and arouse their curiosity regarding science and technology. Teachers should also make full use of new media. The Internet is a part of adolescents' lives and their acquisition of information. Establishing popular science websites that are tailored for and appeal to youths can motivate them to acquire more knowledge and not just entertainment from the Internet (Gee 2003).

Informal science education research in China can be divided into three phases from 1997 to 2012: preliminary, comprehensive, and empirical investigation. Some national scholars, who introduced foreign research results, have tended to divide informal science education into its informal existence in reality and a digitally constructed virtual environment. The former includes an individual's everyday life and family environment, as well as museums, science centers, botanical gardens, zoos, aquariums, and libraries (Ren et al. 2012). A virtual environment, on the other hand, is established through information technology such as the Internet, multimedia and virtual reality technologies, and online games (Rennie 2007). This classification mainly depends on different learning environments. Chinese researchers have also investigated informal science education environments based on community electronic media and popular science sites.

The historical trends and frequencies of the types and foci of research on informal science learning indicate growing interest in practical issues and virtual environments that are likely to set the trend in informal learning research in the near future. Their implications for future research may be in the areas of assessing learning outcomes and exploring the challenges of assessing science learning in informal settings. Informal science learning formats for young people in China can be divided into three categories: media, science popularization activities, and science popularization sites, and their results should be assessed for further development. Chinese researchers should construct evaluation frameworks as a tool for both planning and assessment, identifying areas that can be assessed in these informal settings. At the same time, we need to pay attention to the promotion of students' understanding of nature of science and scientific inquiry in informal science learning environment, and the integration of informal science education resources in different regions of China, and to optimize the educational resources.

References

Bell, P., Lewenstein, B., Shouse, A. W., Feder, M. A. (2009). Learning science in informal environments: People, places, and pursuits. *National Academies Press, 32*(3), 127.

Cheng, D. H. (1989). Out-of-school activities: The road to success. In K. M. Cheng & K. F. Leaning (Eds.), *Popularization of science and technology* (pp. 101–104). Paris: UNESCO.

Dillon, J. (2012). Science, the environment and education beyond the classroom. In B. J. Fraser, K. G. Tobin, & C. J. McRobbie (Eds.), *Second international handbook of science education* (pp. 1081–1096). New York: Springer.

Gee, J. P. (2003). What video games have to teach us about learning and literacy. *Computers in Entertainment (CIE): Theoretical and Practical Computer Applications in Entertainment, 1* (1), 1–4.

Paris, S. (1997). Situated motivation and informal learning. *Journal of Museum Education, 22* (2&3), 22–26.

Polanyi, M. (1967). *The tacit dimension*. New York: Anchor Books.

Ren, F. J., Li, X. J., Zhang, H. L., & Wang, L. H. (2012). Progression of Chinese students' creative imagination from elementary through high school. *International Journal of Science Education, 34*(13), 2043–2059.

Rennie, L. (2007). Learning science outside of school. In S. Abell & N. G. Lederman (Eds.), *Handbook of research on science education* (pp. 125–170). Mawah: Lawrence Erlbaum.

Tal, T. (2012). Out-of-school: Learning experiences, teaching and students' learning. In B. J. Fraser, K. G. Tobin, & C. J. McRobbie (Eds.), *Second international handbook of science education* (pp. 1109–1122). New York: Springer.

Wang, J. Y., & Yuri, O. (2012). The development of science education in P. R. China. *Journal of Science Education, 13*(1), 4–8.

Chapter 14
Effects of Media on Science Learning of Chinese Youths: A Synthesis of Literature and a Case Study

Ling Chen, Yan Yan, and Jie Yuan

14.1 Introduction

The media plays an important role in the public's daily life because of its significant role in society and public life. Household TV ownership in China increased from 83.30 % (National Bureau of Statistics of China 1995) in 1994 to 97.82 % (ChinaIRR 2012) at the end of 2011, which means almost all Chinese minors grow up with a TV in their homes from birth. A similar situation exists with digital media. Internet usage has been reported to be 91.4 % as of October 30, 2011, which means most minors have some experience with Internet usage (Li et al. 2012). The market demand for books aimed at minors is also high in China. Book ownership rates increase constantly. In 2012, minors under age 17 had a book reading rate of 83.1 % (China Publish 2012). The variety of public media has established a ubiquitous informal learning environment, which may have both intended and unintended effects on children and adolescent development. This chapter is concerned with the effect of media on science learning of youths through informal education.

L. Chen (✉) • Y. Yan • J. Yuan
China Research Institute for Science Popularization, No. 86 Xueyuannanlu, Haidian District, Beijing, China
e-mail: chenling@cast.org.cn

L.L. Liang et al. (eds.), *Chinese Science Education in the 21st Century: Policy, Practice, and Research*, Contemporary Trends and Issues in Science Education 45, DOI 10.1007/978-94-017-9864-8_14

14.2 Conceptual Framework

In this chapter, we use the definition of science communication as presented in *Science Communication: A Contemporary Definition*. "Science Communication (SciCom) may be defined as the use of appropriate skills, media, activities, and dialogue to produce one or more of the following personal responses to science (the vowel analogy): *Awareness*, including familiarity with new aspects of science; *Enjoyment* or other affective responses, e.g., appreciating science as entertainment or art; *Interest*, as evidenced by voluntary involvement with science or its communication; *Opinions*, the forming, reforming, or confirming of science-related attitudes; *Understanding* of science, its content, processes, and social factors" (Burns et al. 2003). This definition clarifies the functions of science communication and is convenient for the evaluation of its effect. Similarly, science learning refers to gaining science knowledge; developing science awareness, enjoyment, interest, and understanding; and forming personal viewpoints. In China, media providing informal science education can be divided into three forms: print media, broadcast media, and Internet media. Print media mainly consist of print type (i.e., word typeset on a paper page), nonprint type, and optoelectronic type; broadcast media mainly consist of radio and television programs and advertisements; and Internet media mainly consist of web indexes, network planes, animations, and forums (Baidu Encyclopedia). All of the above mentioned media can be carriers of science, thus fulfilling the above functions of science communication and promoting science learning.

Various forms of science education can improve students' scientific literacy. Different researchers have given different definitions of scientific literacy. In their article, *How literacy in its fundamental sense is central to scientific literacy*, Norris and Phillips (2003) discriminated fundamental senses and derived senses of science literacy. Fundamental senses of scientific literacy refer to the abilities of reading and writing, and derived senses of scientific literacy refer to being knowledgeable, learned, and educated in science. In their point, conceptions of scientific literacy attend to the derived sense but tend to neglect the fundamental sense. Reading and writing are constitutive parts of science. In this chapter, we also attached great importance to the fundamental senses of scientific literacy. As said in the article, "without the expressive power and relative fixity of text; and without the comprehension, interpretive, analytical, and critical capacities we have developed for dealing with texts; then western science as we know it could never have come into being" (Norris and Phillips 2003). Therefore, scientific literacy is comprised of concepts, skills, understanding, and also values generalizable to all reading and knowledge of the substantive content of science. The fundamental sense is central to scientific literacy.

14.3 Previous Research on Effects of Media on Science Learning: A Synthesis

In 2009 the financial giant Morgan Stanley released the report *How Teenagers Consume Media*, which indicated that adolescents primarily consume television and Internet media, while radio and newspapers are unpopular among this age group. The situation in China is similar. Based on related research (CSM 2006; Han and Gao 2010), three types of media, television, books, and the Internet, have high penetration, contain more science education content than other media, and are more accepted and used by Chinese adolescents. These three media types have the highest utilization rates during informal science learning for Chinese adolescents.

14.3.1 Availability of Scientific Content in Science Communication

All types of media consider adolescents a special target audience and pay particular attention to them. There are customized children's TV channels and science education TV channels, as well as websites customized for teenagers. Furthermore, there exist specific popular science websites for teenagers. Children's popular science books are also an important category of books. These are all important carriers of informal youth education. There are a large number of these carriers: there are currently 33 children's TV channels in mainland China, an extra five satellite cartoon channels, and 77 education channels (science education channels included, Wang 2011); there are 34 popular science websites (columns included, Li 2007) with the goal of cultivating youths' innovation and improving scientific literacy in the Chinese mainland population; and there were 3018 new popular science children's books published in 2010 (Open Book 2011).

There is broad content coverage within media communications. In the television realm, while there are only a few customized science education programs for youths (e.g., *Open Sesame*, *Whimsical*, and *The Tree of Wisdom*), there are many general popular science and education programs from both domestic (e.g., *Approaching Science*, *The Light of Science and Technology*, *Invention*, *Natural Story*, *Science Review*, *Marvelous Laboratory*, and *The Charm of Science*) and abroad (e.g., *Discovery*, *National Geographic*, *Massive Nature*, and *Saga*) sources, which provide commentary in many scientific areas, including geography, nature, biology, and chemistry. By browsing and reviewing 34 popular science websites for youths, Li found their contents involved science topics from 11 subject classifications, including biology, aerospace, environmental science and ecology, life science, and ocean science; in general, these sites provided a relatively comprehensive coverage of science knowledge (Li 2007). In response to a survey question asking students whether they would "be able to find required science knowledge on the Internet," 79 % of respondents answered "always be able to" or "be able to at most times,"

18 % answered “only be able to in a small part,” and only 2 % of respondents chose the option “always unable,” which shows that the students perceived there to be abundant science information on the Internet (Su et al. 2009). There is even greater topic coverage in children’s popular science books. Many bestsellers are published in series and cover content in various scientific areas, like the current bestsellers *The Chinese Children’s Encyclopedia*, *Knowledge for Children*, and *My First Scientific Comic Book*.

14.3.2 Effect of Media on Science Content and Process Skills

Currently, only a few studies (Yao 2008; Yang 2007; Liang 2011; Su et al. 2009) have reported evidence that watching science and technology programs contributes to an improvement in grasping, understanding, and applying science knowledge and cultivates the spirit of scientific inquiry and scientific attitudes in youths. Yao’s research was a quantitative study carried out simultaneously in two Beijing area secondary schools. In brief, the television viewing behavior of 345 students was monitored to investigate how often the participants watched science and education programs. After analyzing the questionnaires, Yao found that the majority of students loved to watch science and education TV programs, and these programs had significant influence on the scientific literacy of students from secondary schools. Science and education TV programs could promote the scientific literacy of teenagers. In addition, a Grade 11 student’s experience watching science and education television programs was used as a case study. The result showed that the science and education TV programs had a long-term potential impact for teenagers (Yao 2008). In Yang’s study, the television viewing behavior of 58 middle school students from Guangzhou was investigated. Using attitude scale-based measurement and Chi-square test analysis, the research demonstrated that a series of prescriptive programs called *Intelligent Lighting Control Network* that the researcher produced for the high school students had good teaching effects in that they deepened the students’ understanding and comprehension of science knowledge; stimulated and improved students’ interest, attention, enthusiasm, and motivation to learn science; and promoted students’ understanding of science, technology, and society (Yang 2007). Liang chose 84 Guangzhou middle school students to investigate the learning outcomes of the science and education program *Acid Rain*. The results showed that the majority of the respondents were satisfied with the program and thought the program could improve their enthusiasm for science and deepen their understanding of science (Liang 2011). Su’s study, which consisted of a survey of 700 students, found that the Internet had already become an important channel for youths to acquire science and education knowledge. Youths had a positive attitude to science and education knowledge and participated actively in the Internet science communication. However, they also found that the youths’ sense of trust of the Internet remained low. They suggested that Internet science communication should improve the quality of its content, form, and service (Su et al. 2009).

A study showed that after watching science education TV programs, more than 90 % of high school students said they "have known more" or "have known more deeply" about the topics presented in the program (Yang 2007). More than 80 % of students reported that they "completely understand," "mostly understand," or "partly understand" the program contents after viewing. The data above indicate that science education programs contribute not only to increasing knowledge through science learning but also serve to reinforce students' current understanding of scientific knowledge. Another research study on a Grade 11 student's experience found that science and education programs expanded the student's perspective to science and made her realize that science is amazing, broad, and profound (Yao 2008).

Research on Internet-based popular science resources reported that most respondents agreed that science knowledge on the Internet was helpful in understanding scientific concepts and processes and in daily life and learning (Su et al. 2009).

14.3.3 Effect of Media on Scientific Awareness and Interests

The media has an influence on cultivating scientific awareness and initiating interests in science. Its ability to attract youths to watch and participate proactively varies depending on media format.

Recent research on TV science education programs has shown that most youths love watching TV science education programs and confirmed that science education programs can inspire their interest in science. A survey showed that nearly half of youths loved watching science education programs, while less than 10 % of them did not or never watched, which indicated that current TV science programs were relatively popular with youths (Yao 2008). However, most students said they only watched episodes that they were interested in and did not necessarily watch science programs entirely. This seems to indicate that youths hold some interest in particular topics of scientific contents and also that current science programs were somewhat lacking in audience loyalty. Furthermore, another survey showed that 63.5 % and 11.5 % of respondent students stated that they "love" and "greatly love" youth science education programs, respectively, and that 71.2 % and 28.8 % of students had "interest" and "great interest" in youth science education programs, respectively (Huang 2011). All of these survey results indicate that youths love and are interested in science education TV programs, which was also the conclusion based on the survey data reported in Liang (2011).

As for books, according to currently available statistical data, the annual sales of *Insect* (youths' version), a best-selling children's popular science book in 2009 and 2010, numbered around 30,000. The number of books with a total annual sales volume of more than 10,000 was 31 in 2008, 38 in 2009, and 18 in 2010 (Open Book 2011). This large sales volume indicates, to some extent, that the target

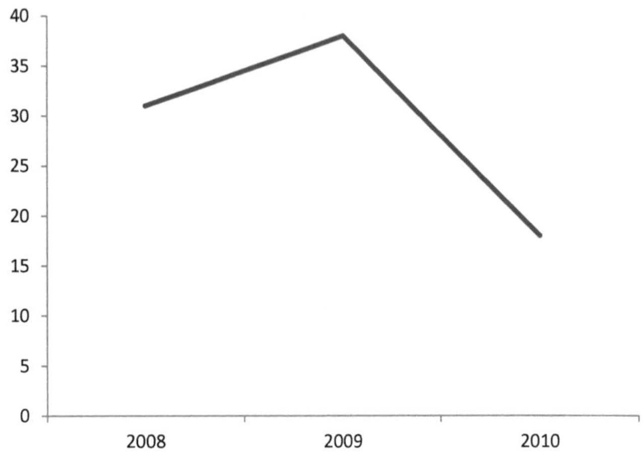

Fig. 14.1 The number of books with total annual sales volume of more than 10,000

Table 14.1 Time spent on popular science websites

Visiting times per month	Rate (%)	Stay hours	Rate (%)
>10	11	>2	9
5–10	15	1–2	11
3–4	36	1/2–1	25
1–2	29	<1/2	45
0	9	0	10

audience for children's popular science books likes the books and is willing to pay to read them (Fig. 14.1).

Consumption of informal science education on the Internet differs from consumption from television and books. In a survey on utilization of popular science resources on the Internet by adolescents, results show the visit frequency of respondents to popular science websites was low. Only 11 % of the surveyed adolescents reported visiting popular science websites more than ten times per month, 15 % 5–10 times monthly, 36 % 3–4 times monthly, 29 % 1–2 times monthly, and 9 % never. The overall time adolescents spent viewing popular science websites was also low. On average, 80 % of visitors stayed on popular science websites or web pages for less than an hour during each visit, and among them, 10 % "almost do not stay," 45 % for "less than 30 minutes," and 25 % for "30–60 minutes." Only 11 % and 9 % of respondents stayed for "1–2 hours" and "more than 2 hours," respectively (Su et al. 2009). The above indicates that the content and design of popular science websites are not attractive enough to youth to make them "forget to leave" (Table 14.1).

14.3.4 Effect of Media on the Formation of Individual Youths' Viewpoints

The media contributes to the formation of individual youths' viewpoints. A survey study showed that when watching *Intelligent Lighting Control Network*, more than half of students thought "seriously" or "highly seriously" about combining their knowledge with that presented in the program. Sixty-three point five percent of students thought that the program content was explained in detail and the case study was clear; thus the program contributed to student thinking. Furthermore, after watching the program, more than 70 % of students reported that they would like to learn more about related subjects by seeking knowledge from books, journals, and web page material. More than 60 % of students thought the program inspired them to learn positively and proactively (Huang 2011). A case study presented in another study also indicated that science and education programs enhance teenagers' understanding of scientific principles, which can be combined with the teenagers' current knowledge to enable development of their own unique viewpoints (Yao 2008).

14.4 A Case Study

We conducted a survey focusing on the effect of media on Chinese adolescents' creative imagination. We were concerned with the effect of different media forms. TV programs, non-textbooks, and the Internet are having different impacts on adolescents' acquisition of scientific knowledge, formation of scientific imagination, and development of scientific awareness. We selected Zhongguancun Middle School in Beijing as a study case. The sample for our investigation consisted of all students from Grades 7 to 10 of this school. In order to avoid the strong homogeneity within each grade, we selected four classes from Grade 7 and three classes from Grade 10 for the final investigation sample. The final sample of respondents includes 146 students from Grade 7 and 120 students from Grade 10.

After giving out questionnaires, we used the analytical tool SPSS 17.0 to get the scores of respondents' creative imagination. Some of the scores illustrate the effect of media science content on youth informal science learning and scientific ability.

14.4.1 Relation Between Watching TV Programs and Science Learning

Television is a main media channel for teenagers to get science information and knowledge in China. The great majority of teenagers watch TV in their time outside school. In our investigation, among the 146 Grade 7 teenagers, 117 students watched less than 7 h television every week, 20 students watched 8–14 h, 7 students watched 15–21 h, and 1 student watched more than 22 h per week. Among the 120 Grade 10 respondents, 99 students watched TV for less than 7 h weekly, 16 students watched 8–14 h, 3 students watched 8–14 h, 3 students watched 15–21 h, and 1 student watched more than 22 h per week. In summary, respondents from Zhongguancun Middle School did not spend much time watching television every week, especially the high school students.

For students from different grades, their time spent on watching television showed different relations to science learning and scientific creative imagination. Grade 10 students who watched TV programs for less than 7 h every week got an average creative imagination score of 43.67. Students who spent 8–14 h on watching TV per week got an average score of 45.57. This suggests that the more time high school students spent on watching television programs, the higher their score of scientific creative imagination was. However, when the time watching TV exceeded a certain point, the score became lower. This result illustrates that watching television programs for a reasonable amount of time was associated with the improvement of adolescents' scientific creative imagination.

Among all of the TV programs, several types of TV programs showed a significant association with students' scientific creative imagination. In our questionnaire, there were five possible answers to the question regarding students' favorite TV program type: cartoon, science documentary, teleplay, arts and entertainment, and science and technology news. Most secondary school respondents

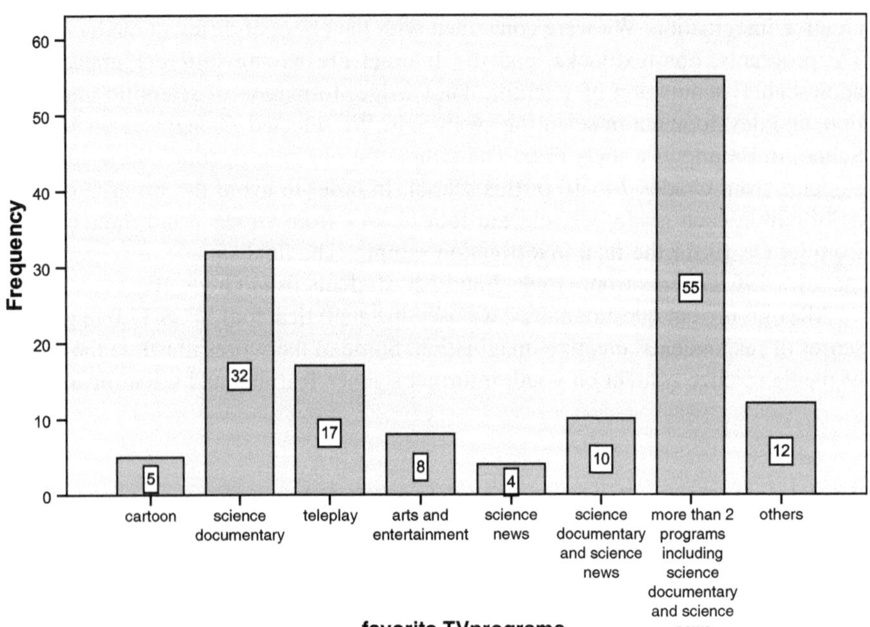

Fig. 14.2 Favorite TV programs of secondary school students

Table 14.2 Cross statistics of the relationship between TV program and students' scientific creative imagination score

Favorite TV programs		Students' scientific creative imagination score range: 0–20	21–40	41–60	61–80	More than 80	Total
Cartoon	Count	0	1	3	1	0	5
	% of total	.0 %	.7 %	2.1 %	.7 %	.0 %	3.5 %
Science documentary	Count	2	3	15	11	1	32
	% of total	1.4 %	2.1 %	10.6 %	7.8 %	.7 %	22.7 %
Teleplay	Count	0	6	6	5	0	17
	% of total	.0 %	4.3 %	4.3 %	3.5 %	.0 %	12.1 %
Arts and entertainment	Count	0	2	3	3	0	8
	% of total	.0 %	1.4 %	2.1 %	2.1 %	.0 %	5.7 %
Science news	Count	0	0	3	1	0	4
	% of total	.0 %	.0 %	2.1 %	.7 %	.0 %	2.8 %
Science documentary and science news	Count	0	2	6	2	0	10
	% of total	.0 %	1.4 %	4.3 %	1.4 %	.0 %	7.1 %
More than two programs including science documentary and science news	Count	3	14	23	14	0	54
	% of total	2.1 %	9.9 %	16.3 %	9.9 %	.0 %	38.3 %
Others	Count	0	3	4	4	0	11
	% of total	.0 %	2.1 %	2.8 %	2.8 %	.0 %	7.8 %
Total	Count	5	31	63	41	1	141
	% of total	3.5 %	22.0 %	44.7 %	29.1 %	.7 %	100.0 %

preferred science documentary programs and science news programs. According to the statistical data, 32 Grade 7 students preferred science documentary TV programs only, 4 students liked watching science and technology news programs the most, and 10 students liked watching both of them. In addition, of the 100 respondents who selected more than two favorite programs, 55 respondents selected at least one of the two types of science-related programs, as shown in Fig. 14.2.

We divided all scores into five equal-interval ranges as shown in Table 14.2. Through analyzing the cross statistics of the relationship between secondary students' favorite TV programs and their scientific creative imagination score, we found that among the students who prefer science-related programs, there were 28 getting their score 61–80. The total number of students who got the score 61–80 was 41. It means that in the students getting higher scientific creative imagination score, there were more than half students like science TV programs. It seems that students who got a high score preferred science-related TV programs.

Although science-related TV programs were the favorite programs of secondary school students, these TV programs showed opposite associations to secondary school students' science creative imagination. For Grade 7 students, the time spent on watching TV was inversely proportional to the scores of their scientific creative imagination. This might be because secondary students, especially Grade 7 students, were not able to distinguish useful and valuable information. They might watch various TV programs, but they did not understand some of the information contained in the shows. This means that some programs entertained these young teenagers without impressing them.

Therefore, watching TV programs is one possible means for high school students to improve their science learning, but it might not be an effective channel for Grade 7 students to get science knowledge. It thus appears that science TV programs contributed to increasing teenagers' science knowledge and understanding of science, but have different effects for teenagers of different ages.

14.4.2 Relation Between Reading Books on Science Learning

According to our statistical data, science books have an obvious effect on youth science learning. Most adolescents like reading in their time outside school. Among the 146 Grade 7 students, 56 respondents read non-textbooks for less than 7 h every week in their time outside school, 49 students read 8–14 h, 25 students read 15–21 h, and 16 students read more than 22 h per week. Of the 120 high school respondents, 76 students spent less than 7 h on reading non-textbooks every week, 30 students spent 8–14 h, 9 students spent 15–21 h, and 5 students spent more than 22 h per week. Accordingly, all these respondents got different scientific creative imagination scores, as shown in Tables 14.3 and 14.4.

According to the cross tabulations above, we proposed a hypothesis that the more time secondary school students spent on reading, the lower their creative imagination scores were. Then we did a correlation analysis about the hypothesis.

Table 14.3 Relation between reading books and scientific creative imagination of secondary school youth

Time spent on reading (hour/week)	Number of respondents	Score of scientific creative imagination
Less than 7 h	56	38.80
8–14 h	49	37.33
15–21 h	25	36.59
More than 22 h	16	34.46

Table 14.4 Relation between reading books and scientific creative imagination of high school youth

Time spent on reading (hour/week)	Number of respondents	Score of scientific creative imagination
Less than 7 h	76	43.55
8–14 h	30	42.98
15–21 h	9	48.76
More than 22 h	5	45.46

Through the correlation analysis of reading hours per week and secondary school students' scientific creative imagination score, we can see that the correlation coefficient of these two variables $r = -0.07$ and the significance value $P = 0.396$. It is concluded that the two variables are not significantly correlated. Therefore, the hypothesis we built at first was not supported. Therefore, the viewpoint that reading hours had a negative correlation with students' creative imagination scores was not reasonable and accurate. On the contrary, reading might be a helpful factor to enhance students' ability of creative thinking. Similarly, from the statistics, we considered that the more books Grade 10 students read, the higher their creative scores were. But if the reading time of high school students exceeded 22 h, their scores became lower. To what extent the reading hours have impact on students' scientific creative imagination, a larger sample survey are needed to be performed in the future.

14.4.3 Relation Between Internet Access and Science Learning

In our investigation, almost every middle school student has used the Internet at least occasionally. The Internet has become a necessary tool of learning, communication, and amusement in the lives of youths. Nearly 80 % of the secondary school respondents used the Internet for less than 7 h a week. Identical to the association of TV watching and non-textbooks, the use of the Internet had the same relation to youths' scientific creative imagination, i.e., the more time secondary school students spent on the Internet, the lower their scores in creative imagination were.

High school students were more sensible and intelligent than secondary school students. They used the Internet for their needs. When they opened a website, they already had a purpose: searching for information, browsing news, communicating with friends, or playing games. Statistical data shows that high school students who used the Internet for less than 14 h a week had an improvement of their creative imagination level with an increase in Internet time. In other words, high school students who spent less than 2 h on the Internet everyday showed an increase in their scores of creative imagination as their Internet time increased

Table 14.5 Relation between Internet access and scientific creative imagination of high school youth

Time spent on the Internet (hour/week)	Number of respondents	Score of scientific creative imagination
Less than 7 h	82	43.67
8–14 h	27	45.80
15–21 h	6	43.40
More than 22 h	5	37.58

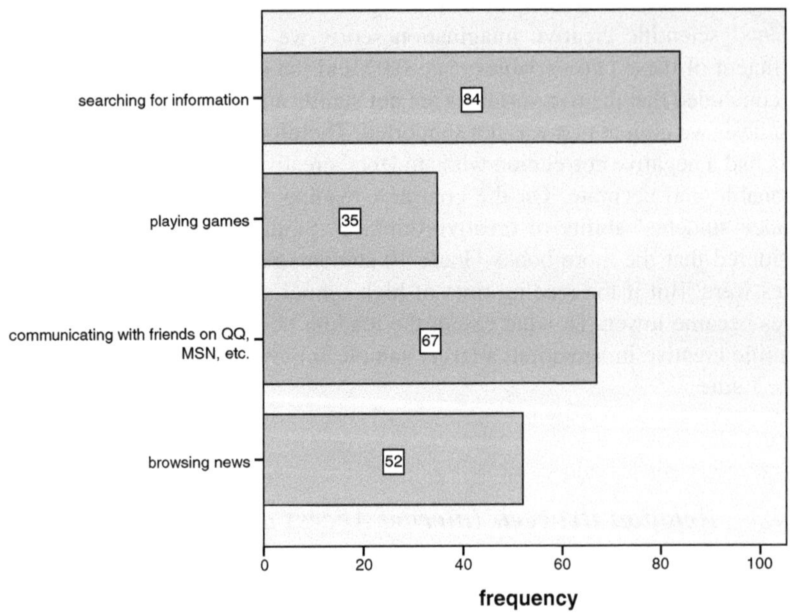

Fig. 14.3 Things high school students do online

(Table 14.5). A moderate amount of Internet use was associated with the improvement of high school students' scientific creative imagination.

How much time spent on the Internet is not the only factor impacting students' science learning; what students do online is a significant factor as well. In our investigation, we designed a multiple-choice question about the things students do online. As shown in Fig. 14.3, high school students most often go online to search for information, followed by communicating with friends using QQ (a messaging service in China), or MSN, and browsing news on the Internet. Thirty-five respondents selected playing games online. When some students searched for information on the Internet, they also communicated with friends online at the same time. Thus it can be seen that high school students use the Internet purposefully and consciously. Learning new things and communicating were high school students' two major Internet activities. The findings also reflect that university entrance exam pressure results in Chinese high school students not having much out-of-class time to do things unrelated to study and school life. The Internet has become a useful learning tool for teenagers and helps them to acquire more knowledge outside of textbooks in a short time. A rich reserve of the most modern and latest knowledge is necessary for teenagers to have higher abilities in science learning and science thinking.

14.4.4 Issues Related to Media's Impact on Youths' Scientific Imagination

Firstly, media need to develop specific and classified science content designed for teenagers. Science programs and science columns should coincide with the scientific interests of adolescents and be easy to understand. Science documentaries and science news are two of teenagers' favorite kinds of media content. Previous studies show that media still lacks specific science works designed for youths. Based on our investigation, we roughly estimate that half of the middle school students do not use media according to their learning needs. On the one hand, this is because some of them do not know how to find and use media content, which will be discussed below. On the other hand, media lack popular youth science resources. The science interests of adolescents are very different from those of adults. Even among adolescents, different age groups have different science interests. Therefore, Chinese media need to develop some excellent and high-quality science resources suitable for teenagers. On this basis, media should continue to promote the development of science series designed for different age groups of youth. The benefit of this would be to enhance the effect of media on adolescent informal science learning.

Secondly, teachers and parents should teach adolescents to use various media forms according to their needs outside of school time. Most teenagers receive different kinds of science content through TV, books, and the Internet every day, but some do not know which content fits their scientific needs. Younger teenagers such as Grade 7 students have just finished their primary school life and do not have the ability to form a more mature cognitive competence. Today, a majority of Chinese teenagers are the only children of their families. Their parents spend a lot of money on and put a lot of effort into their education and are willing to give children effective guidance as much as possible. In view of this situation, parents and teachers should give teenagers proper guidance and make a list of available science content to help them distinguish which kinds of media resources are suitable for them. The media could also design a guidance manual to tell young teenagers how to select their favorite science programs, books, and Internet contents. Meanwhile, formal science learning in school can be combined with the informal science learning from the media to improve students' science literacy. This kind of informal science learning could be also conducive to the reform of the examination-oriented education system in China that will be discussed in Sect. 14.5.

14.5 Discussion

14.5.1 The Quality of Original Chinese Work Needs Improvement

While youths enjoy watching science education programs and reading popular science books, based on the abovementioned previous surveys, youths prefer TV programs from abroad, such as *Discovery*, *National Geographic*, and *Saga*, which makes it hard for most popular domestically originated programs such as *Science Approaching* and *Exploring* to attract audiences. Based on ongoing statistics, there are rarely popular-selling domestic brands (except perhaps encyclopedias such as *The Chinese Children's Encyclopedia* and *Knowledge for Children*) among the best-selling children's popular science books. However, classic works from abroad like *Insect* sell well in every version. Since it was introduced to China in 2005, *My First Scientific Comic Book* has been popular with youths. It can be clearly seen that there is a serious lack of domestic original works in television and print science communication media, which is an indication of poor quality of these works. The quality of original Chinese works needs to be improved.

Innovation in Internet science communication is necessary as well. Internet delivery of science communication differs from passively watching TV or simply reading books, as the Internet supplies a platform for multiple synchronous forms of learning science by adolescents. Specialized science education and popular websites (such as the Science Museums of China, www.kepu.net.cn) supply science knowledge to adolescents; adolescents can use search engines to find answers to science problems; authors of specialized science blogs (like the many science blogs on www.sciencenet.cn) supply professional scientific content for adolescents to learn, interact with, and communicate through; adolescents can also set up blogs (or microblogs) to share their science experiences; online science games provide adolescents with better learning experiences through entertainment; and online digital science and technology museums and libraries avail adolescents to experience science at home. Diverse means, more initiative, and personal applications should make the Internet increasingly attractive to adolescents; unfortunately, these platforms currently fail to be attractive in both content and form. Researchers and practitioners of Internet science communication need to explore ways to make these sites more attractive to adolescents.

14.5.2 The Examination-Oriented Education System Should Be Reformed

In the context of the examination-oriented education system, youths' enthusiasm to study science through media has been limited. *Approaching Science* (a famous TV science education program on CCTV-10) did a 2009 survey called "what is your favorite TV science education program" of 6664 primary and middle school students in 21 cities. In the survey, when respondents answered the question about what their most favorite TV program content was, 53 % chose entertainment, 47 % chose TV drama, 46 % chose science and education, 36 % chose news, 32 % chose cartoons, and 10 % chose other. When asked how much time they spent watching TV every day, 43 % of respondents selected "only on the weekends," 26 % chose "one hour," 21 % chose "half an hour," and 10 % chose "never." Thus, although youths are interested in watching programs with science and education content, their viewing time is very limited. Under the pressure of examination, nearly half of respondents only watch TV on the weekends, and 10 % of them give up watching TV in order to spend time gaining as much knowledge from textbooks as possible.

The statistical data feature of CCTV-10 about Chinese youths' favorite programs and TV time is similar with the results of our survey shown in Sect. 14.4. Besides watching TV, using the Internet and reading non-textbooks are also limited for teenagers, especially high school students. Most of these teenagers study almost every day in their high school life, even during out-of-school and vacation time. Adolescents make all these efforts to pass the university entrance examination so that they can go to the ideal university that they or their parents want. For most youths, university is a relatively fair opportunity to achieve their upward social mobility. However, the examination-oriented education system makes students, parents, and teachers pay too much attention to examination scores. The real academic achievements of adolescents do not only include their scores and records in school but also their various knowledge reserves and their ability to think independently. In order to develop students' abilities of creation and competences of innovation, parents and teachers should give them more time and space to learn things independently. The school education department should encourage schools to diversify curriculums to response to the media development in the digital era. The improvement of informal science learning through media forms should become an important content of exam-oriented education reformation.

References

Burns, T. W., O'Connor, D. J., & Stocklmayer, S. M. (2003). Science communication: A contemporary definition. *Public Understanding of Science, 12*, 183–202.

China IRR. (2012). *The analysis of development of digital TV industry in 2011*. (中国产业研究报告网). Retrieved from http://www.chinairr.org/view/V07/201207/06-103030.html.

China Publish. (2012). *Minors between 0-17 had a book reading rate of 83.1%*. (中国出版网). Retrieved from http://www.chuban.cc/ztjj/yddc/2012yd/201204/t20120426_105903.html

CSM, China Youth & Children Research Center. (2006). *The youth consume market report to media*. (中国青少年研究网). Retrieved from http://www.cycs.org/Article.asp?ID=4614

Han, Y., & Gao, Y. M. (2010). An investigation report on the media contact habit and consumption behavior of middle school student. *Pioneering with Science and Technology Monthly (科技创业月刊), 3*, 130–132.

Huang, J. (2011). *The research on the writing and directing strategies of the science-educational TV programs based on constructivism learning theory (建构主义学习理论指导下的科教电视节目编导策略研究)*. Hangzhou: Zhejiang University.

Li, Z. (2007). Internet – Based science communication for the youth. *Science Popularization (科普研究), 2*(3), 46–50.

Li, W. G., Shen, J., & Ji, W. M. (2012). *Annual report on the new media use by minors in China (2011–2012) (中国未成年人新媒体运用报告 (2011–2012))*. Beijing: Social Science Academic Press.

Liang, J. (2011). To discuss the application of narrative method in the youth science and educational TV programs——take "Acid Rain" for example. *Modern Educational Technology (现代教育科技), 21*(4), 66–71.

National Bureau of Statistics of China. (1995). *The statistics bulletin on national economy and social development*. (中华人民共和国国家统计局). Retrieved from http://www.stats.gov.cn/tjgb/ndtjgb/qgndtjgb/t20020331_15389.htm

Norris, S. P., & Phillips, L. M. (2003). How literacy in its fundamental sense is central to scientific literacy. *Science Education, 87*(2), 224–240.

Open Book. (2011). *The retail marketing analysis report on science popularization books 2006–2010* (2006–2010 年科普类图书零售市场分析报告). Beijing open book information technology co., LTD (北京开卷信息技术有限公司).

Su, B., Shi, Y. M., & Liu, S. (2009). The survey on utilization of science popularization resource on internet by youth——take the city Hefei in Anhui province for example. *JianghuaiTribune (江淮论坛), 3*, 145–153.

Wang, Q. (2011). *The history and status quo of science and education channels*. Retrieved from http://www.csm.com.cn/index.php/knowledge/showArticle/kaid/394

Yang, J. S. (2007). *The research on produce and application of the science and educational TV programs for scientific literacy education of high school students(面向中学生科学素养教育的科教电视节目制作与应用研究)*. Wuhan: South China Normal University.

Yao, W. (2008). *Study on the effects of scientific& educational TV programs on teenagers' scientific literacy (电视科教节目对青少年科学素养形成的影响研究)*. Beijing: Beijing Normal University.

Chapter 15
A Case Study of a Science Teacher in a Science Club Teaching Scientific Inquiry

Xiuju Li

15.1 Introduction

Science fairs give students the opportunity to select interesting problems to solve as science projects. They brainstorm to discuss a variety of topics in order to then identify an exciting idea for their investigation (Barry and Kanematsu 2006). It is recognized by many educators that science fairs help students discover topics of interest, design experiments to investigate science problems, and provide the first steps on the ladder to success (Mann 1984). For example, students who enter the Westinghouse Science Talent Search frequently go on to have careers in scientific fields (Marsa 1993), and students who have participated in science fairs report that this experience influenced their career choices in science (Olson 1985). Many science educators believe that participation in science fairs helps students develop the attitudes, skills, and knowledge that will help them to fit in and be successful in a scientific and technological society.

The first student science fair in the United States was held in 1928. After becoming a national event in 1950, they became more widespread (Bellipanni and Lily 1999). In 1958, the fair became international for the first time, when Japan, Canada, and Germany joined the competition. Today, the Intel International Science and Engineering Fair (Intel ISEF) is the world's largest international pre-college science competition, and it annually provides a forum for more than 1500 high school students from about 70 countries, regions, and territories to showcase their independent research. Every year, millions of students worldwide compete in local and school-sponsored science fairs. The winners of these events go

X. Li (✉)
China Research Institute for Science Popularization, 86# Xueyuan South Road, Haidian District, Beijing, China
e-mail: littleju@126.com

L.L. Liang et al. (eds.), *Chinese Science Education in the 21st Century: Policy, Practice, and Research*, Contemporary Trends and Issues in Science Education 45, DOI 10.1007/978-94-017-9864-8_15

on to participate in ISEF-affiliated regional and state fairs, from which the best wins the opportunity to attend the Intel ISEF. The Intel ISEF unites these top young scientific minds, showcases their talent on an international stage, enables them to submit their work for judging by doctoral level scientists, and provides them with the opportunity to compete for over $3 million in prizes and scholarships. The China Adolescents Science and Technology Innovation Contest (CASTIC) is the national science fair in China, and it is affiliated with the ISEF. CASTIC has had a strong influence on the way students study and select science pathways (Li 2008). Each year, teachers and parents encourage hundreds of thousands of students to enter science fairs, and, according to official figures released earlier this year, 15 million students participate in CASTIC at all levels (CASTIC 2012). Both team projects and individual projects are allowed to participate in the CASTIC. CASTIC provides an annual forum for more than 500 students from middle school to high school to showcase their independent research in 16 categories including biology, chemistry, engineering, and math.

Teachers play a key role in students' research during the course of their science projects. Science clubs give teachers an opportunity to select student researchers and to provide systematic guidance to them regarding appropriate methods of scientific inquiry (Galen 1993). Therefore, encouraging teachers to encourage students who are interested in science to engage in science-fair projects and use science clubs to organize their research can effectively improve the scientific literacy of students.

This study examines the strategies used in a science club by a science-fair-award-winning teacher to instruct talented senior high school students in scientific inquiry and then gives some suggestions to other teachers who want to guide students to participate in the science fair.

15.2 Rationale and Theoretical Framework

The cognitive apprenticeship theory was developed by Collins et al. (1989), who based it on the situated cognition learning theory. Cognitive apprenticeship differs from traditional apprenticeships in that the tasks and problems are chosen to illustrate the power of certain techniques or methods and to give students practice in applying these methods in diverse settings. The complexity of tasks is gradually increased so that component skills and models can be integrated. In cognitive apprenticeship, conceptual and factual knowledge is demonstrated within each specific context. The term *cognitive apprenticeship* refers to the fact that the focus of learning through guided experience is on cognitive and metacognitive development, rather than on physical skills and processes.

Based on the theory of cognitive apprenticeship, a framework is used to describe four dimensions of a learning environment: content, method, sequence, and sociology. Relevant to each of these dimensions is a set of characteristics that should be considered when constructing learning environments. The characteristics of content include domain knowledge, heuristic strategies, control strategies, and learning strategies. The characteristics of methods include modeling, coaching, scaffolding, articulation, reflection, and exploration. The characteristics of sequence include increasing complexity, increasing diversity, and the placement of global skills before local skills. The characteristics of sociology include situated learning, the culture of expert practice, intrinsic motivation, exploiting cooperation, and exploiting competition.

Collins et al. (1989) developed six teaching methods rooted in the cognitive apprenticeship theory and claimed that these methods help students attain cognitive and metacognitive strategies for "using, managing, and discovering knowledge" (Brown et al. 1989). The first three (modeling, coaching, and scaffolding) provide the core of cognitive apprenticeship and help with cognitive and metacognitive development. The next two (articulation and reflection) are designed to help novices with awareness of problem-solving strategies and execution, similar to that of an expert. The final step (exploration) is intended to guide the novice toward independently solving and identifying problems within the domain (Collins et al. 1989).

Collins published a paper about learning environment in 1993. In this paper, he discussed the advantages and disadvantages of modeling, coaching, scaffolding, articulation, and reflection, which are the basic teaching methods (Collins 1993). Collins continued to develop the theory of cognitive apprenticeship. In 1997, Collins published a paper about cognitive apprenticeship and the changing workplace. In this paper, Collins pointed out that students can learn content and skills while conducting a complex task and that the situation based on the goal was also one kind of cognitive apprenticeship (Collins 1997).

In 2006, Collins published a paper on cognitive apprenticeship again in the *Cambridge Handbook of the Learning Sciences*. In this paper, he described the new development of situated learning, community of practice, scaffolding, articulation, and reflection. After continuous research and reflection, Collins revised the framework of the cognitive apprenticeship theory. He deleted exploiting competition from sociology and changed the culture of expert practice into community of practice. Cognitive apprenticeship focuses on four dimensions that constitute any learning environment: content, method, sequencing, and sociology (Collins 2006).

Content means the types of knowledge required for expertise. It includes four parts which are domain knowledge, heuristic strategies, control strategies, and learning strategies. Methods refer to ways to promote the development of expertise. It includes six parts which are modeling, coaching, scaffolding, articulation, reflection, and exploration. Sequencing refers to keys to ordering learning activities. It includes three levels which are increasing complexity, increasing diversity, and global to local skills. Sociology refers to social characteristics of learning environments. It includes four parts which are situated learning, community of practice, intrinsic motivation, and cooperation.

15.3 Research Questions

Within the framework of cognitive apprenticeship developed by Collins et al. (1989), this study intends to answer the following question: What are the guiding strategies of an award-winning teacher to prepare students for CASTIC? How does the teacher organize and manage the science club?

15.4 Methodology

15.4.1 Participants

The subjects of this study were members of a science club, which included the students and the instructor who was a junior high school teacher in Shanghai city. The science club, which was called the "Biology and Environment Science Club," consisted of students who were interested in science and research. Typically, there were 20–25 students in this science club a year. The teacher for this club had 15 years' experience instructing students in developing science-fair projects, and his major was biology and environmental science. He received the national top 10 science teacher title in 2003 in China. His students received many awards, such as the Grand Awards in ISEF and special awards in CASTIC.

15.4.2 Data Collection and Tools

This research employed a case study design. Data were collected by semi-structured interview (Ou 2006) and document analysis (including club diary, research diary, science-fair schedule, and webpage discussion area).

We interviewed the teacher three times. We did the first and the second interview at the Winter Camp of Intel ISEF in China which took place in January 2007. In the first interview, we wanted to collect the teacher's learning experience and his motivation to guide students to develop science-fair projects. Sample questions included "Please describe your learning experience" and "What is your motivation for guiding students to develop science-fair projects?" In the second interview, we wanted to collect as much data as possible about the science club. Sample questions included "How did you arrange your time to guide the students?" and "How did you enable your students to learn the basic skills of scientific research?" We did the third interview during science teacher's training which was organized by the China Association for Science and Technology in March 2007. In the last interview, we wanted to collect information on the teacher's professional development. Sample questions were "Where did you obtain your knowledge of guidance?" and "How did you improve your ability to guide?"

We also collected the documents of the science club in the third interview. The club diaries included materials of the science club basic curriculum which were developed by the case teacher, discussions between students and the teacher, meetings of the science club, and students' feelings about science research. We collected part of the club diaries which were recorded from January 2006 to January 2007.

The research diaries included the progression of students' research projects, such as research questions, methods, and results. The research diaries also record the students' difficulties in their research, the teacher's guidance, and the student researchers' discussions. We collected research diaries of ten students' projects completed from January 2006 to January 2007.

The science-fair schedule included nearly all the information of science fairs in China. The webpage discussion area included all the research papers of the students' projects in the science club and the progression of students' projects. We collected ten students' research papers and the progressions of ten students' projects which were completed from January 2006 to January 2007.

15.4.3 Data Analysis

Firstly, the documents collected by semi-structured interviews were sorted out and analyzed by induction analysis. In order to improve the validity of the context analysis, the results were returned to the case teacher for rechecking.

Secondly, the documents of science club were analyzed by coding, using the framework developed by Collins et al. (1989) for designing an ideal learning environment from cognitive apprenticeship perspectives. We coded the methods the teacher used to guide the students in science fair. We used MO as the code name for modeling, CO as the code name for coaching, SF as the code name for scaffolding, AR as the code name for articulation, RE as the code name for reflection, EX as the code name for exploration, and OT as the code name for other methods. For example, according to the Collins et al (1989) definition of modeling, we selected the modeling content from the documents of the science club and marked them with MO. We then sorted out and analyzed all MO units to conclude the methods. In order to improve reliability and validity, two people from the research group selected the coding content and marked independently. Table 15.1 below lists the coding table of teaching methods.

Table 15.1 The coding table of teaching methods

Coding category	Definition	Examples	Source
MO modeling	1. Operating the experimental instruments	1. Before you do the experiment, you should learn how to use the instruments	Club diaries
	2. Showing how to use the statistical software	2. You can do it according to …	Research diaries
	3. Showing how to write the paper		Science-fair schedule
			Webpage discussion
CO coaching	1. Coaching students for writing papers	1. Students learn how to write a paper from teachers	Club diaries
	2. Providing opportunities of presentation for students	2. We will hold a seminar, please prepare …	Research diaries
	3. Giving guidance and suggestions for students doing science projects		Science-fair schedule
			Webpage discussion
SF scaffolding	1. Providing strategies for students to help them find ideas	1. You can find ideas from observing	Club diaries
	2. Building an expert system for giving students support	2. If you need consulting, you can find experts from our experts system	Research diaries
	3. Proving awarded project papers for students		Science-fair schedule
			Webpage discussion
AR articulation	1. Students upload the research progression to the webpage	1. Please elaborate your progression of projects	Club diaries
	2. Students keep a research diary	2. Do the results from fieldtrips match the experiments?	Research diaries
			Science-fair schedule
			Webpage discussion
RE reflection	1. Students compare their research with teacher's research	Is there any difference between your method and other projects?	Club diaries
	2. Students communicate ideas with experts		Research diaries
	3. Students compare their research with awarded projects		Science-fair schedule
			Webpage discussion

(continued)

Table 15.1 (continued)

Coding category	Definition	Examples	Source
EX exploration	1. Push students into the problem-solving model	With so many materials, you can think …	Club diaries
	2. Encourage students build their own research process		Research diaries
			Science-fair schedule
			Webpage discussion
OT other	1. Give encouragement	You really did a great job!	Club diaries
	2. Arrange schedule		Research diaries
			Science-fair schedule
			Webpage discussion

15.5 Results

15.5.1 Content

Content includes four parts: domain knowledge, heuristics strategies, control strategies, and learning strategies. Domain knowledge includes the conceptual and factual knowledge and procedures explicitly identified with a particular subject matter. These are generally expounded in school textbooks, class lectures, and demonstrations (Collins et al. 1989). In science clubs, the activities undertaken are mainly projects, which differ from the demonstrations in conventional classes. In the science club of this case study, the teacher used a variety of activities to encourage student domain knowledge learning. These activities were core course, research projects, making a science club website, training new science club members, and various community activities.

- The teacher developed a core course to nurture students' basic scientific research skills, designed many activities, and enriched cases and classroom lectures to help students learn and understand basic scientific research skills.
- After students completed the core course, they began to carry out science research projects engaging in a set of research activities. Students selected topics independently and then formed teams according to their interests. Each team was required to report their progress to the teacher at a specific stage and to maintain

a research diary. The teacher encouraged students to make full use of the resources in their school, especially lab resources. During the research project work, the teacher and students formed a learning community and discussed their experiences and ideas in the science club.

- The teacher organized students to collaborate with a technology company to make a website for the "Biology and the Environment Science Club" (the name of the science club; www.bioenv.com). Students were responsible for updating the science club's website including uploading many research project cases providing references for students interested in scientific research.
- New members became familiar with the systems and rules of the science club, so the teacher arranged training for them to help them participate.
- The teacher also used a variety of activities within the communities where students live which served as a special teaching strategy. Combining student projects with communities encouraged students to pay more attention to the environment in which they live and cultivated their ability to apply problem-solving skills in real life. The situated cognition and learning theory emphasizes the creation of learning environments that particularly involve real situations and meaningful activities and the provision of opportunities for understanding and experiencing interactions. The teacher organized the students to participate and engage in community activities consistent with the situated learning theory. This teaching strategy was a win-win solution: many students' projects involved community activities, and the students' research, in turn, improved the communities' environments.
- The natural environment was the best resource for students to carry out research, so the teacher arranged appropriate field observations for students. Because of safety and cost considerations, the places selected for field observations were near the school and students' homes.

The analysis above demonstrates that the teacher used a variety of activities for students to learn and understand domain knowledge. Table 15.2 below lists the type of domain knowledge students can learn from each activity.

Heuristic strategies are generally effective techniques and approaches that teachers use to guide students to accomplish tasks. Heuristic strategies enable students to make inferences from one case to another, thereby effectively obtaining

Table 15.2 Science club teaching activities and domain knowledge

Science club's teaching activities	Domain knowledge
Core course	Conceptual, factual, and procedural
Research projects	Conceptual, factual, and procedural
Making a science club website	Conceptual, factual, and procedural
Training new science club members	Factual and procedural
Activities in the community	Conceptual, factual, and procedural
Field observation	Conceptual, factual, and procedural

twice the result with half the effort (Collins et al. 1989). The teacher in the case study used two heuristic strategies: the first was encouraging student researchers to take part in science competitions in order to improve their ability to conduct scientific research. For both the students and their teacher, science competitions provided a good opportunity to communicate with other competitors and thus gain more experience. The teacher often organized students to participate in CASTIC, Junior Scientist, ISEF, and other science competitions. The second strategy was encouraging students to participate in science competitions at all levels. Entering students into these different levels helped students reflect upon and revise their science projects.

As the name suggests, control strategies control the process of carrying out a task. The teacher in the study used some of these to manage the students in the science club.

- The teacher used science club regulations, which were discussed and endorsed by all science club members, to manage the students. The teacher also tried to create a science club culture of enthusiasm for science and protecting the environment as another way to guide the students.
- Each research team kept a research diary, recording when, where, and how the team members accomplished their scientific research. Students were trained to use self-reflection and self-direction.
- Students were responsible for updating and maintaining the science club's website. Each research team uploaded their research data and results onto this website. This information also provided a good opportunity for students to reflect upon their research. The science club website also had an evaluation system which could track the progress of students' research. When they met with a difficulty or experienced a problem, experts online could give them support.
- Science club members could also discuss problems with their teacher or other experts in the science club. These spontaneous meetings between students and their teacher often encouraged students to continue with their research.

Learning strategies promote any of the content described above, i.e., domain knowledge, heuristic strategies, and control strategies. The students performed their research in groups and selected their own team leader. They worked cooperatively, and in the science club, various activities provided enabled them to improve their abilities. Besides the core course and the research course, the teacher also created other opportunities for students to learn research skills. Students could contact experts for advice when needed. The science club website was also an important place for students to communicate and learn.

These strategies provided full digital and actual learning environments to strengthen students' motivation for learning and improved students' ability to learn domain knowledge, heuristic strategies, and control strategies.

15.5.2 Methods

The teaching methods were designed to give students the opportunity to observe, engage in, invent, or discover new findings (Collins et al. 1989). In Collins et al.'s (1989) theory, teaching methods include modeling, coaching, scaffolding, articulation, reflection, and exploration. We analyzed the teacher's interview data and the science club's materials and summarized the teaching methods of the teacher as follows.

Modeling involves showing an expert carrying out a task so that students can observe and build a conceptual model of the processes that are required to accomplish the task. In cognitive domains, this requires the externalization of usually internal processes and activities, specifically, the heuristics and control processes by which experts make use of basic conceptual and procedural knowledge (Collins et al. 1989). The teacher in the study used four methods: modeling the operation of experimental equipment, modeling the use of statistical software, modeling the writing of a research article, and modeling the research strategies.

Coaching consists of observing students while they carry out a task and offering advice or providing them with scaffolding, feedback, modeling, and/or reminders, which will help them master new tasks and bring their performance closer to that of an expert (Collins et al. 1989). Coaching focuses on the enactment and integration of skills in the service of a well-understood goal through highly interactive and highly situated feedback and suggestions. For the case study teacher, the coaching method was requiring students to summarize published papers: in the core course, students summarized at least ten papers. The teacher used this strategy to train the students to write research articles, providing an opportunity for students to practice an oral presentation and exhibit a project; he also regularly monitored the science club website so that he could give the best advice to each team about their research project.

Scaffolding refers to the support the teacher provides to help a student carry out a task. Fading consists of the gradual removal of this support until students can work independently. The teacher in this case study used the following scaffolding methods:

- Framing a research question at the start of research. Students often met difficulties in framing a research question when starting a research project. The teacher in the study taught students "five-W questioning" to help them improve their ability to formulate good research questions.
- Establishment of an expert support system. The teacher enlisted the assistance of several responsible and highly motivated professors who helped his students with their research projects and promoted their research abilities. He then encouraged his students to contact these experts for support and guidance and modeled how they should do this.
- Provision of research samples. The teacher provided students with some award-winning research papers and diaries from CASTIC and ISEF, anticipating that

they would benefit from comparing their research project to these award-winning papers.

Articulation includes any methods to help students voice their knowledge, reasoning, or problem-solving processes in a domain (Collins et al. 1989). There were several methods of articulation that this case study teacher used. The first involved evaluating the progression of students' research through the information they had loaded onto the science club website, such as their research hypotheses, research designs, and final research reports. The second involved students who keep writing research diaries in order to practice their writing skills. These strategies enabled students to express or articulate their knowledge and improve their writing skills.

Reflection (Brown 1985a, b) involves enabling students to compare their own problem-solving processes with those of experts, other students, and, ultimately, an internal cognitive model of expertise. Reflection is enhanced by the use of techniques for reproducing or replaying the performance of both expert and novice for comparison. Through reflection, the learner can question themselves. In the science club, each research group had two teachers, one from the middle school and the other from the expert support system. The experts helped the young researchers solve challenges, thereby cultivating their abilities through this interaction. Comparing their work with award-winning research projects provided another way for them to reflect upon their endeavors and refine their science projects.

Exploration involves pushing students into a mode of problems solving independently. Encouraging students to explore is critical so students learn how to frame interesting questions or problems that they ultimately can solve. As part of science project guidance, exploration as a method of teaching involves helping students frame a question. Before framing the question, the "five-W questioning" method was taught to the students. The teacher also promoted research exploration among the students by scheduling meeting times and stating exactly when CASTIC was to be held. Students would therefore work hard to complete the research on time.

In conclusion, the case study teacher used operational strategies to help his students learn science process skills. In particular, he used the strategies of modeling, coaching, scaffolding, articulation, reflection, and exploration.

15.5.3 Sequence

Collins et al. (1989) thought that in an ideal learning environment, students would develop their skills better if learning was structured in a particular way: from simple to complex with increasing diversity and from global to local skills. The chief effect of this sequencing principle is to allow students to build a conceptual map before attending to the details of the terrain. The case study teacher arranged the learning activities of the science club in the following way (Fig. 15.1).

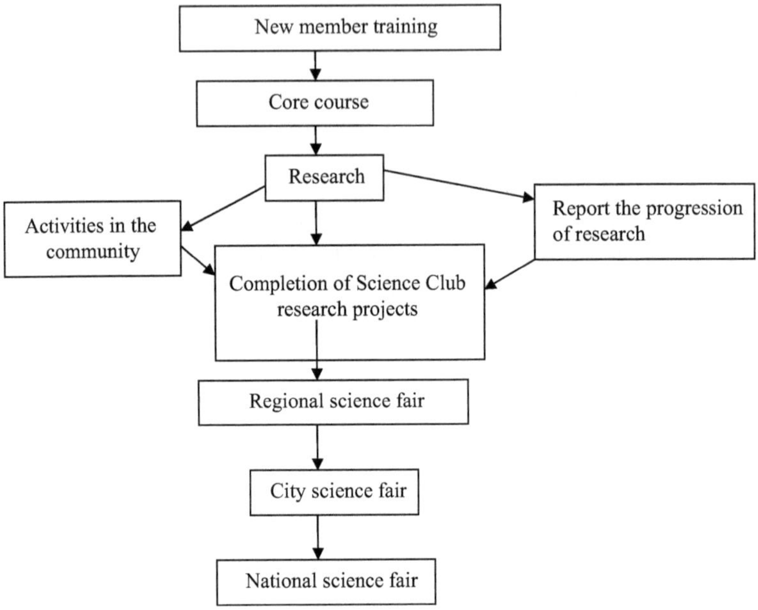

Fig. 15.1 The sequence of learning activities in science club

15.5.4 Sociology

The final dimension in Collins et al.'s (1989) framework is sociology of the learning environment—a critical dimension often ignored in making decisions about curriculum and pedagogical practice. The sociology in the science club is as follows:

A critical element for learning was that students were carrying out tasks and solving problems in an environment that reflected the multiple ways their knowledge would be put to use in the future. In the science club, the case study teacher created numerous activities enabling students to understand and apply their knowledge.

Community of practice refers to the creation of a learning environment in which participants actively communicate about and engage in the skills involved in expertise, where expertise is understood as the practice of solving problems and carrying out tasks in a domain. Completion of research projects can develop students' science process skills. The case study teacher established an expert support system for the science club and encouraged the students to communicate with experts. These strategies prompted students' research and constructed a learning environment and community of practice.

Related to the issue of situated learning and the creation of community of practice is the need to promote intrinsic motivation for learning. Malone (1981) discussed the importance of creating a learning environment in which students perform tasks because they are intrinsically related to an interesting or at least coherent goal, rather than for some extrinsic reason like getting a good grade or pleasing the teacher. The case study teacher strengthened students' intrinsic motivation through these strategies. First, the case study teacher modeled good practice by undertaking the research himself, providing a good example for the students. Second, the case study teacher created a positive culture within the science club to motivate his students. These two strategies motivated students to complete the research because they loved science and not because of parental or teacher pressure. This intrinsic motivation prompted students to overcome the difficulties in research.

Students formed groups according to research direction. In groups, students cooperated to make progress on and complete the research; they helped and learned from each other. The science club also promoted competitive environment by creating a research project competition where groups would vie with each other in order to make progress on their research and achieve better outcomes.

15.6 Discussion/Conclusions

In summary, the case study highlighted the teaching strategies that an award-winning teacher used to help students develop research projects in the context of a science club. The case study teacher designed various activities to organize students into research groups. Based on situated cognition and learning theory, the science club constructed a sociological environment in a science community. The interaction among students, teachers, and experts improved the effects of learning.

In China, students do not have much time to develop science projects because of study pressure, so most students do their science projects in their spare time. Because of a shortage of science project teachers, most students do not participate in science fairs. It is important to organize students to develop science projects using science club model which can cater to and guide more students simultaneously. Science clubs have advantages in terms of organizing students because all members are linked by a common interest in science and research. Science clubs have clearly stated regulations discussed and endorsed by all science club members. The cooperative and competitive environment in science clubs also motivates students to complete research to the best of their abilities.

Most science-fair projects come from science club sand are team projects, which is a typical characteristic of the science club model. The cooperative structure has shown to be preferred in inquiry learning (Tjosvold et al. 1977), promote positive attitudes (Humphreys et al. 1982), and be the most effective structure for the majority of students when paired with competition (Okebukola and Ogunniyi 1984). It also has been shown that the individualistic structure is very effective in competitions, especially for highly capable students (Michaels 1977). The individualistic structure may also be desirable if one of the goals of the school science fair is to select a winner to progress to another competition that accepts only individual efforts. Since the cooperative structure is the most effective for the majority of students, the science club model is useful for most schools to nurture students developing science-fair projects. Although the science club model can help students finish science projects, the key to the science club model is in the design of content and activities and the use of teaching strategies. The case study teacher designed a core course, instituted various activities, and used effective strategies to encourage students in their research.

Acknowledgments I would like to thank Professor Enshan Liu for his patient guidance and valuable suggestions. I also want to give my sincere appreciation to teacher Zhang who is the teacher in this study, for his cooperation and all his assistance.

References

Barry, D., & Kanematsu, H. (2006). International program to promote creative thinking in chemistry and science. *The Chemist, 83*(2), 10–14.

Bellipanni, L. J., & Lilly, J. E. (1999). What have researchers been saying about science fairs? *Science and Children, 99*, 46–50.

Brown, J. S. (1985a). Process versus product: A perspective on tools for communal and informal electronic learning. *Journal of Educational Computing Research, 1*, 179–201.

Brown, J. S. (1985b). Idea-amplifiers: New kinds of electronic learning. *Educational Horizons, 63*, 108–112.

Brown, J. S., Collins, A., & Duguid, P. (1989). Situated cognition and the culture of learning. *Educational Researcher, 18*, 32–42.

CASTIC. (2012). *Introduction of China adolescents science and technology innovation contest.* Retrieved from http://castic.xiaoxiaotong.org/intro.aspx?ColumnID=1011000000

Collins, A. (1993). Design issues for learning environments. In S. Vosniadou (Ed.), *Psychological and educational foundations of technology-based education.* New York: Springer.

Collins, A. (1997). *Cognitive apprenticeship and the changing workplace.* Proceedings of 5th annual international conference on post-compulsory education and training. Queensland.

Collins, A. (2006). Cognitive apprenticeship. In R. K. Sawyer (Ed.), *The Cambridge handbook of the learning sciences* (pp. 47–60). New York: Cambridge University Press.

Collins, A., Brown, J. S., & Newman, D. (1989). Cognitive apprenticeship: Teaching the craft of reading, writing and mathematics. In L. Resnick (Ed.), *Knowing, learning, and instruction: Essays in honor of Robert Glaser* (pp. 453–494). Hillsdale: Lawrence Erlbaum Associates.

Galen, D. (1993). Science fair: A successful venture. *The American Biology Teacher, 55*(5), 465–467.

Humphreys, B., Johnson, R. T., & Johnson, D. W. (1982). Effects of cooperative, competitive and individualistic learning on students' achievement in science class. *Journal of Research in Science Teaching, 19*(5), 351–356.

Li, X. (2008). *Studies on school education environment of high-quality middle school CASTIC' projects* (中学创新大赛项目学校教育环境研究). Beijing: Beijing Normal University.

Malone, T. (1981). Toward a theory of intrinsically motivating instruction. *Cognitive Science, 4*, 333–369.

Mann, J. Z. (1984). *Science day guide.* Columbus: Ohio academy of science.

Marsa, L. (1993). Do high school science competitions predict success? In C. M. Czerniak (Ed.), *Predictors of success in a district science fair competition: An exploratory study* (pp. 21–22). Alabama: School Science and Mathematics Association.

Michaels, J. (1977). Classroom reward structures and academic performance. *Review of Educational Research, 47*(1), 87–98.

Okebukola, P., & Ogunniyi, M. B. (1984). Cooperative, competitive, and individualistic interaction patterns effects on students; achievement and acquisition of practical skills. *Journal of Research in Science Teaching, 21*, 875–884.

Olson, L. S. (1985). *The North Dakota science and engineering fair-its history and a survey of its participants*. Fargo: North Dakota State University.

Ou, Y. (2006). *Exploration of cognitive apprenticeship perspectives in case study of a science-fair awarded teacher's application of science club to instruct students in progress of science inquiry* (以认知学徒制观点探讨科展优秀教师应用科学社团指导学生进行科学探究活动的个案研究). Taiwan: Taiwan Normal University.

Tjosvold, D., Marino, P. M., & Johnson, D. W. (1977). The effects of cooperation and competition on student reactions to inquiry and didactic science teaching. *Journal of Research in Science Teaching, 14*, 281–288.

Chapter 16
Students' Views of Science Learning During Visits to Science Museums: A Case Study

Lihui Wang

16.1 Introduction

Learning in museums is an important way to promote the understanding of science. It is evident that as a learning venue, the museum has a positive role in promoting the active learning process to students (Allen 1997; Crowley et al. 2001; Bell et al. 2009). The constructivist model (Hein 1998) of the museum-based learning approach emphasizes that learners construct new knowledge through active engagement with their own experience and existing knowledge. Museum-based learnings can improve students' interest in science. A study on the long-term effects of learning in museums conducted by Falk and Dierking (1997) found that field experience in museums produced a lasting influence on pupils' lives. In another study, Jarvis and Pell (2005) found that girls 10–11 years old who lacked confidence in science learning thought this kind of visit was "real and very interesting." After having joined a challenging visit to the UK Science Centre, their anxiety about science learning was also reduced and their confidence strengthened. Many teachers and science educators also believe that learning in museums can help students gain new knowledge and develop their skills.

Since the passing of *the Law of the People's Republic of China on the Popularization of Science and Technology* (2002) and the *Outline of the National Scheme for Scientific Literacy (2006–2020)* (2006), the establishment of museums has progressed rapidly in China and may have already reached its peak. By the end of 2013, there were 678 science and nature museums and 380 science centers in China. The number of science and nature museums was 9.5 % higher than that of 2011.

L. Wang (✉)
China Research Institute for Science Popularization, No. 86 Xueyuan South Road, Haidian District, Beijing 100081, China
e-mail: wanglihui1001@sina.com

L.L. Liang et al. (eds.), *Chinese Science Education in the 21st Century: Policy, Practice, and Research*, Contemporary Trends and Issues in Science Education 45, DOI 10.1007/978-94-017-9864-8_16

More than 50 % of science and nature museums were located in eastern part of China. Some museums such as China Science and Technology Museum and Guangdong Science Center are very famous to Chinese students. The number of visitors to science and technology museums in 2013 went up to 98 million (Ministry of Science and Technology MOSAT 2014). To improve access to museums, the museums managed by the Culture and Heritage Department of the government have been open to the public for free since 2009, and from 2014 all science and technology museums offer free entry to young people. A survey conducted by China Research Institute for Science Popularization in Beijing at Zhongguancun Middle School showed that most students had visited science museums. Twenty-four percent of the students often visited science museums during weekend and vacations and 62.7 % of the students hoped to take part in projects and events offered at science museums.

With the developments of science and technology museums, Chinese scholars, who had been mainly concerned with science learning in schools, began to focus on the museum-based approach to learning science (Bai and Chen 2005; Xu 2007). Special attention has been paid to the informal learning which occurs in science museums. Some scholars became interested in using museum science learning as a supplement to school-based learning to promote science learning (Xie and Shang 2009). There are some studies from the constructivist perspective designed to understand the role of the museum as an avenue of science education (Wu et al. 2009; Chen 2003). However, there have been very few in-depth studies on the features of museum-based learning; most current studies are theoretical introductions. This chapter reports a case study of students' views of visits to museums in China. I intend to answer the following questions: What are the students' perceptions of the features of museum-based science learning? What is the perceived relationship between museum-based learning and school learning?

16.2 Theoretical Framework

Constructivism argues that humans generate knowledge and meaning from an interaction between their experiences and their ideas. This means that learners construct knowledge both individually and socially. The principles of constructivism give us fresh perspectives on museum learning (Falk and Storksdieck 2009). According to the constructivism theory, the museum can be seen as an experienced and perceived environment where the audience will be modeled as an active learner. That is, the learner can access knowledge according to their own interests in the museum instead of it be instilled in the classroom.

The theoretical framework of science and technology museum-based learning mostly adopts learning models based on constructivism. Based on this theory, Falk and Dierking (2000) propose the contextual model of learning. This framework draws from constructivist, cognitive, and sociocultural theories of learning, and the key feature of this framework is the emphasis on context. They believe that learning is achieved through three contexts. First, the students are prompted to awaken their existing knowledge. Second, they are provided with new information. Third, they work with the new and existing knowledge and apply it to address new problems or issues (Atiomo 2009).

Compared with other educational forms, museum-based learning is quite unique. A museum is an informal learning space. Effective museum-based learning means that at the end of the visit, a visitor has a greater knowledge and understanding of a theme than at the beginning. When a visitor comes into an exhibition space, the learning activities mainly include reading text, listening to explanations, observing exhibits, watching videos, experiencing, and experimenting. In the museum, each exhibit provides the context for understanding the exhibit. When the exhibit is in a specific form, its knowledge and its relative significance are exposed.

16.3 Methodology

16.3.1 Questionnaire-Based Interviews

In this study I used a questionnaire-based interview of students. There were 24 questions in the questionnaire, and it was initially reviewed by two youth science learning research experts.

Based on the context model of museum-based learning, visitors, as a part of the museum, and their interaction with the exhibition and other visitors are an important way to learn. In the questionnaire, this study collected the influential factors of the pre-visit and in the middle of the visit. The questionnaire had five sections: (1) number of visitors and duration of visits to the science and technology museum; (2) the recognition/popularity of the science technology museum in school science teaching; (3) who was responsible for deciding to visit the museum; (4) the impact of the museum on the student's learning of science knowledge, including the motivation for learning and how to master the scientific method; and (5) the impact of the museum on the student's individual cognitive development. The reliability of the scales in the final questionnaire, based on Cronbach's alpha coefficients of internal consistency, was 0.703 and above.

16.3.2 Data Collection

The survey was administered on a weekend in December to all students who visited the China Science and Technology Museum and the Beijing Museum of Natural History who arrived and departed between 10 AM and 3 PM. Except for holidays, during the week, students in China are often organized to visit the science museums by schools, and the visiting time is limited. This study focused on individuals during the weekend. The respondents' answers to the questionnaire were recorded by the investigators, each taking 5–10 min to complete. Of the 142 completed questionnaires, 140 were valid. There were 66 responses from the China Science and Technology Museum and 74 from the Beijing Museum of Natural History. I combined the data from the two museums for analysis. Of the visitors, 51.3 % were boy students and 48.7 % were girl students, while 86.7 % were primary school students and 13.3 % were junior high school students.

16.3.3 Data Analysis

In the survey, the investigators were coded as A, B, C, and D. Before the interview, the respondents were asked whether they had been interviewed. Each respondent was coded in accordance with A1, A2, A3...B1, B2, B3, and so on. The data were input into excel and analyzed by SPSS version 16.0. Frequency analysis was used to describe the data.

16.4 Results

16.4.1 Students' Views on Features of Museums as Science Learning Venues

Museum-based science learning is connected to learners' primary knowledge, and interaction is important in obtaining new knowledge. The analysis of students' views on science museums is as follows:

1. Relevance to prior knowledge. Prior knowledge is the origin of new knowledge. The students' prior knowledge mostly was learned from school. In response to the question "During your visit, did you use knowledge gained in the classroom to help understand the exhibits?" 63.5 % of the students answered that they did. However, there was a large portion of the students who gave a negative or an uncertain response. This means that those students were unable to link classroom learning with learning in the science museum. When asked "Did the visit to the museum increase your knowledge?" 91.3 % of them responded positively.
2. Emotional response to visits. Visiting a museum may result in pleasure, fun, and surprise, as experienced by 98.6 % of the students. Among the students, 53.5 % liked the exhibits and exhibition areas and had always been interested in science. The opportunity for a "hands-on experience" was liked by 13.2 % of the students, while 12.4 % of the students wanted an in-depth understanding of the knowledge learned in the classroom.

A specific scientist/astronaut/computer professional/biologist was admired by 93.4 % of the students, and more than half of them thought they would join the profession that related to the exhibits they liked in the museum.

3. Promoting exploratory learning. In response to the question "During your visit to the museum, have you tried to observe the scientific aspects of the exhibits from different angles?" 46.6 % of the students said that they had certainly been able to do so; however, 39.1 % of the students were unsure.

A large number (87.4 %) of the students who visited the museum thought the "do-it-yourself" tips helped them understand the scientific aspects of the exhibits. The question "Through visiting the museum, did you understand the science-related processes and methods of scientific inquiry?" received a 56.2 % positive response, but 38 % of the students said they were unsure.

4. The cognitive impact of the visits. Data concerning the students' individual cognitive investigations were collected mainly through the following four questions: (a) During your visit to the museum, did you find that the exhibits were logically organized? (b) Did looking at the exhibits/multimedia/touch-screen displays leave a clear impression in your mind? (c) During the museum visit, did you find that some of the exhibits made you think of learning certain concepts and how they worked? and (d) During the museum visit, did vivid and colorful presentation of images stimulate you to imagine and design something new? Positive responses from students to these four questions were 37.8 %, 91.9 %, 50.0 %, and 77.4 %, respectively.

According to the answers to above questions, it is possible that the exhibits helped students to form a three-dimensional impression in their minds and inspire them in the practical design of new things. But understanding the link between the exhibits and the principles behind them was difficult for a large proportion of students, and this is precisely the point of linking classroom learning with technol ogy through visiting a museum.

16.4.2 Time Spent During Museum Visits

The main model of science learning in science and technology museums in China is that weekday visits to the museum are organized by schools, and weekend and holiday visits are arranged by parents or others. Organizing student groups to visit is the main way for most schools to use the museum; however, a quick "play" is not conducive to the learning experience of students in the museum. It is worth noting that this was the first visit for 59.4 % of the children and only 14.1 % had visited more than three times. It is evident that not many students visit the popular science venues frequently.

The behavior patterns of audiences in science venues vary depending on the composition of the audiences. Although the survey was conducted in winter when

Table 16.1 Duration of student visits to the museums

Visiting time	Frequency	Percent
1 h	4	2.9 %
2 h	25	17.9 %
3 h	44	31.4 %
4 h	35	25.9 %
5 h or more	28	20.2 %
Missing	4	2.9 %
Total	140	100 %

the weather was colder and daylight shorter, 57.3 % of the students spent 3–4 h (including eating and resting) visiting the museum, while 20.8 % of students spent less than 2 h. This indicates that the students who came to these venues stayed a considerable period of time to fully understand the layout of the venues and explore exhibits they wanted to learn from. Table 16.1 presents time spent in the museums.

Spending more time in museums and repeated visiting museum undoubtedly can be helpful to students' science learning, so parents and teachers should spend more time guiding students to learn from science museums.

16.4.3 Museum-Based Learning and School Learning

Museum-based learning and school learning are both important venues for science learning. In this investigation, I found museum-based science learning should be paid more attention:

1. Only 18.4 % of the students' schools had ever organized a visit to a science museum before. In the survey, I found that for science-related courses, only 29.8 % of teachers asked students to visit science and technology museums. However, 79.1 % of the students wanted to visit the museum and learn there. As a guide for the students in the classroom, the science teacher plays an important role in promoting informal science learning. If the teacher mentioned content in the classroom related to the experience of visiting the museum, either before or after visiting the museum, the students can combine their museum experience with classroom learning. This enables them to understand the relevant information better and faster and expands the breadth and depth of their knowledge.
2. Museum-based visits and students' interest in science. Museum learning takes place without any pressure; it can maximize the motivation of students toward learning and enhance individual interest in science. Based on our survey, I found that students became more interested in science. Among the students, 53.5 % liked the exhibits and exhibition areas and had always been interested in science. The reason for liking the exhibits for 13.2 % of the students was the "hands-on experience," and for 12.4 % of the students it was the knowledge: "I learned in

Table 16.2 Reasons for liking the exhibition

Selection	Frequency	Percent
Was always interested	69	53.5 %
Hands-on experience	17	13.2 %
Learned related knowledge on class	16	12.4 %
Wanted to have in-depth understanding	9	7.0 %
Others		
Missing	31	13.9 %
Total	140	100 %

the classroom, but I wanted to have an in-depth understanding." Table 16.2 presents the findings.

3. Cooperation between schools and museums. In the survey, nearly 80 % of the students wanted to have classes in museums, but there were few science museums that could provide science courses to students. One student was once a member of a "biology class taken in a museum" conducted by the Beijing Museum of Natural History, in which the class was co-taught by school teachers and museum experts, introducing courses about "biology and the environment" including animals, botany, paleontology, and anthropology. This student said that while learning in the museum, he could find the real exhibition corresponding to the pictures on the book and understand the theory more easily. His classmates and he all liked studying in the museum. There is still a lot of work required to build a bridge between museum learning and classroom learning in China, and there is also the lack of evaluation of learning effects on the museum audience.

16.5 Discussion

This chapter reports on students' perceptions of museum-based learning. In China, most parents and teachers recognize that museum-based science learning is important.

This study is about students' attitude of visiting science museums, and there are some limitations in the research. (a) In the questionnaire, interview questions might be leading respondents to certain answers due to the design; (b) the interaction with the exhibits, other visitors, and their parents in the museum is the important aspect of learning; because of the large number of subjects, there was no observation of visitors interacting with the exhibits; and (c) the comparison of the behavior before and after the activity can understand the learning effect of students, but this study did not choose specific science topics, so there was no knowledge of what students had learned in their classrooms before visiting the museum.

Constructivists describe learning as the process of constructing new knowledge by linking the already-mastered knowledge and experience of the learners with new information. Exhibits showing daily life can help visitors understand scientific ideas, emphasizing the significance of already-mastered knowledge. Museum-based learning experience stimulates the inner motivation of students and helps them construct new knowledge using their cognition; as a result, students can understand new scientific concepts and knowledge. However, since families can have an important impact on students, parents should study with students and guide students' learning.

It is important to develop school-based learning enhanced by museum-based learning and to promote the integration between museum learning and school learning. School-based curriculum can be developed in response to schools' particular situations and needs. An excellent school-based curriculum can then be extended into informal learning places to cater to the individual needs of students. Based on the general characteristics of all museums, a school-based curriculum can be developed for primary and secondary students. The content of a school-based curriculum in science and technology should reflect the resources of the available museums. The curriculum may be designed to improve students' scientific literacy based on their mastered knowledge and experience, while recognizing their interests, and should take into account both the logical progression of knowledge development and the character of museum exhibits.

Exhibition is an important strategy to facilitate museum science learning. In order to enhance the ability of museums to provide primary, junior high, and high school education programs, museums need to strengthen exhibits' designs. Most students in our survey study thought exhibits and exhibitions were too general and lacked detailed information or were too abstract. The integration of exhibits and textbook scientific knowledge is necessary for students' science learning. In order to do this, museums need support from the formal educational sectors.

References

Allen, S. (1997). Using scientific inquiry activities in exhibit explanations. *Science Education, 81* (6), 715–734.

Atiomo, W. (2009). A constructivist strategy for medium/large student groups- the contextual learning model. *The Open Medical Education Journal, 2*, 1–9.

Bai, Z. X., & Chen, J. R. (2005). Discussion on the social educational function of the Sci-tech Hall (科技馆的社会教育功能). *Sci/tech Information Development & Economy, 3*, 176–177 (In Chinese).

Bell, P., Lewenstein, B., Shouse, A. W., & Feder, M. A. (2009). *Learning science in informal environments: People, places, and pursuits* (p. 35). New York: National Academies Press.

Chen, W. P. (2003). Constructivism and museum education (建构主义与博物馆教育) [in Chinese]. *Chinese Museum, 2*, 18–23.

Crowley, K., Callanan, M. A., Jipson, J., Galco, J., Topping, K., & Shrager, J. (2001). Shared scientific thinking in everyday parent- child activity. *Science Education, 85*(6), 712–732.

Falk, J. H., & Dierking, L. D. (1997). School field trips: Assessing their long-term impact. *Curator, 40*(3), 211–218.

Falk, J. H., & Dierking, L. D. (2000). *Learning from museums: Visited experience and the making of meaning*. New York: Alta Mira Press.

Falk, J. H., & Storksdieck, M. (2009). Using the contextual model of learning to understand visitor learning from a science center exhibition. *Science Education, 89*(5), 744–778.

Hein, G. E. (1998). *Learning in the museum*. London: Routledge.

Jarvis, T., & Pell, A. (2005). Factors influencing elementary school children's attitudes toward science before, during, and after a visit to the UK National Space Center. *Journal of Research in Science Teaching, 42*(1), 53–83.

Ministry of Science and Technology (MOSAT). (2014). *China Science Popularization Statistics* (中国科普统计). Beijing: Scientific and Technical Documentation Press (In Chinese).

Wu, X. C., Zeng, Z., Xie, J., & Kang, C. Y. (2009). The essential characteristics of museum- based science learning and its influential factors (场馆科学学习:本质特征与影响因素). *Journal of Beijing Normal University (Social Sciences), 215*(5), 13–19 (In Chinese).

Xie, J., & Shang, X. Q. (2009). Some thoughts about the development of educational activities in Science & Technology Museums (关于科技博物馆教育活动开发的若干思考). *Science Popularization, 3*, 39–43 (In Chinese).

Xu, S. Y. (2007). Some thoughts about the development tendency and characteristic of Science and Technology Museums (关于科技馆发展趋势和特点的思考). *Science Popularization, 4*, 15–20 (In Chinese).

Part VI
Science Teacher Education

Ling L. Liang

Editor's Introduction: Part VI

As we all probably agree, to successfully implement any new reforms as envisioned in the science education standards documents, there must be coordinated changes in curriculum, instruction, assessment, and teacher preparation and development. This section focuses on Chinese science teacher education, in which both preservice and in-service teacher development programs are presented.

What does a teacher need to know in order to teach science effectively? What type of knowledge distinguishes a novice from an expert teacher? Since the 1980s, research on science teacher education and professional development has shifted from a *training* perspective focusing on the trainees' performance of observable behaviors to a *learning* perspective that seeks to understand teachers' development of knowledge and beliefs (Bryan 2012). Schulman (1986, 1987) first conceptualized that teacher knowledge consists of subject matter knowledge (SMK), pedagogical content knowledge (PCK), and curricular knowledge. According to Schulman, what distinguishes novices from expert teachers is "the capacity of a teacher to transform the content knowledge he or she possesses into forms that are pedagogically powerful and yet adaptive to the variations in ability and background presented by students" (Shulman 1987, p. 15). This type of professional knowledge is PCK—a term coined by Shulman.

In the past two decades, PCK has been studied and investigated by science education scholars around the world and has been viewed as the central element of teacher professionalism (Park and Oliver 2008). Building upon Grossman's and Tamir's work (Grossman 1990; Tamir 1988), Magnusson et al. (1999) had further conceptualized PCK as a tool to better understand science teacher development, defining it as an integration of five components: (1) *Orientation to Teaching*

The original version of this part front matter was revised. An erratum can be found at http://dx.doi.org/10.1007/978-94-017-9864-8_20

Science, i.e., the teachers' knowledge and beliefs about the purposes and goals of teaching science at a particular grade level; (2) *Knowledge of Science Curriculum*, i.e., teachers' knowledge about mandated goals and objectives, and the specific curriculum programs and materials available. This knowledge enables teachers to identify core concepts, modify activities, and eliminate aspects judged to be peripheral to the targeted conceptual understandings; (3) *Knowledge of Students' Understanding of Science*, which includes knowledge of students' conceptions of particular topics, learning difficulties, motivation, and diversity in ability, learning style, interest, developmental level, and need; (4) *Knowledge of Instructional Strategies*, including subject-specific and topic-specific strategies. Subject-specific strategies are general approaches to instruction that are consistent with the goals of science teaching in teachers' minds, such as learning cycles, conceptual change strategies, and inquiry-oriented instruction. Topic-specific strategies refer to specific strategies that apply to teaching particular topics within a domain of science (e.g., how to teach Newton's Laws) and (5) *Knowledge of Assessments of Science Learning*, which is comprised of knowledge of the dimensions of science learning that are important to assess and knowledge of the methods by which that learning can be assessed.

More recently, in a multiple case study of three high school chemistry teachers conducted by Park and Oliver (2008), teacher efficacy emerged as an affective affiliate of PCK. It is suggested that teachers who have higher levels of science teaching efficacy and strong PCK would more likely enact appropriate PCK in classroom teaching.

Teacher learning and development of PCK begin at preservice education and continue throughout a teaching career. Not all professional development programs are equally effective in terms of initiating fundamental changes in how teachers view learning and teaching and in changing practices. Literature in teacher development has suggested that a targeted, sustained, and coherent teacher development model is far more effective than a short-term, disjointed, "one-size-fits-all" type of program. Some principles of effective teacher development have been summarized in different sources:

1. Teacher educators should treat teachers as they expect teachers to treat students, i.e., teachers need to construct their own knowledge as active learners.
2. Learning is situated. Therefore teacher development should be situated in classroom practice.
3. Teachers need room for reflection in order to understand the emerging patterns of change. Sufficient time should be given for teachers to adjust to the changes made.
4. Teachers need opportunities for action to test what works in classrooms, as well as a community to share experiences (Loucks-Horsley et al. 1998; Hoban 2002).

Effective teacher professional development programs should be coherent, focus on content knowledge, provide opportunities for collaboration, and incorporate adequate support for individual teachers (Luft and Hewson 2014). Other promising strategies for professional development that have been suggested and implemented include providing collaborative structures and immersion experiences to engage teachers in scientific research activities or activities designed for student learners; examining teaching and learning through reflection; using action research and case discussion (using text-based or video cases); and practicing teaching, including coaching, mentoring, and demonstration lessons (e.g., Simon and Campbell 2012).

In China, the majority of preservice elementary science teachers, including those in early childhood education, are still prepared in secondary level normal schools (usually 3-year) or 4-year normal colleges, while some are prepared in 4-year normal universities. On the other hand, most secondary preservice science teachers are prepared in normal universities or 4-year colleges. More recently, borrowing the models from developed countries such as the United States, schools of education have been established in some major research universities in China to attract more students with strong academic backgrounds into the teaching career. These non-normal colleges and/or research universities offer their graduates with opportunities to complete additional required coursework related to education and earn teaching certificates.

Traditionally, secondary preservice science education programs in China focus on developing teacher candidates' subject matter knowledge. Both preservice and in-service teacher education also emphasize on the development of teachers' subject-specific and topic-specific pedagogical content knowledge (PCK). While working in K-12 schools, all teachers are expected to participate in continuous professional learning and development at various levels (e.g., school, district, or national level). Teachers are promoted along a career ladder moving from a "junior rank" toward an "intermediate" and a "senior" rank. A small fraction of the teacher population may be honored with the "master-teacher" rank (or 特级教师).

Chapter Introductions

In Chap. 17, C. Liu and E. Liu unpack the typical preservice science teacher preparation programs in China. To help readers better understand the nature of secondary science teacher education programs in China, a simple comparison between a sample Chinese preservice biology teacher preparation program and a US counterpart at a large public research university is presented in Table VI.1.

In the United States, there are various teacher education models, ranging from 4-year undergraduate programs to 1- or 2-year postbaccalaureate graduate studies to alternative teacher certification programs. However, the curricula of preservice science teacher education are remarkably similar. Besides typical courses in the foundations of education, educational psychology, education diversity or special education, and student teaching, usually there are only one or two courses addressing general science teaching methods.

Compared to their US counterparts, Chinese prospective biology teachers are required to take more courses in general education and biology and related subject areas. Usually, the undergraduate biology curriculum taken by biology teacher

Table VI.1 A comparison of sample preservice biology teacher training programs (4 years): China vs. United States

Category of course			Credits required (an example: China) On average, 1cr. = 16 class hr.	Credits required (an example: United States)
General courses, such as history and foreign language			44	27
Science content courses	Courses in related subject area, such as chemistry and physics		33.5	24
	Biology courses	Basic courses in biology	33.5	13
		Specialized or advanced courses in biology	25	12
Teacher education courses	*Required courses related to teaching		16	31
	*Education research courses		4	0
	Teaching practice/student Teaching		11	12
	Field experience		Incorporated into teaching practice (1)	4
	Professional integrity		2	1
*Thesis			6	0
Total			175	124

candidates in many Chinese Normal Universities is of the same level of rigor as that of those who will go on to do graduate work in biology. To develop pedagogical content knowledge, Chinese preservice biology teachers take multiple subject-specific and topic-specific courses, including required courses such as Pedagogy of Biology (three credits), Comprehensive Experiments in Biological Education (four credits), and Teaching Skills Training (two credits). By contrast, their US counterparts take two required courses related to teaching science ("Exploring Secondary School Science Teaching" and "Methods of Teaching Junior/Middle/ Senior High School Science"), neither is biology or topic specific. In addition, Chinese preservice teachers have education research requirements, while their US counterparts have none. Teacher research has been deemed an important aspect of teacher learning and a cornerstone of educational reform (Wallace and Loughran 2012). The educational research course and thesis requirement have the potential to develop more reflective and inquiry-oriented teachers who better understand the nature of science and scientific knowledge development. Overall, the Chinese preservice teacher education program appears to be more content intensive, especially in subject matter knowledge and topic-specific instructional strategies in PCK.

Based on the literature on teacher knowledge domains, it appears that the current Chinese preservice teacher education model would produce teachers with relatively strong subject matter knowledge and PCK who are more ready to teach some common topics in the discipline upon graduation. However, compared to their US counterparts, the Chinese preservice teachers have fewer field experience requirements and less required coursework on general knowledge related to education. For instance, the US biology teacher candidates are required to take eight 3-credit courses related to general education topics prior to student teaching, including Teaching in a Pluralistic Society, Learning Theory into Practice, Adolescents in a Learning Community, Using Computers in Education, Legal and Ethical Issues in Education, Introduction to Educational Thought, Education and American Culture, and Teaching Students with Special Needs: Secondary Classrooms. By contrast, Chinese prospective teachers only need to take three courses: Educational Psychology, Pedagogy, and Educational Technology. Consequently, Chinese teacher candidates may have very limited knowledge of students as diverse learners, the students' cognitive, social, and emotional development as adolescents, etc. With limited required field experience, Chinese teacher candidates appear to have fewer opportunities to develop their knowledge of educational contexts.

The authors of Chap. 17 also describe the four most widely adopted in-service science teachers' professional development models in mainland China, i.e., Online Teacher Training Programs, Model Lesson Study, Distinguished Teachers' Workshops, and Collective Lesson Preparation. Again, all of the above PD activities focus on developing topic-specific instructional strategies for PCK. These models provide systemic support for teacher learning and development throughout their professional career by incorporating principles such as teacher ownership, focus on practice, self-reflection, coherence, and collegial discourse within multilevel professional learning communities. Apparently, these strategies used in Chinese teacher professional development are all in line with the existing research literature. However, empirical research on the effects of the teacher professional development models is scarce in the Chinese context. For instance, to what extent have the teacher education and professional development programs actually impacted prospective or in-service teachers' classroom practices in China? To what extent have the teacher education and professional development programs actually impacted Chinese teachers' beliefs about the nature of scientific knowledge and inquiry and their conceptions of learning and teaching science? Answers to these questions remain largely unknown due to a lack of empirical studies in these areas.

Around 1995, Gao conducted a study on conceptions of teaching held by high school physics teachers (Gao and Watkins 2002). Based on both interviews and survey data, Gao identified five common teaching conceptions held by Chinese teachers: knowledge delivery, exam preparation, ability and development, attitude development, and cultivation of good conduct. Gao further combined the five teaching conceptions into two overarching categories: *molding* orientation and *cultivating* orientation. The core of the molding orientation is to mold students to meet expectations either in knowledge accumulation or exam performance. For the teachers with a molding orientation of teaching, the teaching process is mostly one-way where students are viewed as passive receptors, whereas the teachers with a cultivation orientation are those who see the purpose of teaching as development of student cognitive ability or attitude or cultivation of good conduct. Teachers with a cultivation orientation tend to perceive students as active learners and teachers as facilitators or role models. Gao's research suggests that Chinese teachers may hold one or more competing concept of teaching simultaneously, but the majority of Chinese teachers seemed to view student performance on external exams as the most important indicator of effective teaching and successful schooling. The recent survey results of physics teachers in secondary schools (as presented in Chap. 5) seem to indicate that this is still the case today. For any fundamental changes in classroom practices to occur in China, the external exam system needs to be reformed. In addition, teacher educators and teacher professional development designers need to pay more attention to addressing preservice or in-service teachers' beliefs about science knowledge and process, as well as beliefs about science learning and teaching.

Chapter 18 reports an investigation into secondary science teachers' subject matter knowledge in China. Specifically, Liu and Li examine the relationship between teachers' and students' misconceptions about photosynthesis and respiration. Although the study of teachers' SMK through understanding their conceptions in science has gained attention in the field since the 1980s, it is still a relatively new research area for many Chinese science teacher educators. Consistent with the findings in the existing literature on teachers' SMK (Abell 2007), the study by Liu and Li reveals that some Chinese junior secondary school biology teachers have similar misunderstandings about the key biological concepts as their students do, and the teachers' misconceptions have likely contributed to their students' misconceptions. These findings may surprise many Chinese science educators given the rigorous science coursework required of teacher candidates in China. It is now suggested that those Chinese teachers' preconceptions or misconceptions may have never been confronted during their own professional learning process. This further indicates the limitations of traditional, direct science instruction with regard to concept learning and adds to the argument that university science teacher educators need to treat teacher candidates as they expect the teachers to treat their students in the schools. When the teachers and/or teacher candidates have productive experience in inquiry-based, constructivist instruction in their own professional learning, they are more likely to use the same strategies in their own classrooms to strategically address student misconceptions and enhance student learning and understanding of science. Future studies on teachers' SMK may go beyond exploring teachers' understanding of key science concepts by addressing teachers' understanding of relationships among concepts, theories, and scientific laws, as well as their understanding of the process and nature of science and technology (Carlsen 1999).

In the international science teacher education literature, the use of video cases or text-based cases has been widely adopted as one of the promising strategies for both preservice and in-service teacher development at elementary and secondary levels (Martin and Siry 2012). In Chap. 19, Lu and Kwan describe and evaluate the use of video case instruction designed for preservice chemistry teachers. Based on a self-reported survey, at the conclusion of the intervention, the majority of the preservice teachers viewed video case instruction as valuable and useful for enhancing their competence in instructional design, instructional practice, and problem-solving skills. The findings indicate that the use of video case instruction for preservice Chinese teachers shows some of the common benefits as found elsewhere, such as those reported in Germany, the United States, and Hong Kong (e.g., Martin and Siry 2012; Seidel and Prenzel 2005; Wong et al. 2006). In each normal university in China, there may be hundreds of teacher candidates in each science subject (chemistry, physics, and biology) every year, making it difficult to schedule extended field placements and to find many exemplary classrooms that incorporate standard-based and inquiry-oriented teaching. The video case studies provide a cost-effective way for preservice teachers to observe and reflect on authentic examples of "best practices." Future research may go beyond the use of simple self-reported surveys and investigate the effects of video case instruction in developing teacher candidates' personal theories or beliefs about teaching science (e.g., Abell et al. 1998), as well as its impact on teacher candidates' classroom practices during or after student teaching.

References

Abell, S. K. (2007). Research on science teacher knowledge. In S. K. Abell & N. G. Lederman (Eds.), *Handbook of research on science education*. Mahwah: Lawrence Erlbaum Associates, Inc., Publishers.

Abell, S. K., Bryan, L. A., & Anderson, M. A. (1998). Investigating pre-service elementary science teacher reflective thinking using integrated media case-based instruction in elementary science teacher preparation. *Science Education, 82*(4), 491–510.

Bryan, L. A. (2012). Research on science teacher beliefs. In B. J. Fraser et al. (Eds.), *Second international handbook of science education* (pp. 307–321). New York: Springer.

Carlsen, W. (1999). Domains of teacher knowledge. In J. Gess-Newsome & N. G. Lederman (Eds.), *Examining pedagogical content knowledge: The construct and its implications for science education* (pp. 133–144). Boston: Kluwer.

Gao, L., & Watkins, D. A. (2002). Conceptions of teaching held by school science teachers in P.R. China: Identification and cross-cultural comparisons. *International Journal of Science Education, 24*(1), 61–79.

Grossman, P. L. (1990). *The making of a teacher: Teacher knowledge and teacher education.* New York: Teachers College Press.

Hoban, G. (2002). *Teacher learning for educational change*. Buckingham: Open University Press.

Loucks-Horsley, S., Hewson, P. W., Love, N., & Stiles, K. E. (1998). *Designing professional development for teachers of science and mathematics.* Thousand Oaks: Corwin Press.

Luft, J. A., & Hewson, P. W. (2014). Research on teacher professional development programs in science. In N. G. Lederman & S. K. Abell (Eds.), *Handbook of Research on Science Education* (Vol. II, pp. 889–909). New York: Routledge.

Magnusson, S., Krajcik, J., & Borko, H. (1999). Nature, sources, and development of pedagogical content knowledge for science teaching. In J. Gess-Newsome & N. G. Lederman (Eds.), *Examining pedagogical content knowledge: The construct and its implications for science education* (pp. 95–132). Boston: Kluwer.

Martin, S. N., & Siry, C. (2012). Using video in science teacher education: An analysis of the utilization of video-based media by teacher educators and researchers. In B. J. Fraser (Ed.), *Second international handbook of science education* (pp. 417–433). New York: Springer.

Park, S., & Oliver, J. S. (2008). Revisiting the conceptualization of pedagogical content knowledge (PCK): PCK as a conceptual tool to understand teachers as professionals. *Research in Science Education, 38*, 261–284.

Seidel, T., & Prenzel, M. (2005). How to run a video study. In *Technical report of the IPN video study* (pp. 54–61). Kiel: Waxmann Verlag GmbH.

Shulman, L. S. (1986). Those who understand: Knowledge growth in teaching. *Educational Researcher, 15*(2), 4–14.

Shulman, L. (1987). Knowledge and teaching: Foundations of the new reform. *Harvard Educational Review, 57*(1), 1–22.

Simon, S., & Campbell, S. (2012). Teacher learning and professional development in science education. In B. J. Fraser et al. (Eds.), *Second international handbook of science education* (pp. 307–321). New York: Springer.

Tamir, P. (1988). Subject matter and related pedagogical knowledge in teacher education. *Teaching & Teacher Education, 4*, 99–110.

Wallace, J., & Loughran, J. (2012). Science teacher learning. In B. J. Fraser (Ed.), *Second international handbook of science education* (pp. 307–321). New York: Springer.

Wong, S. L., Yung, B. H. W., Cheng, M. W., Lam, K. L., & Hodson, D. (2006). Setting the stage for developing pre-service teachers' conceptions of good science teaching: The role of classroom videos. *International Journal of Science Education, 28*(1), 1–24.

Chapter 17
An Overview of Professional Preparation for Preservice and In-Service Science Teachers

Cheng Liu and Enshan Liu

17.1 Introduction

In the past decades, which have featured rapid Chinese economic, social, and educational growth, Chinese students, representing a large source of science and engineering talent in K-16 education both in China and all over the world, are gaining increased recognition. Some recently published studies have made information about Chinese science education available to the international community, including student achievement such as PISA in 2009, the history and recent reform of science education in China (Liu et al. 2012), and the career hierarchy of K-12 mathematics teachers (NRC 2010). However, less is known about science teacher education and professional development for K-12 science teachers in mainland China. This chapter aims to provide an overview of various professional development models for Chinese science teachers, including typical training programs for preservice teachers in normal universities and four representative models of professional development programs for in-service teachers (i.e., online teacher education programs, model lesson study, workshops for distinguished teachers, and collective lesson preparation).

17.1.1 Background

Beginning in 2000, a new round of education reform began with the notion that education must be oriented toward modernization, the world, and the future. Within

C. Liu (✉) • E. Liu
The College of Life Sciences, Beijing Normal University, 19 Xin jie kou wai Street, Beijing 100875, China
e-mail: liucheng@bnu.edu.cn

L.L. Liang et al. (eds.), *Chinese Science Education in the 21st Century: Policy, Practice, and Research*, Contemporary Trends and Issues in Science Education 45, DOI 10.1007/978-94-017-9864-8_17

1 year, the new science curriculum standards for Grades 1–9 were released by the MOE (Ministry of Education [MOE] 2001). Two years later, high school science curriculum standards were released (MOE 2003a, b, c). In January 2012, the MOE issued the revised edition of science curriculum standards for Grades 1–9.

Along with the implementation of these science curriculum standards in most schools throughout the country was a shift in emphasis from the transfer of knowledge to the development of students' scientific literacy. In particular, the development of a deeper understanding of significant scientific concepts through inquiry-based teaching is becoming more highly emphasized in Chinese science education (MOE 2011; Liu 2011, 2012).

Only when individual teachers are able to understand these new science curriculum standards and use their own pedagogical content knowledge to teach in a manner consistent with the beliefs and goals of this new approach is it possible for classroom teaching practices to meet these standards. In other words, teachers' professional development is a key to ensuring the success of the curriculum reform efforts.

There are over 1.25 million K-12 science teachers in mainland China, with a broad range of academic backgrounds. Table 17.1 presents the number of teachers and students by science subject and by grade level. According to the MOE's educational statistics (MOE 2015), 476,290 full-time science teachers are employed at senior secondary schools to teach separate science subjects. At junior secondary schools, science is taught as a separate science curriculum nationwide but as an integrated subject in a few provinces. Junior secondary schools employ 597,778 science teachers to teach 43,846,297 students in Grades 7–9, and 184,967 science teachers are employed at primary schools to teach integrated science in Grades 1–6. The student-teacher ratio varies in different areas of the country due to differences in the economic, social, and educational conditions between different

Table 17.1 The number of teachers and students by science subject and by grade level

	Primary school		Junior secondary school		Senior secondary schools	
	Full-time teachers	In-school students	Full-time teachers	In-school students	Full-time teachers	In-school students
Physics	Not applicable[a]		236,822	43,846,297	145,594	24,004,723
Chemistry			151,323		140,126	
Biology			143,084		101,897	
Geology			135,442		88,673	
Integrated science	184,967	94,510,651	31,107		Not applicable[a]	
Total	184,967	94,510,651	597,778	43,846,297	476,290	24,004,723

MOE (2015)

Data is cited from year 2014 Statistics Report published online by MOE in 2015. The data show the number of teachers and students in year 2014

[a]Data is not applicable because physics, chemistry, biology, and geology subjects are not taught separately in primary schools, and science is not taught in the integrated way in senior secondary schools

schools, school districts, cities, and provinces. Some teachers teach fewer than 30 students per class, while some teachers in other provinces teach more than 60 students per class.

In addition to the student-teacher ratio, teachers work in different provinces and settings (urban, suburban, or rural) and are diverse not only in terms of their dialects, cultures, and economic conditions but also in terms of their teaching beliefs, instructional goals, skills, and teaching experience. In mainland China, there is a clear career hierarchy with a series of titles representing in-service teachers' professional development progression from "junior rank" to "intermediate rank" and on to "senior rank" and finally "master teacher" (i.e., 特级教师 in Chinese). Teachers must meet specific criteria to be promoted to a higher rank; for example, to reach higher ranks, teachers must conduct research, publish their findings in scholarly journals (in Chinese), and participate in teaching skill contests.

In the context of the new round of education reform and the large number and great diversity of teachers, initiating a standards-based professional development community for science teachers is crucial to fulfill the intentions and goals of the new science curriculum standards.

17.1.2 Rationales for Teachers' Professional Development Models

Professional development models designed for Chinese science teachers are built on the theory of pedagogical content knowledge (PCK) established by Lee Shulman (Shulman 1986, 1987). It is clear that teacher educators and researchers have identified pedagogical content knowledge as a critical component of the knowledge needed to teach (Gess-Newsome 1999). Based on Shulman's work, many different definitions and frameworks of PCK have been proposed. However, among all the various conceptualizations of PCK, there is always essential elements:

> At the heart of PCK lies what teachers know about how their students learn specific subject matter or topics and the difficulties or misconceptions students may have regarding this topic related to the variety of representations (e.g., models, metaphors) and activities (e.g., explications, experiments) teachers know to teach this specific topic. These components are mutually related: The better teachers understand their students' learning difficulties with respect to a certain topic and the more representations and activities they have at their disposal, the more effectively they can teach about this topic. (Van Driel et al. 2014, pp. 849)

PCK could be defined as "the knowledge that a teacher uses to provide teaching situations that help learners make sense of particular science content" (Loughran et al. 2001, p. 289). The development of a preservice teacher education program in mainland China is an application of the theory of PCK, which is embodied in the integration of studying science knowledge with the application of knowledge about learning, pedagogy, and students to actual teaching practice. The teaching practicum is emphasized in the development of teacher education program to help teacher integrate the content knowledge and pedagogical knowledge into topic-specific PCK. In sum, the notion of PCK is embedded in in-service teachers' professional development models in either explicit (online teacher education programs) or implicit ways (other in-service teacher education programs).

The second rationale behind the development of science teacher education programs is the need for engaging both prospective and practicing teachers in professional learning communities, in which teachers could work collaboratively among peers. A professional learning community has been defined as a school environment where teachers work collaboratively in purposefully designed groups to improve student achievement within a structure of support provided by the school administrator (Ontario Principals' Council 2008). Collaboration is important in successfully supporting the use of new practices among teachers (Luft and Hewson 2014). Facing the great diversity of science teachers in mainland China, professional learning communities are established not only at the school level (collective lesson preparation) but also at the school district (model lesson study and workshops for distinguished teachers) as well as provincial or national levels (online teacher education programs). These professional learning communities are based on the belief that PCK could be best understood and used through critical reflection among peers with similar professional experiences.

Finally, teaching is itself complex, requiring constant learning and reflection. New knowledge, skills, and strategies for teaching come from a variety of sources, including research, new materials and tools, descriptions of best practices, colleagues, supervisors, self-reflection on teaching, and reflection on the learning of students in the classroom. It is very important that teachers continually consider and contribute to the advances in knowledge regarding teaching and learning (NRC 1996). Loughran (2014) pointed out that learning about science teaching through reflection on experience is a necessary and effective way for teacher education. He summarized many professional development programs to illustrate how important both student teachers and experienced teachers need real opportunities to learn from their experiences, not to just have experiences. Therefore, providing opportunities for teachers to engage in both self-reflection and collegial reflection about the clinical experience or practicum of teaching is another priority for teacher education programs in mainland China.

17.2 K-12 Preservice Science Teacher Education

According to the Teacher's Law of the People's Republic of China issued in 1993 (Standing Committee of National People's Congress 1993), a teaching license is required for employment as a teacher in the K-12 education system. Both education experience with specialization in a certain science discipline (studying science content knowledge) and a certification of pedagogical training (studying pedagogical knowledge) are necessary to apply for a science teaching license. In alignment with the requirement of getting a teaching license, training programs for preservice science teachers are designed to help teacher candidates obtain a deeper understanding about both science content and pedagogical knowledge and apply this understanding to their lesson plans and teaching practice.

17.2.1 Institutions Responsible for K-12 Preservice Science Teacher Preparation

The education system in mainland China begins with kindergarten. Chinese students attend school from kindergarten through 12th grade, in primary schools (grades 1–6) and secondary schools (grades 7–12). Tertiary education refers to grade 13 and upward and is also known as higher education or college or university education. Children are enrolled in the education system when they are 5 or 6 years old. Teacher education programs described in this chapter involve only the professional development of teachers who currently or will soon hold teaching positions in kindergarten, primary, and secondary schools. Teachers holding positions in tertiary education are not considered.

Institutions approved to provide education for K-12 teachers are incorporated into the secondary and the tertiary education level. At the secondary education level, only certain types of vocational schools, called early childhood normal schools and secondary normal schools, are specialized in preservice science teacher education programs. All of the programs held at early childhood normal schools are designed specifically for preparing kindergarten teachers. However, graduates from secondary normal schools may be allowed to teach at both elementary school and kindergarten. Candidates enrolled in these institutions always spend approximately 3 years developing their PCK specific to teaching in kindergarten and primary school. All graduates have over 12 years of education experience before becoming a science teacher in kindergarten or elementary school. Graduates from other secondary schools without further education experience are not allowed to teach. At the tertiary education level, only normal university and certain types of junior colleges (called specialized higher normal schools or normal colleges) are eligible to offer preservice science teacher education programs. Both graduation certificates and academic degrees could be awarded by normal universities, whereas junior colleges offer graduation certificates only. Students enrolled in preservice science teacher education programs in normal universities may obtain teaching certifications as soon as they are qualified to graduate from a normal university. Graduates from other colleges or universities with 4 years' schooling or more are also allowed to apply for a teaching license if they submit a certification of credit hours in education and psychology courses and pass the national teachers' qualification examinations. Today, the blending of the aforementioned vocational schools into the secondary education level occurs primarily in rural areas. Almost all preservice teacher education programs in urban areas, such as in Beijing, are offered in colleges or universities.

The grades that preservice teachers are allowed to teach depend on the levels and types of the vocational schools, colleges, or universities from which these teachers have graduated. According to the Teachers Law of the People's Republic of China issued in 1993:

> (1) To obtain qualifications for a teacher in a kindergarten, one shall be a graduate of an infant normal school or upwards. (2) To obtain qualifications for a teacher in a primary school, one shall be a graduate of a secondary normal school or upwards. (3) To obtain qualifications for a teacher in a junior middle school, or a teacher for general knowledge courses and specialized courses in a primary vocational school, one shall be a graduate of a specialized higher normal school, or other colleges or universities with two or three years' schooling or upwards. (4) To obtain qualifications for a teacher in a senior middle school, or a teacher for general knowledge courses and specialized courses in a secondary vocational school, technical school or a vocational high school, one shall be a graduate of a normal college or other colleges or universities with four years' schooling or upwards. (Standing Committee of National People's Congress 1993)

17.2.2 Admission into Preservice Science Teacher Preparation Institutions

Admission into a science teacher preparation institution is determined based on both candidates' university applications and their scores on entrance exams (paper and pencil tests, i.e., 高考 in Chinese). As they approach the end of their K-12 school education, candidates need to take entrance exams for further education and apply to institutions in which they are interested. And the institutions admit students based on their scores and grade history.

At many of the institutions at the tertiary education level cited above, some special financial support is available to recruit qualified preservice science teachers from rural areas. The dominant source of support is called the "Government-Sponsored Normal Students Program," which was instituted in 2007. Under this program, students from rural areas are enrolled tuition free into preservice teacher education programs in normal universities. Once they graduate from a normal university, they must return to their home rural area to teach for 10 years in a K-12 school. The government will guarantee their teaching position in the rural area.

Every student recruited into preservice science teacher preparation programs in normal universities has shown good performance in the entrance exam, as have students from rural areas enrolled in the "Government-Sponsored Normal Students Program." In other words, the admission into preservice science teacher education programs is based on the applicants' academic performances, regardless of their financial situations.

17.2.3 *Typical Preservice Science Teacher Preparation Programs at Normal Institutions*

The teacher preparation programs at all of these institutions are typically 2–4 years in length and focus on both the theory and practice of science content and pedagogical knowledge. The graduates will obtain a teaching license without extra exams so long as they accumulate enough credits in both the science subject areas and pedagogy domains and are qualified to graduate. Graduates from nonteacher preparation colleges or universities—that is, not the normal institutions mentioned above—may also apply for teaching licenses if they pass additional qualifying exams to obtain pedagogy certification.

Table 17.2 below presents a typical preservice biology teacher education program designed by one of the top six normal universities in China. The teacher candidates must earn 175 credits (1 credit represents 1 class hour per week) to graduate and obtain the teaching license during their 4-year tertiary education experience.

All preservice teachers are selected from different provinces by national entrance exams for higher education. Teacher candidates are required to take three types of courses, including "general courses," "professional courses," and "teacher education courses," to complete their training program.

A total of 44 credits are allotted for general courses, in which participants learn the arts and humanities (e.g., literature, history, and philosophy) and fulfill physical education requirements. Teacher candidates must complete all of these general courses by their third year.

Taking the biology program as an example, to gain a deep understanding about biology content and be proficient in biological experimental skills, students must take professional courses, starting from some courses in related subject areas during

Table 17.2 A 4-year preservice biology teacher education program in normal university A

Category of course			Credits	
General courses, such as arts and history			44	
Professional courses	Courses in a related subject area, such as chemistry and physics		33.5	92
	Biology courses	Basic courses in biology, such as microbiology and genetics	33.5	
		Specialized courses in biology, such as animal behavior and ornithology	25	
Teacher education courses	Compulsory courses, such as educational psychology		16	39
	Education research courses, such as practice in biology education research		4	
	Teaching practice		11	
	Professional integrity		2	
	Thesis		6	
Total			175	

the first 2 years, such as undergraduate mathematics, organic chemistry, physical chemistry, and basic physics. In the second and third years, students focus primarily on basic biology courses, such as general zoology, general botany, biochemistry, molecular biology, microbiology, genetics, and cell biology. In addition to taking basic biology courses, participants are required to choose some specialized courses (e.g., animal behavior, genetic engineering, or immunology) to specialize in certain biological subfields.

During the third and fourth years, teacher candidates focus on teacher education courses, including compulsory courses such as educational psychology, pedagogy of biology, comprehensive experiments for biology education, and teaching skills training. In these teacher education courses, general pedagogical knowledge, curriculum knowledge, and knowledge about learners are taught. For example, in the pedagogy of biology course, preservice biology teachers must be aware of students' alternative, prescientific ideas about K-12 biology topics and learn how to apply conceptual change models and other constructivist strategies in science teaching. In addition, knowledge about the K-12 biology and science curriculum reform in both mainland China and all over the world will be taught in order to help preservice biology teachers better understand the context of the biology curriculum. In the meantime, preservice teachers must learn teaching strategies to initiate and support students' scientific inquiry, knowledge about performance and summative assessment, and so on, in the context of the K-12 biology domain. Also, they are required to earn four credits in education research courses, such as practice in biology education research and reform trends of middle school biology education, to familiarize themselves with reform trends and standard educational research methods.

Teacher candidates must then participate in a teaching practicum for at least 6 weeks in a school where they have the opportunity to apply their understanding of biology and education in an actual classroom teaching setting. Preservice teachers involved in the Government-Sponsored Normal Students Program will practice teaching for almost half a year in K-12 schools, considering that these candidates will certainly become teachers as soon as they have graduated. During the teaching practicum, every trainee will be assigned to a K-12 school as a student teacher to teach in a real classroom setting under the supervision of a certified teacher. Each student teacher must have two certified teacher supervisors. One of these supervisors is an experienced science teacher, who allows the inexperienced student teacher to observe teaching in the classroom and then helps the student teacher revise their teaching plan and provides onsite support. This experienced science teacher usually also invites the student teacher to join the in-service collective lesson preparation community at the school level (see Sect. 17.3.4 in this chapter). Another certified teacher is always a homeroom teacher who has the responsibility for taking care of students' in-school lives. The student teacher is also required to act as the assistant to the homeroom teacher throughout the day, observing how to manage students' affairs and assisting in addressing students' problems.

The last step for teacher candidates to apply for a diploma is to design and implement a study and to write a thesis. Usually, these studies focus on biology research topics, such as cellular and molecular domains, ecology, and zoology. Through the thesis study, preservice teachers can not only learn more about certain biological topics but also understand more about real scientific inquiry in the natural science domain. Along with ongoing science education reform that started in 2000, some teacher candidates focus their thesis on the K-12 biology education domain. They do literature and descriptive research about what students know about certain biology topics, what their prescientific concepts are, what good inquiry teaching and learning is from the teachers' perspectives, what characteristics different textbooks have to support conceptual and inquiry teaching and learning, and so on. Through the process of designing and implementing these kinds of thesis studies, preservice teachers can not only develop their understanding and abilities about science education research but also learn more about their students, colleagues, and instruction materials.

Similar to the above preservice biology teacher education program, Table 17.3 below presents a typical preservice physics teacher education program designed by the same normal university. Participants must earn 160 credits to graduate and obtain a teaching license during their 4-year tertiary education experience.

This training plan for physics preservice teachers has a credit distribution that is very similar to that of the biology preservice teacher education program. The main differences are the courses related to the specific science disciplines. In addition, the programs designed for chemistry and geology preservice teachers at the same university have similar course structures to the one described above. Teacher education programs at different universities vary in course offerings and credit requirements. Table 17.4 below shows a sample biology preservice teacher education plan from another of the top six normal universities.

In this case, teacher candidates are required to earn fewer credits, and all courses and credits are categorized differently from those in university A. However, the

Table 17.3 A 4-year preservice physics teacher education program in normal university A

<table>
<tr><th colspan="3">Category of course</th><th colspan="2">Credits</th></tr>
<tr><td colspan="3">General courses, such as arts and foreign language</td><td colspan="2">44</td></tr>
<tr><td rowspan="3">Professional courses</td><td colspan="2">Courses in related subject area, especially mathematics</td><td>18</td><td rowspan="3">80</td></tr>
<tr><td rowspan="2">Physics courses</td><td>Basic courses in physics, such as mechanics, thermodynamics, and electromagnetics</td><td>43</td></tr>
<tr><td>Specialized courses in physics, such as general relativity and general astronomy</td><td>19</td></tr>
<tr><td rowspan="5">Teacher education courses</td><td colspan="2">Compulsory courses, such as educational psychology</td><td>12</td><td rowspan="5">36</td></tr>
<tr><td colspan="2">Education research courses, such as research on teaching physics experimentation in grades 7–12</td><td>2</td></tr>
<tr><td colspan="2">Teaching practice</td><td>11</td></tr>
<tr><td colspan="2">Professional integrity and scientific research training</td><td>3</td></tr>
<tr><td colspan="2">Thesis</td><td>8</td></tr>
<tr><td colspan="3">Total</td><td colspan="2">160</td></tr>
</table>

Table 17.4 A 4-year preservice biology teacher education program in normal university B

Category of course		Credits	
General courses, such as history and foreign language		30	
Basic courses	Compulsory courses, such as inorganic chemistry, organic chemistry, zoology, botany, psychology, and lesson planning	35	47
	Elective courses, such as philosophy of education and curriculum design and evaluation	12	
Specialized courses	Compulsory courses, such as molecular biology, cell biology, and ecology	22	30
	Elective courses, such as developmental biology, immunology, and parasitology	8	
Innovation	Research project or publication	3	3
Practice	Teaching skills training	2	20
	Teaching practice and field experience	10	
	Thesis	8	
Total		130	

pattern of focus on both a specific discipline and related science content knowledge and pedagogical knowledge is similar. For example, during the courses of zoology, botany, molecular biology, and so on, Preservice teacher needs to get deeper understanding about discipline and content knowledge to differentiate between the misconception and scientific conception. In the meantime, during the courses of psychology, lesson planning, teaching skill training, and so on, they need to know better about students' learning difficulties with respect to a certain biology topic and the more activities that might help students change their prior understanding. In addition, the teaching practicum is included as an indispensable component based on the same concern that preservice science teacher education should not only help participants obtain a deeper understanding of the content and pedagogical knowledge but also apply their knowledge to actual classroom teaching.

Although the structure of training programs in different institutions appears slightly different, they share two common characteristics. First, the training programs for preservice science teachers are designed in alignment with the PCK perspective too. These programs are conducted primarily by science departments comprising both scientists and education specialists. Candidates are trained not only in pedagogical knowledge and teaching practices by educators but also in science and scientific research by scientists from respective science departments. Such programs thereby make preservice teachers well-rounded in terms of their science backgrounds. Second, credits for general courses, such as arts and history, are required in all of the preservice science teacher's preparation programs. This common feature indicates that training programs are designed to develop not only teachers' PCK with regard to a certain science subject area but also their spirit and literacy in humanities. The full development of teachers' characters is of greater consideration than the sole development of technical teaching skills.

17.3 Professional Development for In-Service Science Teachers

After graduating from teacher's preparation programs, preservice teachers obtain a job and enter the professional development system, becoming in-service teachers. Teachers are granted a title based on their academic degree and teaching experience in the field. For example, new secondary school science teachers who have master's degrees can obtain a junior rank at the end of their first full year of teaching, whereas at least 3 years of teaching experience is required for new teachers with bachelor's degrees to obtain the same rank. In the meantime, the supervision of a new teacher's full year of teaching is part of the responsibilities of teachers with intermediate or higher ranks. Junior-ranking teachers must accumulate more teaching experience and a certain number of credits in in-service professional development every year, participate in teaching skill contests, and investigate some teaching problems to be promoted to the intermediate ranks. To earn a senior rank, a teacher must have a deep understanding of science content and pedagogical knowledge, be able to implement a sound science teaching research program, and make an important contribution in training lower-ranking teachers. The requirements to earn a master teacher rank are the most rigorous. This honorary title is awarded only to teachers who have significant ideas about both the science discipline and educational theory and practices in K-12, are able to initiate and conduct an educational research program, publish enough papers in science teaching journals (in Chinese), are advocated for by most peer teachers, and make outstanding contributions in training lower-ranking teachers.

The evaluation criteria for promoting teachers from one rank to the next may differ slightly between schools, schools districts, cities, and provinces. The description above is the general situation across mainland China. The salary and available resources for teachers are associated with their title (except for the master teacher). For example, an intermediate-ranking teacher would be paid higher than a junior-ranking teacher if they get exactly the same job in the same school. However, the pay scale varies depending on the financial and economic development status of the school district or local community. The title of master teacher is only honorary and has no differences in salary and recourses. This type of title promotion may represent an incentive for teachers to participate in in-service teacher professional development as shown below.

In-service teachers may earn credits either as full-time (or part-time) teachers or while on a leave of absence from work with approval by the MOE. A national in-service teacher education program has recently begun to enroll science teachers from all over the country. Participating teachers are allowed to take a leave of absence from their work to participate in more intensive face-to-face training workshops. Their work is taken over by preservice science teachers who have already earned all of the necessary credits in science subject content and education theory and who are ready to become student teachers. This policy offers in-service teachers more time for their professional development while affording preservice teachers greater opportunities to integrate theory with classroom practice. The face-to-face professional development programs, combined with online teacher education, are funded by the MOE with approximately 500 million Chinese yuan financial support, including payments for participating teachers and preservice teachers.

However, many teacher education programs are also designed for science teachers with full-time or part-time positions who are unable to be absent from their work. Four types of these programs are described in the following sections: online teacher education programs, model lesson study, workshops for distinguished teachers, and collective lesson preparation.

17.3.1 Online Teacher Education Programs

With the new round of education reforms beginning in 2000, the new national science curriculum standard turned its focus from emphasizing knowledge to emphasizing the development of students' scientific literacy through inquiry-based teaching. This standard requires that in-service teachers' training be developed in a manner consistent with this shift to ensure that the curriculum reform efforts will be effectively transferred into real classroom teaching. However, engaging in face-to-face teacher education poses a practical problem due to the huge number of teachers with diverse backgrounds and a lack of teacher educators. Online teacher education programs have been designed and conducted to solve this problem.

From 2007 to 2011, a nationwide standards-based online professional development program, funded by the MOE, was initiated for in-service science teachers throughout the country. The program was developed by the same team who designed the national curriculum standards. This program focused on helping science teachers translate the aims and goals of the national curriculum standards (e.g., "developing student scientific literacy," "science for all children," and "teaching science through inquiry"), into their daily classroom practice.

Under this model, teachers may access any online community they like, but they are preassigned to an online community of teachers based in their own province. In this program, participants come from urban, suburban, or rural areas with great diversity in not only their demographic background, comprising their dialects, culture, and economics, but also their teaching beliefs, skills, and experience. This high level of diversity made the program quite challenging but also highly rewarding, as teachers were able to learn from each other.

A panel of teacher education experts was assembled, including the designers of national curriculum standards, master teachers, teacher educators, and science education researchers from universities. They all had access to an online community to provide feedback and suggestions to teachers. In the meantime, every participating teacher had free access to online videos and was able to submit his assignments, check peer and trainer reviews about his/her assignments, make

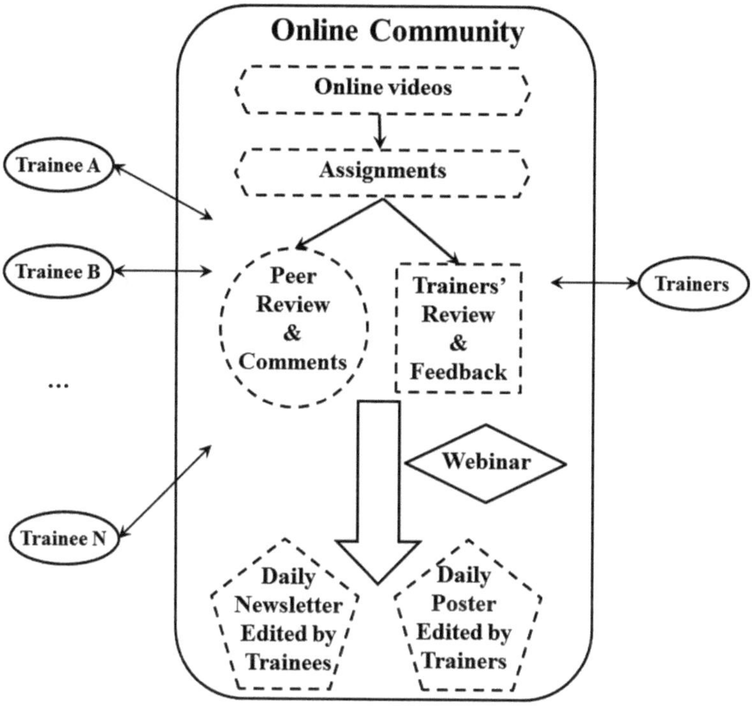

Fig. 17.1 Work pattern for online teacher education programs

comments on others' assignments, participate in webinars, and follow daily newsletters and daily posts from his online community (See Fig. 17.1).

The typical duration of the program was ten consecutive days during summer vacation. Every morning, participants logged onto their own unique online account containing their personal academic record and watched online videos concerning the panel's comments and suggestions regarding a special topic related to a real classroom teaching situation. At the end of the videos, a few reflective questions by the trainee panel were presented. Participating teachers were required to use their own real teaching experiences as examples to illustrate their answers to these questions. In the afternoon, teachers began to write down their answers and upload them to the online community through their accounts. In the meantime, teacher educators reviewed teachers' answers and gave brief feedback. Teachers could also read others' answers and make comments to each other. In the evening, a daily post summarizing the key points produced from the entire day of online training was published by teacher professional development leaders to all of the participants in order to review the most important concepts. In addition, participating teachers made similar summaries for themselves by editing and publishing a daily newsletter. In addition, participants were required to join in the online webinar every 3 or 4 days, communicating with the science educator about any difficulties they faced when trying to apply the new standards to real classroom teaching. Participating

Table 17.5 An outline of the national online professional development program for grades 10–12 biology teachers

Topics	Questions addressed and discussed in the video
Topic 1: the rationale for changing the biology curriculum and implementing the principle of the curriculum into real classroom teaching	What is the difference between the new biology curriculum standards and the old ones?
	How should teachers understand the principles of the new curriculum?
	What knowledge and skills are necessary for teachers to translate the principles of the new curriculum into real classroom teaching?
Topic 2: the content changes in biology curriculum and its challenges	What challenges will students encounter while implementing the new curriculum?
	How can teachers manage the new content knowledge?
	How can teachers teach effectively in response to the changes in curriculum?
Topic 3: achieving the goal—"Teaching the Big Ideas of Biology"	What is your opinion of the "big ideas" required in the national biology curriculum standards?
	How is it possible to select big ideas from given content?
	How can one teach big ideas effectively in a biology classroom?
Topic 4: achieving the goal—"Cultivating Students' Active Learning"	What characteristics does the inquiry-based teaching approach have?
	What content is adequate for inquiry-based teaching?
	How does one teach biology through inquiry in a classroom with a large student-teacher ratio?
Topic 5: achieving the goal—"Teaching Effectively with a Lecture and Demonstration Approach"	What is your opinion about the effectiveness of the use of a lecture and demonstration approach in biology classroom?
	What content is better suited for lecture than for inquiry?
	How does one teach biology in an effective way by using demonstration?
Topic 6: teaching compulsory module 1—"Molecules and Cells"	What issues should be noticed while teaching this module?
Topic 7: teaching compulsory module 2—"Genetics and Evolution"	Which parts are the most important or difficult to teach in this module?
Topic 8: teaching compulsory module 3—"Stability and Environment"	How does one teach the big ideas in this module?
Topic 9: teaching elective module 1—"Practice of Biotechnology"	How does one propose a feasible and affordable list of new equipment for the biology labs in your school?

(continued)

Table 17.5 (continued)

Topics	Questions addressed and discussed in the video
	How does one choose the biotechnological activity available in the curriculum given the current condition of your school?
	How can the success rate of biotechnological practice be improved?
Topic 10: assessments in biology teaching	What principles and assessment methods are consistent with the requirements of the new curriculum standards?
	What are the assessment strategies for classroom teaching and how should they be used?
	What are the assessment strategies for lab-based teaching and how should they be used?
	How does assessment encourage students' active learning?

teachers could post their questions through an online forum. Science educators checked the messages and questions posted by participants and chose the most typical questions to make comments on and suggestions for all participants through the forum system.

For example, Table 17.5 is an outline of this program for training biology in-service teachers for grades 10–12. Typically, ten topics were given, including the understanding of the beliefs and goals of new biology curriculum standards, analysis of PCK (containing biological subject knowledge, general pedagogical knowledge, curricular knowledge, knowledge about learning, and knowledge about educational contexts), and the means of incorporating these new standards into real classroom teaching practices.

From 2007 to 2011, science teachers from all provinces in mainland China participated in such online program. This professional development model has been used for both elementary and secondary teachers not only at the national or provincial level but also at the county or town level. More and more websites are being built to integrate multiple online resources for in-service teachers' learning and practicums. This program made its contribution to ensure the success of the national curriculum reform amid the large number of teachers with great diversity and a lack of teacher educators.

17.3.2 Model Lesson Study

Model lesson study is a type of professional development workshop for in-service teachers at the school district level. This program provides on-site peer review for teachers, who demonstrate their classroom teaching in front of other teachers, and it provides opportunities for teacher observers to reflect on what they have observed and make connections to their own classroom practices. In every school district, there is a special occupational position called the "subject coordinator" (教研员in Chinese), who is responsible for initiating in-service teacher education in a certain subject area. Experienced senior-ranking teachers are eligible for this type of position.

The subject coordinator first identifies a teacher (either an experienced teacher or a new faculty member at a school) who has creative ideas for teaching certain science content in nontraditional ways. The coordinator then calls all of the teachers in this subject from the same school district to observe this teacher's model lesson together. After the classroom observation, the subject coordinator initiates a workshop featuring peer review and discussion about this model lesson. The teacher who designs this lesson is also invited to participate in the workshop to explicitly illustrate her/his creative ideas. The subject coordinator and other teachers then provide their opinions and further suggestions.

During peer review and discussion, collegial reflections from the subject coordinator and other teachers may help the model lesson designer to self-reflect in order to improve his/her teaching practice. Meanwhile, observing teachers also benefit from self-reflection regarding how to incorporate the best ideas from this model lesson into their own class. These mutual benefits between the model lesson designer and other teachers derive from the similar teaching environments they share within the same school district.

In addition, exemplary model lessons are invited for exhibition at annual provincial or national conferences on this subject area. It is in this manner that creative ideas about science teaching can be spread over different cities or provinces and impact actual classroom teaching throughout a province or the entire country.

17.3.3 Workshops for Distinguished Teachers

This professional development program was designed specifically for teacher leaders who are responsible for leading and developing other teachers. This program, which is usually initiated and funded by the District Board of Education and aims to help experienced teachers to be more effective leaders of the professional learning community at either the school level or the school district level. Only teachers who have made important contributions to the school district may be enrolled in this program.

Participants have access to several integrated resources, such as one-on-one interactions with mentors who are distinguished master teachers in the same subject area or play a key role within the science education system. They also get ample opportunities to observe their mentors' or other expert teachers' real classroom teaching, free access to book review salons and academic forums, an open platform to design and implement science education research, and different ways to reflect

Table 17.6 Sample schedule of Haidian Cadreman Teachers Academy

Session	Assignments
Phase I: developing self-plan and attending multifaceted activities (6 months)	Teachers share their own ideas and situations focusing on professional development orientation and goals with their colleagues and mentors
	Then, a series of workshop and routine activities will open for each participant, such as book review salons, academic forums, and classroom teaching observations and review (e.g., "model lesson study")
	Finally, teachers must design specific professional development plans for themselves with clear statements about what they really need and how they want to develop
Phrase II: fulfill the self-plan through designing and implementing science education research supported by collaboration (24 months)	Teachers continue to attend book review salons, academic forums, and classroom teaching observations and review
	In the meantime, teachers have discussions with other program participants on the topics they are interested in. Participants find a way to perform a collaborative study as a team
Phrase III: summary and reflection (6 months)	Teachers publish and communicate their achievements and findings with other members in this academy. They must also reflect on how to initiate professional development in their own communities, either at the school level or at the school district level

on their own decision making in teaching and studying. Participants must dedicate themselves to this program for at least 3 years. Table 17.6 above shows an example model called the Haidian Cadreman Teachers Academy.

This model differs substantially from other professional development models mentioned in this chapter because it aims to cultivate teacher leaders of the professional learning community. This program is designed for experienced teachers who have the ability and ambition to make more significant contributions for their school district or school—that is, it is not intended for new teachers or ordinary teachers.

17.3.4 Collective Lesson Preparation

This is a professional development model for in-service teachers at the school level. Within a school, teachers who teach the same science subject for the same grade always belong to the same collective lesson preparation community. Student teachers (aforementioned in Sect. 17.2.3) are always invited to join in the community during their teaching practicum in a K-12 school.

In this community, teachers share their opinions and comments about the unit plan. First, they hold discussions on the objectives, curricular schedules, the most important ideas that should be taught, the most challenging concepts that students will encounter, teaching procedures, students' preconcepts, learning activities, series of questioning, and means of evaluation, among others. A senior-ranking teacher or a master teacher always hosts this discussion and leads the community to achieve some common goals.

Second, teachers within the same community of collective lesson preparation work together to turn their idea into a real teaching plan. The lead teacher splits the entire unit into specific lessons and assigns them to individuals. For example, teacher A takes the responsibility for designing all materials for lesson 1, including the lesson plan, PowerPoint documents, resources (i.e., pictures, animations, and website links), and assessment tools, while teacher B takes lesson 2, and teacher C takes lesson 3.

Usually a week later, teachers share all of the lesson materials with one another. The designers of each lesson must explain the rationale and principles behind the students' learning activities they created. In the meantime, other teachers review the designs, make comments, and discuss with each other about how to improve them. Each lesson designer later revises his original design based on the peer reviews and shares the new version of all of his teaching materials with everyone else.

The teachers then receive all of teaching materials for these units from the other teachers and make their own revisions to meet the requirements of the different

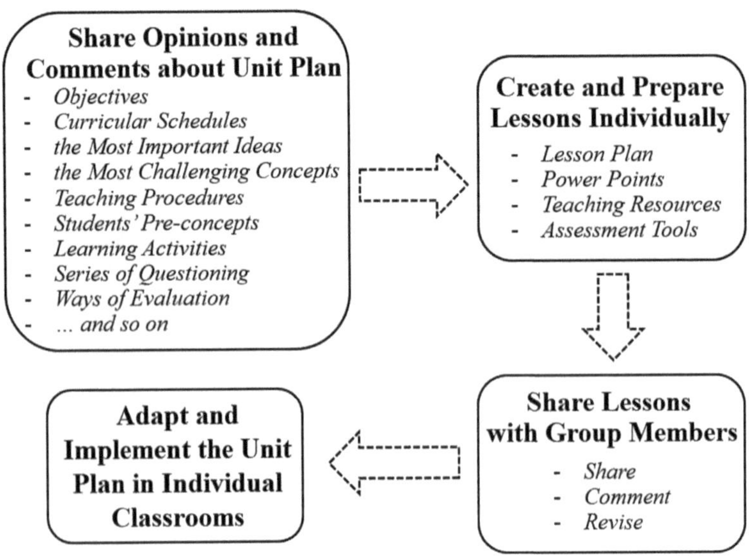

Fig. 17.2 Collective lesson preparation community

students they teach. Finally, teachers from the same collective lesson preparation community will incorporate the shared teaching unit plan into their diverse classrooms based on their own decisions and revisions (See Fig. 17.2).

Although they teach different students in different classes, teachers within a collective lesson preparation community share a common teaching philosophy and goals, a synchronous or parallel teaching schedule, and almost identical teaching policies and equipment. This similarity helps them gain rapid professional development, especially for the new faculty members, who have a relative lack of teaching ideas, instructional materials, and classroom experience.

17.4 Summary

In summary, the system for the professional development of science teachers in mainland China is distributed from the national or provincial level to the school district level and finally to the school level, designed respectively or integrated for all teachers, including inexperienced preservice teachers, new teachers, experienced (intermediate or senior-ranking) teachers, and teacher leaders (See Fig. 17.3). Preservice science teachers may receive training at the national or provincial level by focusing on the theory and practice of science and science education through the approved educational institutions. After they graduate, these teachers may develop their PCK through online teacher education programs (at the national or provincial level), model lesson study (at the school district level), and

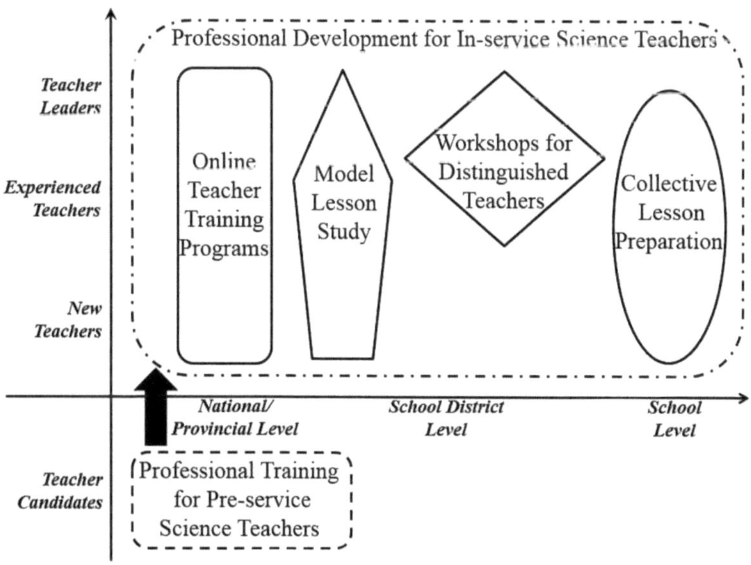

Fig. 17.3 The science teacher professional development system in mainland China

collective lesson preparation communities (at the school level). New teachers will gain rapid professional development through online teacher education programs, model lesson study, and collective lesson preparation communities. Experienced teachers play demonstrational roles in model lesson studies and collective lesson preparation communities. Some experienced teachers can also get further development through workshops for distinguished teachers, enabling them to be better leaders of their school districts or schools.

The current science teacher professional development system has trained millions of science teachers to meet the demands of a huge of population. It has been shown that by establishing professional learning communities that emphasize curriculum standards, PCK perspectives, and self- and collegial reflections, the preservice and in-service teacher development models described above appear to be working well in mainland China, especially in the context of involving an extremely large number of teachers and students.

However, as Loughran (2014) said, "A great challenge for teacher education programs is to help student teachers see beyond their own experiences of teaching and find new ways to engage them in conceptualizing practice as something more than how they themselves were taught" (p. 812). Teachers are required to teach in a manner that emphasizes the beliefs and goals, for example, teaching and learning in the way of scientific inquiry, advocated by the educational reform and its new science curriculum standards. However, most of science teachers in mainland China do not have the experience of doing real authentic scientific investigation. Therefore, it is really a big challenge for teacher education programs in mainland China to help teachers change their beliefs and behavior to a manner that they themselves were never taught.

In the meantime, the teacher education programs in mainland China are in dire need of more qualified and expert science teachers and educators who are able to do science education research on translating the new science curriculum ideas into K-12 science classroom teaching practices. Along with the gradual implementation of the science curriculum reform started in 2000 into more and more schools, there have been many experience-based articles published in science teacher journals (in Chinese) to share interesting teaching ideas, experiences, and so on. However, in the new round of science education reform, using only experience-based teaching decisions is not sufficient for K-12 science teachers to take the various challenges they are going to encounter. More research-based or evidence-based teaching decisions are also necessary to design and implement effective teaching for a large student population with different backgrounds and experiences. In sum, more empirical-based studies about science teacher professional development, for both preservice and in-service science teachers, are necessary.

Acknowledgments Sincere thanks go to Teacher Lehe Xiao, Prof. Hong Cui, Prof. Yuying Guo, Prof. Min Wang, and Dr. Rui Wei for providing the actual K-12 science teacher professional development models to support this chapter. We also gratefully acknowledge the contributions of Prof. Ling Liang and other reviewers who provided valuable comments and suggestions to help revise the draft manuscript.

References

Gess-Newsome, J. (1999). Pedagogical content knowledge: An introduction and orientation. In J. Gess-Newsome & N. G. Lederman (Eds.), *Examining pedagogical content knowledge: The construct and its implications for science education* (pp. 3–17). Dordrecht: Kluwer Academic.

Liu, E. (2011). Prospection on the next ten years of science curriculum development in China (基础教育理科课程改革未来十年展望). *General Education Curricula, 12*, 17–18 (in Chinese).

Liu, E. (2012). Deliver key concepts through inquiry based approaches in school biology—what's new in the 2011 national biology curriculum standard of junior secondary school (在教学中实现主动探究学习与凸显重要概念传递的对接—《义务教育生物学课程标准》修订思路和要点). *Bulletin of Biology, 47*(3), 33–36 (in Chinese).

Liu, X., Liang, L., & Liu, E. (2012). Science education research in China: Challenges and promises. *International Journal of Science Education, 34*(13), 1961–1970.

Loughran, J. J. (2014). Developing understandings of practice. In N. G. Lederman & S. K. Abell (Eds.), *Handbook of research on science education (volume II)* (pp. 811–829). New York: Routledge.

Loughran, J., Milroy, P., Berry, A., Gunstone, R., & Mulhall, P. (2001). Documenting science teachers' pedagogical content knowledge through PaP-eRs. *Research in Science Education, 31* (2), 289–307.

Luft, J. A., & Hewson, P. W. (2014). Research on teacher professional development programs in science. In N. G. Lederman & S. K. Abell (Eds.), *Handbook of research on science education (volume II)* (pp. 889–909). New York: Routledge.

Ministry of Education. (2001). *National science curriculum standards of the compulsory education (grades. 7–9) (trial version)* (义务教育科学课程标准(实验)). Beijing: Beijing Normal University Press (in Chinese).

Ministry of Education. (2003a). *National biology curriculum standards for high school (trial version)* (普通高中生物课程标准(实验)). Beijing: People's Education Press (in Chinese).

Ministry of Education. (2003b). *National chemistry curriculum standards for high school (trial version)* (普通高中化学课程标准(实验)). Beijing: People's Education Press (in Chinese).

Ministry of Education. (2003c). *National physics curriculum standards for high school (trial version)* (普通高中物理课程标准(实验)). Beijing: People's Education Press (in Chinese).

Ministry of Education. (2011). *National biology curriculum standards of the compulsory education* (义务教育生物学课程标准). Beijing: Beijing Normal University Press (in Chinese).

Ministry of Education. (2015). *Educational statistics in 2014*. Retrieved from http://www.moe.gov.cn/s78/A03/moe_560/jytjsj_2014/2014_qg/

National Research Council. (1996). *National science education standards*. Washington, DC: The National Academies Press.

National Research Council. (2010). *The teacher development continuum in the united states and china: Summary of a workshop*. Washington, DC: The National Academies Press.

Ontario Principals' Council. (2008). Building professional learning communities. In Ontario Principals' Council (Ed.), *The principal as professional learning community leader* (pp. 5–16). Thousand Oaks: SAGE.

Shulman, L. S. (1986). Those who understand: Knowledge growth in teaching. *Educational Researcher, 15*(2), 4–14.

Shulman, L. S. (1987). Knowledge and teaching: Foundations of the new reform. *Harvard Educational Review, 57*(1), 1–22.

Standing Committee of National People's Congress. (1993, October). *Teachers law of the People's Republic of China* (中华人民共和国教师法). Retrieved from http://www.moe.gov.cn/publicfiles/business/htmlfiles/moe/moe_2803/200907/49852.html (in Chinese).

Van Driel, J. H., Berry, A., & Meirink, J. (2014). Research on science teacher knowledge. In N. G. Lederman & S. K. Abell (Eds.), *Handbook of research on science education (volume II)* (pp. 848–870). New York: Routledge.

Chapter 18
Enhancing Science Teacher Professional Development: Lessons from a Study of Misconceptions of Junior Secondary Biology Teachers

Enshan Liu and Mingyu Li

18.1 Introduction

Many research studies show that students have ideas about many scientific phenomena or scientific processes before they enter the science classroom. Most of these ideas come from children's daily lives, and many of them are incomplete ideas or misconceptions. It is not easy for students to give up these misconceptions (Abell and Lederman 2007; Vosniadou 2008). Misconceptions can be a substantial barrier for students in learning scientific concepts. If these ideas are not changed during the processes of school learning, they will remain in the student's mind for a long time.

Several theoretical frameworks of how students change their misconceptions about the natural world have been developed over the past three decades since 1982, when Posner and his colleagues proposed the conceptual change model (Posner et al. 1982). Within each of these frameworks, there are three essential perspectives of conceptual change learning related to epistemology, ontology, and affective/social/learner characteristics. Although these perspectives have different explanations about conceptual change, there is a consensus that conceptual change approach always means science teachers use different ways to communicate ideas or concepts with students by presenting these ideas or concepts either externally—taking the form of spoken language (verbal), written symbols (textual), pictures, physical objects, or a combination of these forms—or internally when thinking about them. Usually researchers who use conceptual change approaches in their classroom-based studies report that their approach is more effective than traditional

E. Liu (✉) • M. Li
The College of Life Sciences, Beijing Normal University, 19 Xin jie kou wai Street, Beijing 100875, China
e-mail: liues@bnu.edu.cn

L.L. Liang et al. (eds.), *Chinese Science Education in the 21st Century: Policy, Practice, and Research*, Contemporary Trends and Issues in Science Education 45, DOI 10.1007/978-94-017-9864-8_18

ones dominated by the transmissive view of teaching and learning (Tobin et al. 2012).

However, the research evidence shows that even though students receive science education in school, they might also retain their misconceptions, and the impact of conceptual change instructional practices in real classroom situations tends to be associated with various teacher factors. One of the most important factors is whether teachers hold scientific ideas or misconceptions that are similar to those of students (e.g., Fulmer 2013; Sadler et al. 2013). It is irrational to expect teachers who hold the same misconception as students to present scientific idea or concepts externally or internally in a proper way. Therefore, research on both students' and teachers' misconceptions is important and a prerequisite to improving the efficacy of conceptual change instructional practices in school education.

18.2 Research on Misconceptions and Teachers' Impact on Students' Misconceptions

Various terms have been used for students' misconceptions, such as misconceptions, preconceptions, and alternative conceptions. Driver and Easley (1978) believe that different terms reflect the viewpoint of the scholars. The term "misconception" is used in this chapter because the current study focuses on the students' ideas about science concepts after classroom learning.

To address students' misconceptions with conceptual change strategies, science teachers must have related knowledge and teaching skills. Helping students reconstruct their scientific concepts is not simply a task of using a concept-changing checklist. Teachers' content knowledge and understanding of students' misconceptions are the key elements in implementing conceptual change in the classroom (Diakidoy and Iordanou 2003; Gomez-Zweip 2008). Teachers' impacts on students' misconceptions are classified into the following four categories:

1. The teacher holds misconceptions. The teacher cannot be aware of students' misconceptions and unconsciously delivers misconceptions to students in the teaching process. Research shows that science teachers hold some misconceptions. Furthermore, there are similarities in teachers' and students' misconceptions (Burgoon et al. 2009; Kruger 1990; Yip 1998).
2. The teacher does not hold misconceptions but also does not fully understand the negative impact of misconceptions on teaching and learning. Because the teacher does not care about the students' misconceptions, the chance to reconstruct the students' ideas decreases during the classroom teaching. In this way, students may retain misconceptions after they complete the school education (Morrison and Lederman 2003).
3. The teacher realizes that the misconceptions have negative effects on students' learning. Due to limited knowledge and skills in diagnosing and changing

students' misconceptions, the teacher cannot effectively convert students' ideas to scientific conceptions. Related research studies show that teachers pay limited attention to students' misconceptions and know little about diagnosing children's ideas and concept-changing strategies (Gomez-Zweip 2008; Morrison and Lederman 2003; Li 2007).

4. If teachers do not hold misconceptions and students do not have misconceptions before they enter the science classroom, students may hold misconceptions after the learning process due to a teacher's deficits in pedagogy and teaching skills (Chi et al. 1994).

To date, most of the research studies have focused on diagnosing students' misconceptions, and much less research has examined teachers' misconceptions. One of the reasons may be that researchers had a hypothesis that most teachers are well prepared in content knowledge due to their pre-service education and have sound content knowledge before they start teaching (Yip 1998). However, some researchers had claimed that teachers' content knowledge is not sufficient (Li and Liu 2010).

Based on the literature review, the current research focuses on the following questions:

- After 30 years of related research studies, what is the current situation concerning teachers' misconceptions in the junior secondary biology teaching force?
- Is there a similarity in students' and teachers' misconceptions in mainland China, as the previous international research suggests?
- What type of relationship characterizes this similarity?

18.3 Methodology

18.3.1 Sampling

Samples were taken from a capital city of a northern province in mainland China. In general, the education level of this city is at the average position in China. There are 90 junior secondary schools located in this city. Based on the standardized entrance test scores, those schools can be classified into three levels: high-score, intermediate-score, and low-score levels. A total of 11 sample schools were randomly selected from each level. From each sample school, two or three biology teachers and their corresponding classes were selected as the sample. There are 40–50 students per class. Altogether, 30 biology teachers and 30 classes including 1442 students comprised the sample.

Table 18.1 Content knowledge detected by the questionnaire

Code	Aspects of the concept of photosynthesis and respiration
PR	Photosynthesis vs. respiration
FN	Factors necessary for respiration
MER	Matter transformation and energy transfer during respiration
MEP	Matter transformation and energy transfer during photosynthesis
IF	Factors impacting the rate of photosynthesis reaction
AT	Photosynthesis and autotrophy

18.3.2 Instrument

The instrument used in this survey is a questionnaire with two-tier test items about photosynthesis and respiration developed by Haslam and Treagust (1987). Translation and back translation of the two-tier items in questionnaire were done to make sure the Chinese version is accurate and understandable. The original reliability, using Cronbach's coefficient alpha, was 0.72 tested by Haslam and Treagust in 1987. The reliability based on the sample of this study is 0.84. There are 13 test items in the questionnaire, with 2 points for each item, resulting in a total of 26 points. Among the 13 items, the biology content knowledge refers to six aspects of the concept of photosynthesis and respiration based on the previous research (see Table 18.1) (Canal 1999; Haslam and Treagust 1987; NRC 1996; Waheed and Lucas 1992).

The questionnaire was first translated into Chinese. Then, six bilingual bio-educators and four biology teachers reviewed this questionnaire to check the translation's accuracy in both content and language. They also evaluated the instrument's content validity. Finally, the instrument was slightly modified based on the reviewers' suggestions for a better fit into the Chinese context.

One of the purposes of this study is to examine students' misconceptions after biology classroom teaching. The survey was administered after students learned about photosynthesis and respiration in their biology course. This arrangement fits Haslam and Treagust's (1987) recommendation for the use of the questionnaire. Following the survey, both the teachers' and students' questionnaires were collected for analysis.

18.4 Results

The data from the questionnaires were collected and analyzed using the statistics software SPSS. Based on their performance on the survey, teachers were divided into two groups, the misconception-free (TMF) group with 13 teachers, and the misconception-holding (TMH) group with 17 teachers. For detailed analysis of teacher's impact on student learning, numbers were used as codes for teachers based on their number of wrong answers on the questionnaire; namely, the teacher

Table 18.2 The rate of incorrect answers from teachers and their students

Item code	Measured content	Proportion of teachers' incorrect answers (%)	Proportion of students' incorrect answers (%)
1	Photosynthesis and respiration	13	53.6
2	Photosynthesis and respiration	30	79.9
3	Photosynthesis and respiration	20	48.7
4	Photosynthesis and respiration	16.6	39.9
5	Factors necessary for respiration	3.3	58.4
6	Matter transformation and energy transfer during respiration	6.6	41.6
7	Matter transformation and energy transfer during respiration	16.6	75.2
8	Factors necessary for respiration	10	51.3
9	Matter transformation and energy transfer during respiration	20	84.9
10	Matter transformation and energy transfer during photosynthesis	13.3	60.4
11	Factors impacting the rate of photosynthesis reaction	16.6	67.9
12	Photosynthesis and autotrophy	20	68.3
13	Photosynthesis and respiration	10	66.6

with the lowest score was given number 1 and the teacher with the second lowest score with number 2. Teachers coded with numbers 1–17 belonged to the TMH group, whereas those with numbers 18–30 belonged to the TMF group. The students were divided into two groups as well. Of the classes, the 13 taught by TMF teachers were grouped into the STMF group and the remaining 17 classes were grouped into the STMH group.

The data show that after students learned photosynthesis and respiration in the biology classroom, they retained misconceptions. Approximately 94.3 % of students had one or more misconceptions. With a total of 26 points for test items in the questionnaire, the students' mean score is 14.1, resulting in 54 % accuracy (14.1/26 = 0.54). Approximately 56.7 % of teachers held one or more misconceptions. Teachers' mean score is 22 points (of 26 total points) resulting in 85 % accuracy (22/26 = 0.85). Table 18.2 shows the percentage of teachers and students holding misconceptions for each test item.

As shown in Fig. 18.1, the distribution patterns of incorrect answers for most of the test items are similar for teachers and students, except items 5 and 13. As the questionnaire is a diagnostic instrument for misconceptions on photosynthesis and respiration, an incorrect answer demonstrates that the teacher or student holds a poor understanding or misconception. The result shows that for the items for which teachers hold a misconception, their students would have a greater chance of holding a misconception. This finding is consistent with the previous research studies that reflect the teacher's impact on student learning as the first category mentioned above. Items 5 and 13 seem to fit the second, third, or fourth categories

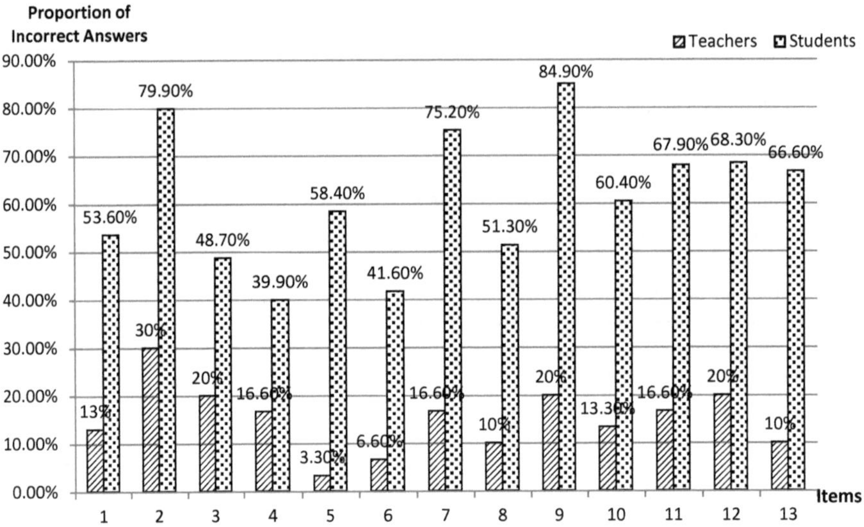

Fig. 18.1 Comparison of the rate of incorrect answers on each item among teachers and their students

Table 18.3 Comparison of scores between the two student groups taught by misconception-holding and misconception-free teachers

	Num.	Mean	Std. deviation	Std. error mean
STMF	13	12.33	6.18	1.71
STMH	17	8.22	4.36	1.06

of teachers' impact on students' misconceptions described above. To study the similarity in the misconceptions of teachers and their students, statistical analysis using SPSS was conducted on the data of 11 items, excluding test items 5 and 13. The results show that there is a significant positive correlation between teachers' misconceptions and those of students ($p = 0.035$; Pearson's $r = 0.64$).

As previously mentioned, 30 teachers were divided into the group of TMF or TMH, and their students were divided into two groups, STMF and STMH. Two-Sample t-Test was used to analyze the data of the two groups of students' performance. The result shows a significant difference in misconceptions between the STMF group and STMH group ($p = .042$; Cohen's $d = 0.77$). Table 18.3 shows the average scores and standard deviations of the two student groups.

A comparison of the ranks of a class's average score between the TMF and TMH groups also demonstrates the teacher's impact on student misconception. In Fig. 18.2, all of the 30 teachers' codes are displayed on the horizontal axis. The teachers' codes in this figure are consistent with the codes mentioned above, i.e., teachers in the TMH group are labeled from 1 to 17 (the teacher who received the lowest score is coded as 1 and the teacher who received the lowest number of wrong

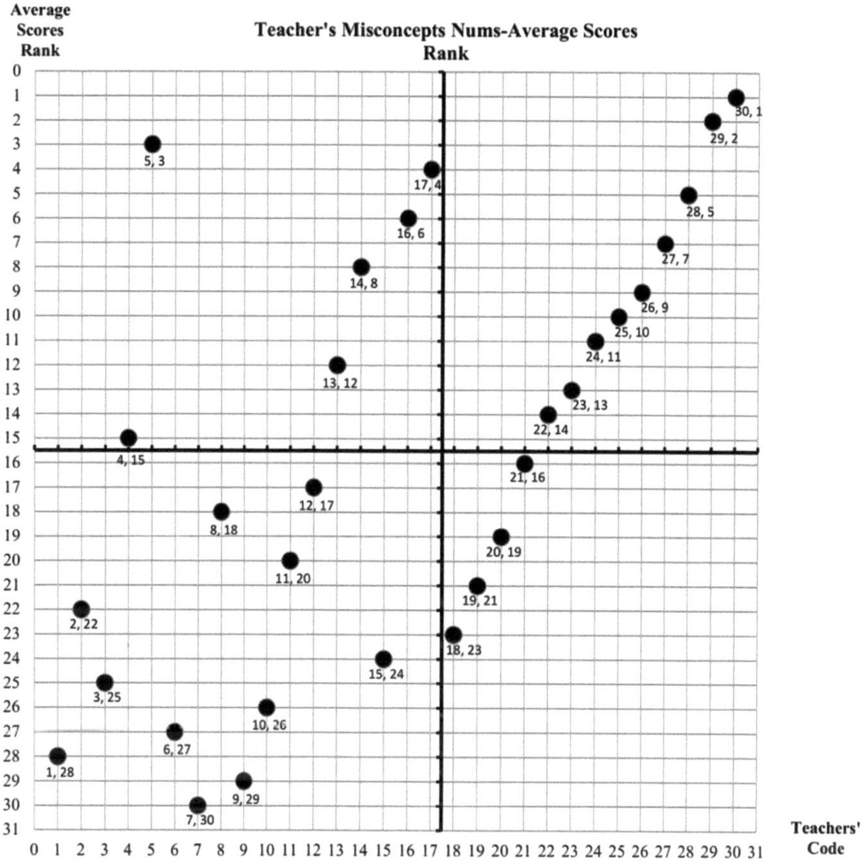

Fig. 18.2 The relationship between students' average score ranking and the number of misconceptions that teachers hold

answers is coded as 17), and the teachers in TMF group are labeled from 18 to 30 (none had a misconception diagnosed by this questionnaire). Therefore, in Fig. 18.2, spots that represent the TMH group fall on the left-hand side and spots that represent the TMF group fall on the right-hand side. The vertical axis shows the rank of the class's average score. For example, the spot "29,2" means that the class is taught by the teacher coded as 29, who is misconception free and the class's test score is at the second top position. The majority (69 %) of the STMF classes rank in the top 15. Furthermore, most of the STMH classes rank below the top 15. The data show that students taught by teachers without misconceptions displayed better performance than those taught by teachers who hold misconceptions. This could be a strong indicator of the impact of teachers' misconception on students' biology concept learning.

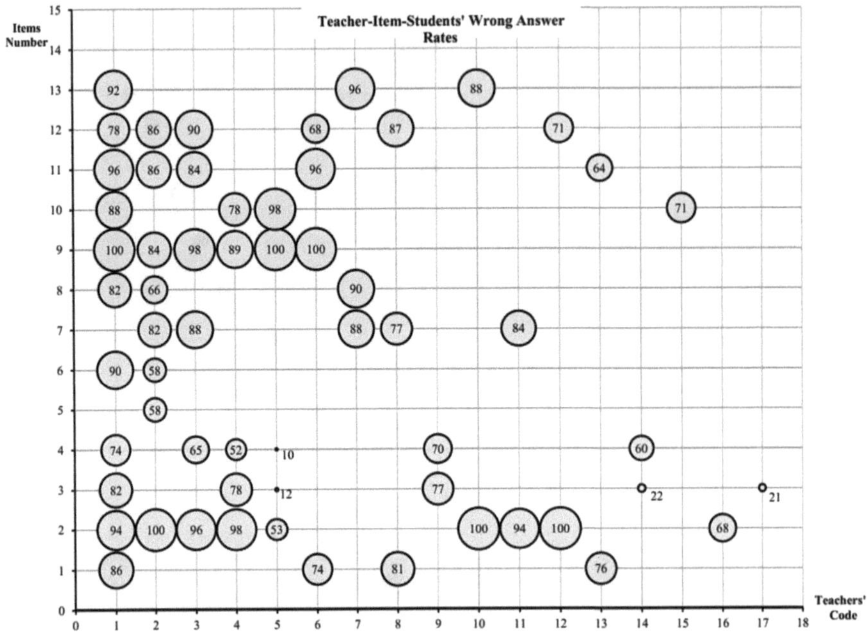

Fig. 18.3 The distribution and percentage of students' incorrect answers on the items for which their teachers held misconceptions

To analyze the relationship between teachers' misconceptions and those of their students in detail, the study also focused on individual teachers of the TMH group and their students. For the teachers with misconceptions, we matched teachers' performance on particular test items with those of their students, showing the relationship between them in a bubble diagram (see Fig. 18.3).

In Fig. 18.3, the horizontal axis refers to the teachers' codes and the vertical axis marks the test item number. Each bubble represents a teacher's incorrect answer (misconception) on a test item. The area of each bubble, i.e., the number inside each bubble, represents the percentage of students' incorrect answers on the corresponding item. Figure 18.3 shows that when teachers hold a misconception on a certain item, a high percentage of their students provides incorrect answers on the same item. For items 2, 7, 9, and 13, approximately 80 % of students answered incorrectly when their teacher held a misconception about the same item. For items 2, 3, 4, 9, and 12, there are at least six bubbles, indicating that these items are challenging for both teachers and their students. Among them, test items 2, 3, and 4 refer to the concept of "relationship between photosynthesis and respiration," item 9 refers to the concept of "matter transformation and energy transfer during respiration," and item 12 refers to the concept of "photosynthesis and autotrophy." Thus, the diagram details the type of similarities in teachers' misconceptions and those of their students. It also indicates the pitfalls in misconceptions about photosynthesis and autotrophy for both teachers and learners.

18.5 Conclusion and Discussion

18.5.1 Conclusion

Based on the survey, several conclusions can be made. First, after students learned photosynthesis and respiration in the biology classroom, many of them retained misconceptions. Approximately 94.3 % of students have one or more misconceptions. Some of their biology teachers also hold misconceptions on photosynthesis and respiration.

Second, there is a similarity in the teachers' misconceptions and those of their students. This result is consistent with the previous research in this area (Burgoon et al. 2009; Kruger 1990; Diakidoy and Iordanou 2003; Gomez-Zweip 2008; Yip 1998; Nancy et al. 2005). This finding does not indicate that students with teachers who have few misconceptions will hold few misconceptions. On the contrary, the data from test items 5 and 13 show that a high proportion of students maintain misconceptions, while more than 90 % of teachers hold scientific ideas. This finding is consistent with the previous research that the poor teaching strategies and skills of the teacher may result in student misconceptions (Morrison and Lederman 2003; Meyer 2004; Li 2007; Knuth et al. 2005). This indicates that students' conceptual understanding might be influenced by their teachers' general pedagogical knowledge or conceptual teaching knowledge and skills (i.e., the second, third, and fourth category), in addition to teachers' content knowledge or scientific understandings (i.e., the first category). The high correlation between the students' and their teachers' misconceptions, based on 11 of the 13 items, might suggest that teachers' misconceptions are a stronger source of students' misunderstandings of science in Chinese schools.

Third, further analysis of the similarity of teachers' misconceptions and those of their students reveals that: (a) There is a significant positive correlation between the teachers' misconceptions and those of their students for most of the test items ($p = 0.035$) except items 5 and 13, (b) The scores of students with misconception-free teachers are significantly higher than those of the students taught by a teacher with misconceptions ($p = 0.042$), and (c) The classes taught by the teachers with sound biology knowledge generally show better performance than those taught by teachers with misconceptions.

Based on detailed analyses of teachers' incorrect answers on a test item and their students' performance on the same item, it is clear that when teachers incorrectly answered an item, their students had a high chance of answering incorrectly. This provides a picture of the type of similarity in the misconceptions of teachers and those of their students. It also indicates that three aspects of photosynthesis and respiration are the most challenging for both biology learners and teachers, namely, the concepts of "photosynthesis and respiration," "matter transformation and energy transfer during respiration," and "photosynthesis and autotrophy."

18.5.2 Discussion

The survey results show that both biology teachers and their students in mainland China have misconceptions. Teachers' misconceptions might have a negative impact on students' learning of science concepts. The results also provide a detailed picture of the similarity in teachers' misconceptions and those of their students. This conclusion is consistent with previous research that found that teachers might be a factor in student misconceptions (Li and Liu 2010; Sanders 1993). After 30 years of research on misconception and conceptual change, the current research serves as a reminder that misconceptions held by science teachers remain a problem in junior secondary biology classrooms. More importantly, the problem found by this study is only representative for large cities such as those in the sampling area; it may be even worse in schools in rural areas.

Previous research studies revealed that students thought about science phenomena and held preconceptions before they entered the science classroom. The basic tenet of conceptual change instruction is that students' misconceptions are resistant to change, and explicit instruction that confronts student misconceptions is likely to be more successful. "Conceptual change approaches" are generally acknowledged as being more successful in classrooms (Duit and Treagust 2003). With these constructivist-based approaches, teachers need to explicitly connect with students' misconceptions, providing learning experience for students to actively explore and analyze evidence conflicting with their existing misconceptions. As advocated by the Chinese National High School Biology Curriculum Standards, inquiry-based teaching is seen as a promising strategy for teaching biology toward conceptual change. In the inquiry-oriented classroom, teachers are expected to engage students and directly challenge their misconceptions to promote conceptual change.

In future studies, we need to examine classroom practices have led to more successful student learning outcomes. One limitation of this research is that we did not distinguish the students who had misconceptions before they entered the biology class from those who developed misconceptions during the classroom teaching. That is to say, did some of the students develop misconceptions directly from their teacher during the classroom learning? If so, how many of the students developed misconceptions in this way? The current research provides new ideas and questions for further study. The current research evidenced another interesting phenomenon. In Fig. 18.3, there are some abnormal points, including the point labeled 10 for item 4 and points labeled 12, 22, and 21 for item 3, compared with other points. These points indicate that only approximately 10–20 % of students held misconceptions that their teacher also held. Therefore, 80–90 % of students had the correct scientific understanding. This result might be due to conceptual learning opportunities beyond the teachers' classroom teaching or other reasons that should be investigated.

The assumption that teachers obtained "perfect" content knowledge in pre-service education has been long-standing. The current survey demonstrated that this assumption is not accurate! The findings of this study provide some insights into teacher education, both pre-service and in-service. In the past decades, many efforts were made to enhance pre-service teachers' content knowledge by reviewing the latest development in biology and increasing lab experience (Chen et al. 2012). However, there is almost no emphasis on conceptual change in teacher training programs. Diagnosing student misconceptions and helping them to construct scientific concepts is not only the task of science teachers in general education but also the mission of the lecturers and professors involved in science teacher training programs. To rebuild the teacher training program with an emphasis on trainees' conceptual change, research on the following questions should be conducted to support the reform: What are the science teachers' perceptions, knowledge, and skills related to student misconception and conceptual change? What are the teacher candidates' perceptions and knowledge about misconception and conceptual change? To what extent are conceptual change theory and skills embedded in science teacher training programs? With the shift from the "top-down" decision-making model toward an "evidence-driven" decision-making model in curriculum development, such research will provide meaningful information for the design of preservice and in-service teacher training programs in China.

Acknowledgments We thank Dr. Cheng Liu for his ideas and help with data processing and revising the draft manuscript and Ph.D. candidate Wenyuan Yang for help with data processing.

References

Abell, S. K., & Lederman, N. G. (Eds.). (2007). *Handbook of research on science education*. Mahwah: Erlbaum.

Burgoon, J., Heddle, M., & Duran, E. (2009). Re-examining the similarities between teacher and student conceptions about physical science. *Journal of Science Teacher Education, 21*, 859–872.

Canal, P. (1999). Photosynthesis and "inverse respiration" in plants: An inevitable misconception? *International Journal of Science Education, 21*, 363–371.

Chen, J., Peng, A., & Xiang, B. (2012). Discussion on the construction of the practice teaching system and platform in biology teacher education specialty (生物学教师教育专业实践教学体系与平台建设). *China Modern Education Equipment* (中国现代教育装备)*, 3*, 70–74 (In Chinese).

Chi, M. T. H., Slotta, J. D., & De Leeuw, J. (1994). From things to processes: A theory of conceptual change for learning science concepts. *Learning and Instruction, 4*(1), 27–43.

Diakidoy, I. A., & Iordanou, K. (2003). Preservice teachers and teachers' conceptions of energy and their ability to predict pupils' level of understanding. *European Journal of Psychology of Education, 23*, 357–368.

Driver, R., & Easley, J. (1978). Pupils and paradigms: A review of literature related to concept development in adolescent science students. *Studies in Science Education, 5*, 61–84.

Duit, R., & Treagust, D. P. (2003). Conceptual change: A powerful framework for improving science teaching and learning. *International Journal of Science Education, 25*(6), 671–688.

Fulmer, G. W. (2013). Constraints on conceptual change: How elementary teachers' attitudes and understanding of conceptual change relate to changes in students' conceptions. *Journal of Science Teacher Education, 24*, 1219–1236.

Gomez-Zweip, S. (2008). Elementary teachers' understanding of students' science misconceptions: Implications for practice and teacher education. *Journal of Science Teacher Education, 19*, 437–454.

Haslam, F., & Treagust, D. T. (1987). Diagnosing secondary students' misconceptions of photosynthesis and respiration in plants using a two-tier multiple choice instrument. *Journal of Biological Education, 21*, 203–211.

Knuth, E. J., Alibali, M. W., McNeil, N. M., Weinberg, A., & Stephens, A. C. (2005). Middle-school students' understanding of core algebraic concepts: Equality and variable. *Zentralblatt fur Didaktikder Mathematik. International Reviews on Mathematical Education, 37*, 1–9.

Kruger, C. (1990). Some primary teachers' idea about energy. *Physics Education, 25*, 86–91.

Li, G. (2007). A survey on junior secondary physics teachers' perception and practice in classroom (初中物理教师针对前概念的教学现状调研). *Journal of Qiqihar Teachers' College, 2*, 120–121 (In Chinese).

Li, G., & Liu, E. (2010). Research on the origin of junior secondary students' preconceptions (初中生生物学前科学概念来源的研究). *Bulletin of Biology, 45*(2), 37–39 (In Chinese).

Meyer, H. (2004). Novice and expert teachers' conceptions of learners' prior knowledge. *Science Education, 6*, 970–983.

Morrison, J. A., & Lederman, N. G. (2003). Science teachers' diagnosis and understanding of students' preconceptions. *Science Education, 87*(6), 849–867.

Nancy, J. P., et al. (2005). Prevalence of blood circulation misconceptions among prospective elementary teachers. *Advances in Physiology Education, 29*(3), 172–181.

National Research Council. (1996). *National science education standards*. Washington, DC: National Academy Press.

Posner, G., Strike, K., Hewson, P., & Gertzog, W. (1982). Accommodation of a scientific conception: Toward a theory of conceptual change. *Science Education, 66*, 211–227.

Sadler, P. M., Sonnert, G., Coyle, H. P., Cook-Smith, N., & Miller, J. L. (2013). The influence of teachers' knowledge on student learning in middle school physical science classrooms. *American Educational Research Journal, 50*, 1020–1049.

Sanders, M. (1993). Erroneous ideas about respiration: The teacher factor. *Journal of Reserch in Science Teaching, 30*(8), 919–934.

Tobin, K. G., Fraser, B. J., & McRobbie, C. J. (Eds.). (2012). *Second international handbook of science education*. Dordrecht: Springer.

Vosniadou, S. (Ed.). (2008). *International handbook of research on conceptual change*. New York: Routledge.

Waheed, T., & Lucas, A. (1992). Understanding interrelated topics: Photosynthesis at age 14+. *Journal of Biological Education, 26*, 193–199.

Yip, D. (1998). Identification of misconceptions in novice biology teachers and remedial strategies for improving biology learning. *International Journal of Science Education, 20*, 461–477.

Chapter 19
Video Case Instruction: A New Approach to Instructional Design and Practice for Preservice Chemistry Teachers

Zhen Lu and Lyna Kwan

19.1 Introduction

Since the beginning of the twenty-first century, China has instituted comprehensive curriculum reform efforts to improve students' scientific literacy. New curriculum standards and teaching materials including textbooks, teacher training, and teaching technology have been introduced and implemented in junior and high schools. Consequently, preservice teacher education also needs to be reformed. Fully preparing preservice teachers for the many requirements of their position is difficult to achieve (Lu and Cheng 2005). In order to cultivate in preservice chemistry teachers the new instructional design ideology and practical teaching capability, some new and effective teaching methods must be explored.

To address this need, we have spent the last 3 years designing a program called *video case instruction* for theoretical and practical study and implementation. We conducted an experiment using video case instruction with 29 senior student teachers majoring in chemistry education at Nanjing Normal University from October 2007 through January 2008. In this chapter, we present some background on video case instruction, describe our use of video cases for preservice chemistry teachers, and then present results on the preservice teachers' perception of the video case instruction.

Z. Lu (✉)
Second Affiliated High School, School of Teacher Education, Nanjing Normal University, Jiangsu 210097, China
e-mail: zhlu@njnu.edu.cn

L. Kwan
Office of Education Research, National Institute of Education, 1 Nanyang Walk, 637616 Singapore, Singapore
e-mail: lyna@nie.edu.sg

L.L. Liang et al. (eds.), *Chinese Science Education in the 21st Century: Policy, Practice, and Research*, Contemporary Trends and Issues in Science Education 45, DOI 10.1007/978-94-017-9864-8_19

19.2 Review of Literature on Video Case Instruction

Video case instruction as a method of teacher education and professional development has been the subject of significant previous research. Videos of exemplar instruction provide opportunities for teachers to recognize classroom interaction patterns in what van Es and Sherin (2002) call *learning to notice*. This process of learning to notice can have significant impact on preservice teachers' ability to recognize and interpret observations of real classrooms (Santagata and Angelici 2010). Seidel and Prenzel (2005) made tremendous progress in the field of video instruction. In a study of German PISA, they compared research of video instruction cases from different countries. They explored the standard, the features, and the forms of the outstanding instruction cases. Furthermore, they developed analytical software called Videograph, providing the functionality to analyze aspects of the video, such as the characteristics of the teaching process and the modes of tasks in class. These proposed models of expert teachers' teaching experience have boosted German science and mathematics education reform. In the same vein, Wong et al. (2006) conducted a study in Hong Kong through the use of classroom videos of exemplary science teaching in order to facilitate student teachers to analyze, understand, reflect, and express their views on good science teaching in some detail. They found that the use of exemplary cases had a positive impact on their preservice teachers, and authenticity has been enhanced by capturing video of real-world science classroom dilemmas and discussing them with the teachers.

19.3 The Development of Video Case Instruction in China

A teaching case is a teaching story, which reflects the problems and strategies a teacher goes through. It demonstrates how to solve real-world teaching problems and provides answers to "what" and "why" by describing an actual situation. The use of video case instruction has some unique advantages, as videos (1) allow access, before and after classroom instruction, to a variety of information and include many instructional perspectives; (2) have a stronger connection to research, highlight the topic, and focus on a combination of a variety of teachers' and researchers' (and even students') views; (3) allow for reanalysis of original material from classroom instruction; and (4) provide a rich variety of media materials where viewers can approach the case content by way of personal preferences and multiple angles.

In April 2004 the Chinese National Academy of Teacher Education launched an action research: the First Classroom Video Case Display and Rating in Primary Schools and Middle Schools. It started with five domains: mathematics in primary school, Chinese in primary school, Chinese in junior high school, mathematics in junior high school, and science in junior high school. In May 2005 the committee developed 555 good classroom instruction video cases and published a book:

Selected Works of Classroom Video Case Display and Rating in National Primary Schools and Middle Schools.

Furthermore, Chinese researchers Bao et al. (2005) published a book entitled *Focusing on Classroom—Researching and Producing Classroom Teaching Case Video*, which is an academic book systematically introducing video case research. It describes the fundamental system of video case research and the design and implementation of a video case. The video cases mainly consist of five parts: background, subject, teaching methods, classroom cognition level, and emotion and attitude.

With the growth of research in video case instruction, further studies based on video instruction have been conducted. For instance, a case-based instruction skill training system which infuses video technique and microteaching in instruction skill training has been developed by Zhu (2004). Microteaching is a classroom instruction training technique which is used in preservice teacher education, where the preservice teachers teach a small portion of a lesson to their classmates to foster the development of multiple instructional skills through the practices and immediate feedback from their classmates and mentor (Kilic 2010). To make sure that the preservice teachers are capable and competent to teach, microteaching provides rigorous training. Figure 19.1 illustrates a case-based instructional skill training system.

Because of the multiple dimensions of cases, the case-based instructional skill training system is useful for teachers. It also shows the teacher's comments about the case by experts, references to theory related to the case, and an evaluation of their overall performances. Case studies, which are recorded onto CDs, can play an important role in the system as excellent teaching cases can provide models for preservice teachers.

Video case instruction offers several instructional functions as follows:

– Video case instruction as a way of learning from experts

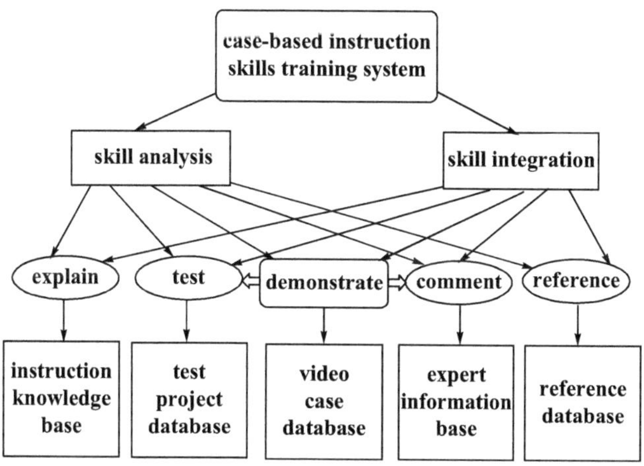

Fig. 19.1 Case-based instructional skill training system

Learning from expert teachers is an important part of professional development for preservice teachers. However, the knowledge of expert teachers tends to be highly situational and personal; hence, such knowledge would be difficult to show. By using video case instruction, the expert teachers have at least two ways to share their expertise: (1) through some "demonstrations" of the teaching of the subject and (2) through some "comments" on the case study as a viewer (and reviewer). Either way is useful for preservice teachers to learn from experts.

- Video case instruction as a tool for teacher professional development and training

The video case can accurately record a typical, real case for teachers' analysis, comparison, and professional development. Video case studies of teacher training create a platform for preservice teachers to express their understanding of the case, reflect on the theory applicable to the case, and transform it into their instructional practice. Such a learning process not only improves the quality of preservice teachers in terms of educational theory but also brings together theory and practice.

- Video case instruction as a tool to better adapt to the new curriculum

The development and implementation of new curriculum resources are big challenges in the new curriculum reformation process. A new curriculum requires significant adjustment since traditional teaching methods and educational philosophy often do not match the requirements of new reforms. Showing a real-world classroom situation for teacher discussion by using video case instruction allows teachers to better adapt to a new curriculum.

- Video case instruction as rich resources for teaching and research

Compared to text case and videotape, video case instruction has greater potential in the construction of related resources. Discussion and reflection around the case can improve the teaching skills of teachers and, at the same time, video case resources can also provide strong theoretical support for research.

- Video case instruction as a reflection of authentic classroom situations

Compared with text cases, video case instruction provides more authentic real-world classroom situations "without modification" and contains rich information about teaching and learning processes. Hence, it helps viewers to form their own perspectives from what they see on the video and link it to their existing knowledge. The inherent characteristics of video case instruction make it a more effective learning tool than text for preservice teachers. In short, in the process of developing preservice teachers' teaching skills, video case instruction is effective to reflect real-world classroom situations, instead of providing unrealistic scenarios to be discussed. Video case instruction allows preservice teachers to have deeper understanding and appreciation of the theories of teaching. The analysis of typical cases from real-world classroom situations enriches student teachers' understanding of teaching pedagogy and helps them master related teaching skills and strategies.

19.4 The Methodology of Video Case Instruction

Having been shown authentic, extensive, and complex teaching situations, student teachers can then discuss and reflect on these outstanding cases with a teacher in order to improve their teaching design and practices. The preservice teachers are expected to further understand the "subject-based pedagogy" and to better integrate the theory and practice, so that their teaching design and practice will be improved through authentic teaching and lesson analysis. We selected four examples of outstanding teaching videos and three examples of instructional design and used information technology and digital technologyto make standardized instruction case courseware. The four steps of using video case instruction are as follows:

19.4.1 Selecting Cases That Demonstrate Excellence

The selected teaching video cases require clear video information (voice, images, etc.) and all ideas, processes, and models have to be realistic and applicable. Cases should contain instructional design, students' learning processes, courseware, expert comments, and other materials. These can assist the learners in constructing their thoughts about teaching design and practice. The selected cases should emphasize core ideas of chemistry teaching theories, outstanding teaching models of the new curriculum and teaching materials, methods of inquiry and demonstration, as well as the issues and trends in chemistry education in order to broaden the student teachers' perspectives on the objective, comprehensive, and dialectical analysis from different aspects.

19.4.2 Learning from a Video Case

The teaching design philosophy, teaching design thinking, and teaching models, as well as the important and challenging points in teaching particular topics, are explained. The video case is shown and the process and models of teaching are described. Teachers of outstanding sample videos and experts are invited to the class to exchange and discuss ideas, to introduce their design concepts, and to share their experiences and critical thinking.

19.4.3 Discussion Among Teachers

The preservice teachers give their opinions about what is worth learning and what can be further improved in the case they have just watched. They mainly discuss "case issues." The core issues usually include problems of teaching, cognitive level, emotions, and attitude. During the discussion, they use some related references such as "experts' reviews" and "students' feedback." When they have different ideas during the process of discussion about one teaching link, they can play the segment back, review it carefully, link it to the theories learned, and criticize and improve these theories through case discussion.

19.4.4 Microteaching Training

We show the preservice teachers different kinds of practical teaching situations through the first four links of case teaching. By analyzing the situations, they are trained to solve problems that may occur during the instructional process. The next stage is microteaching training.

We adopt the following four steps: (1) microteaching design; (2) micro-training (video); (3) panel discussion, which involves replaying video, self-analysis, and discussion and evaluation; and (4) summing up, where the preservice teachers apply the knowledge they have learned. We select some typical micro-training videos as an example and play them for analysis and discussion. As a loop system, the close

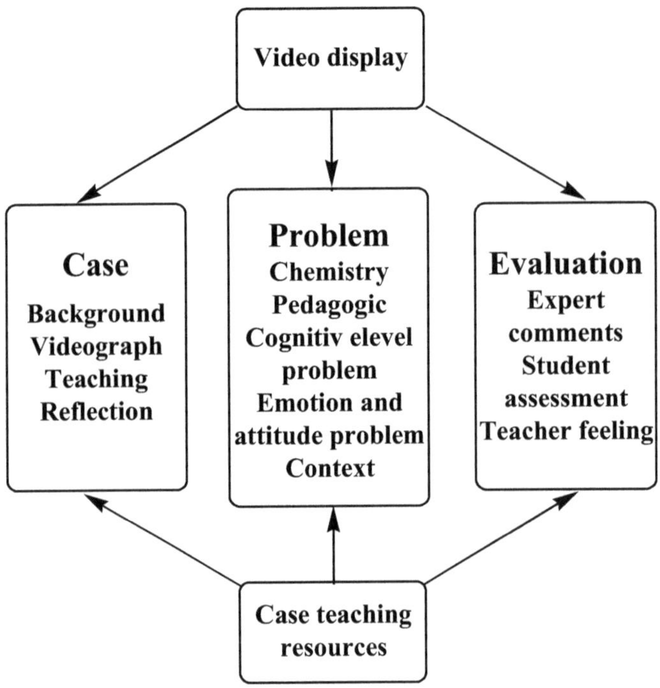

Fig. 19.2 Video case instruction chart

connection of case teaching and microteaching helps students resolve the problems they encounter during the instructional practice. Figure 19.2 presents the chart of video case instruction.

Through video case instruction, preservice teachers achieve deep understanding of excellent teaching cases which help them apply their personal experience and understanding to teaching design so that they can be independent in material analysis, method selection, and design of the teaching process. In this way, their rational design and ideas will be transferred into their teaching practice.

Below is an example of video case instruction on the topic "Electron Configuration of Atoms." The topic was chosen because the concepts are abstract and complex because the three different levels of representation—macroscopic, symbolic, and submicroscopic—are directly related to each other and are therefore difficult to teach and learn (Treagust et al. 2003).

19.5 The Use of "Electron Configuration of Atoms" Video Case Instruction

"Electron Configuration of Atoms" is a further study on the structure of atoms which requires comprehensive conceptual knowledge of the upper atomic structure. The students can have a deeper understanding of the structure of matter through a clear and accurate understanding of the relevant concepts of electron configuration. The video case shows the preservice teachers the process of design and practice in teaching and learning the electron configuration of atoms in chemistry class (Ma et al. 2014). We describe the use of the "electron configuration of atoms" video case instruction below.

19.5.1 Learning from the Case

Case teaching is divided into two parts based on the text content. The first part introduces the topic using the story of the scientists who explored the atomic structure and the history of atomic structure theory. Knowing the history of science helps the students realize that people's understanding of the nature of science is a gradual process of discovery, experimentation, and evolution toward the truth. The second part integrates the knowledge of the motion states of electrons and of atomic structure, enhancing knowledge of the motion of electrons through discussion. The three laws of electron configuration are used for elements 1–18. The teacher should trigger a cognitive conflict among students through the analysis of the experimental phenomenon and results so as to inspire students to think about the rules of sublevel movement patterns of the electron. After the process of lectures, discussions, and exploration, conclusions are drawn. In this way the instructional objectives of moving toward the more advanced "structure principle" are achieved.

Subsequently, the preservice teachers are required to watch the case video, focusing on the features of the textbook, teaching process, and structure of teaching model and teaching important and difficult points of the curriculum.

19.5.2 Discussion Among Teachers About the Case

After watching the case, the preservice teachers share and discuss their thoughts. They analyze the lesson's structure and come up with three key points:

1. Important and difficult points in teaching: understand the concepts of energy sphere and level and understand the characteristics of atomic structure from the perspective of energy.
2. Teaching model: heuristic-mastery mode.
3. Instructional objectives:

Knowledge (concepts to be constructed): (1) Understand the structure theory of atomic structure, know the distribution of atomic energy levels of the outer electrons, and explain the arrangement of the outer electrons of common elements (1–36). (2) Understand the lowest energy principle, know the ground state and excited states, and know that the electrons transition to atomic spectroscopy under certain conditions. (3) Understand the state of the outer electrons' motion, know the electron cloud and the atomic orbital, master the modern model of atomic structure, and understand the difference and links between macro- and microparticles.

Process (process skills used): (1) Through the history of atomic structure, students experience the whole processes of scientific methods and research. (2) By means of multimedia teaching, the teacher helps students understand abstract concepts and at the same time trains students' awareness and ability to use information technology.

Affective (science attitudes to nurture): (1) Let students feel the harmonious beauty of science by demonstrating the model of the atomic structure. (2) Develop students' interest in exploring the microscopic world.

19.5.3 Instructional Practice Using Video Case Instruction

The theory of atomic structure is difficult for students to understand because it is abstract. It is also difficult for knowledge transfer and direct experimental exploration because of its difference from the macroscopic motion. Instead of telling the students directly, the teacher should encourage student-centered learning and scaffold students' thinking by posing critical questions from the experiments to trigger cognitive conflicts, inspire the students to discuss and explore together, and enhance the students' critical thinking. This practice helps students build the spatial imagination ability, analysis, and microscopic abstract reasoning, which helps them construct a more advanced knowledge of atomic structure.

19.5.4 Instructional Design and Microteaching Training

After case teaching, the preservice teachers are assigned to select part of the subject "ionic bond" to carry out a teaching design by applying the heuristic-mastery mode. They then carry out microteaching training and watch and discuss their videos in order to improve and get familiar with the teaching mode and teaching skills.

19.6 Reform of Teachers' Pedagogical Content Knowledge

The concept of pedagogical content knowledge (PCK) was first introduced by Shulman (1986). It involves the transformation of teachers' subject knowledge and pedagogical knowledge into instruction and reinforces the interaction between content, teacher, and student (Shulman 1986). Shulman (1987, p. 8) stated that:

> ... pedagogical content knowledge is of special interest because it identifies the distinctive bodies of knowledge for teaching. It represents the blending of content and pedagogy into an understanding of how particular topics, problems, or issues are organized, represented, and adapted to the diverse interests and abilities of learners, and presented for instruction.

Accordingly, in order to leverage their professional development, teachers are required to keep their knowledge updated and reform their view of the four aspects: students, teaching, education aims, and teaching process. As case teaching involves integrated knowledge of many areas, teachers must have a wide range of knowledge in order to be competent for teaching. Using the case teaching method, teachers should also master classroom management and questioning techniques. They should give students enough time and space to freely express their opinions, work in collaboration with others, and encourage students to ask critical questions and solve scientific problems.

A change in the teaching role is the premise of organizing and implementing the case teaching methodology, which requires the teacher to work together with the students in creating a classroom atmosphere conducive to learning. Case teaching changes the teacher's role completely from the single traditional role as a "preacher" to a multidimensional role: from dominator to organizer, facilitator, and mentor and from knowledge provider to students' learning partner. Teachers' professional skills have also changed from the knowledge providers to curriculum designers and developers. They should inspire and develop students' creative thinking. Teaching methods must be changed from the teacher-centered approach to student-centered and inquiry-based approach, which empowers students to take ownership of their learning and become active learners.

19.7 Preservice Teachers' Perception of Video Case Instruction

After teaching practice, we administereda questionnaire to the 29 preservice teachers to elicit their perception of video case instruction. The findings are presented in Figs. 19.3, 19.4, and 19.5.

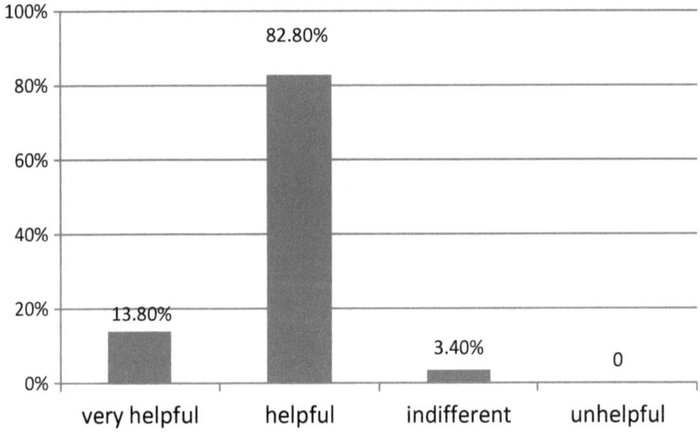

Fig. 19.3 Preservice teachers' perception of the helpfulness of video case instruction

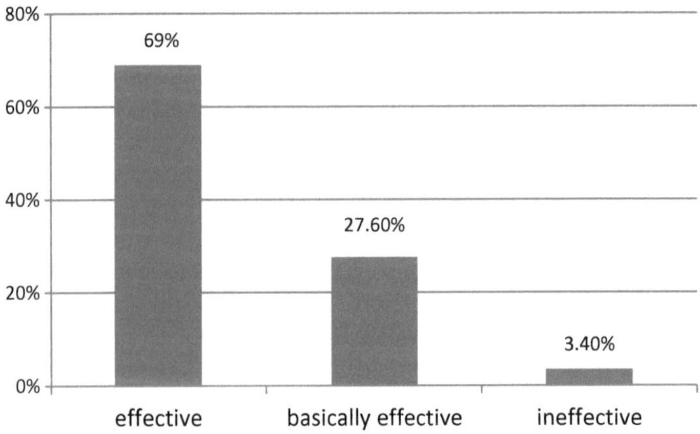

Fig. 19.4 Preservice teachers' perception of the effectiveness of case teaching on teaching design

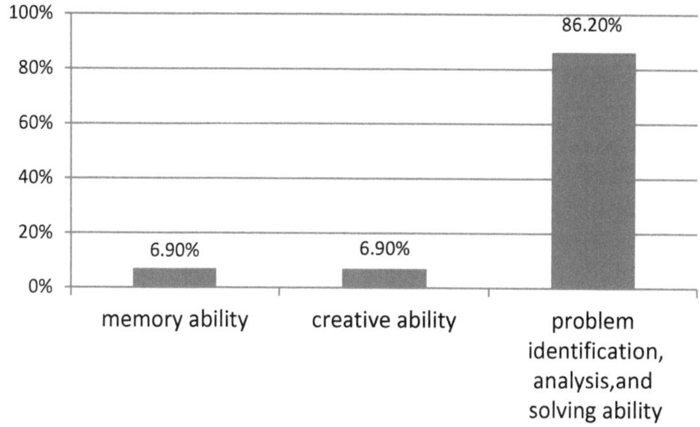

Fig. 19.5 Preservice teachers' perception of the usefulness of video case practice

As seen in Figs. 19.3, 19.4, and 19.5, 13.8 % of the preservice teachers perceived that case teaching was very helpful, 82.8 % perceived that it was helpful, 69 % perceived that the case teaching was effective in enhancing their competence in teaching design, and 86.2 % perceived that video case practice has improved their competence in problem-solving skills.

19.8 Conclusion

There are growing expectations that students should leave school with high thinking, learning, and communication skills to solve complex problems, respond to scientific issues, and expand capacity and productivity. To meet these expectations, schools must be transformed in ways that will enable students to acquire the sophisticated thinking, problem-solving, reasoning, collaboration, and communication skills they will need to be successful in work and life (Binkley et al. 2012). Consequently, reform in curriculum and instructional practice is of high importance. Reforming such instruction requires investment in tools and materials that afford opportunities for teachers, whether preservice or in service, to learn about reformed teaching approaches and consider how to adopt them in class. Video case instruction is one such approach that can allow teachers to observe master teachers, learn from comments from experts, and see connections between theory and practice (e.g., Bao et al. 2005; van Es and Sherin 2002). By using video case instruction, teachers are expected to develop their skills in managing instructional activities by understanding the unique interplay between subject knowledge and pedagogy.

This study aims to improve the quality of curriculum materials and instructional practices through the use of video case instruction. After 3 months of case teaching training, the majority of the preservice chemistry teachers perceived that case teaching was effective in enhancing their competence in instructional design, pedagogy, and conceptual knowledge. This study offers insight into the usefulness of video case instruction in preservice chemistry teacher training, which can serve as one way to accomplish the necessary transformation and reformation of science education in China. Findings indicate that the use of video case instruction for preservice chemistry teachers shows some of the common benefits that have been found in previous work in other countries and education systems, such as Germany, Italy, the United States, and Hong Kong (Santagata and Angelici 2010; Seidel and Prenzel 2005; van Es and Sherin 2002; Wong et al. 2006). Therefore, the present results not only support this work in mainland China, the findings point to a promising commonality in teacher education between China and other education systems.

References

Bao, J. S., Gu, L. Y., & Wang, H. (2005). *Focusing on classroom—Researching and producing classroom teaching case video* (课堂教学视频案例的研究与制作) (pp. 180–190). Shanghai: Shanghai Education Press (in Chinese).

Binkley, M., Erstad, O., Herman, J., Raizen, S., Ripley, M., Miller-Ricci, M., & Rumble, M. (2012). Defining twenty-first century skills. In P. Griffin, B. McGaw, & E. Care (Eds.), *Assessment and teaching of 21st century skills* (pp. 17–66). New York: Springer.

Kilic, A. (2010). Learner-centered micro teaching in teacher education. *International Journal of Instruction, 3*(1), 77–100.

Lu, Z., & Cheng, P. (2005). The design and fabrication of chemistry case teaching resources database under new curriculum reform (新课程改革背景下的化学案例教学资源库的设计与制作). *Chemical Education, 25*(6), 37–38 (in Chinese).

Ma, L. P., Wu, D. Q., & Lu, Z. (2014). Preliminary discussing on teaching structure model of high school chemical theory based on the video cases studied. (化学理论课堂教学结构模型构建初探—基于课堂教学视频案例分析的研究). *Chemical Education, 35*(7), 35–38 (in Chinese).

Santagata, R., & Angelici, G. (2010). Studying the impact of the lesson analysis framework on pre-service teachers' abilities to reflect on videos of classroom teaching. *Journal of Teacher Education, 61*(4), 339–349.

Seidel, T., & Prenzel, M. (2005). *How to run a video study* (Technical report of the IPN video study). Kiel: Waxmann Verlag GmbH.

Shulman, L. S. (1986). Those who understand: Knowledge growth in teaching. *Educational Researcher, 15*, 4–14.

Shulman, L. S. (1987). Knowledge and teaching: Foundations of the new reform. *Harvard Educational Review, 57*, 1–21.

Treagust, D., Chittleborough, G., & Mamiala, T. (2003). The role of submicroscopic and symbolic representations in chemical explanations. *International Journal of Science Education, 25*(11), 1353–1368.

Van Es, E. A., & Sherin, M. G. (2002). Learning to notice: Scaffolding new teachers' interpretations of classroom interactions. *Journal of Technology and Teacher Education, 10*(4), 571–596.

Wong, S. L., Yung, B. H. W., Cheng, M. W., Lam, K. L., & Hodson, D. (2006). Setting the stage for developing pre-service teachers' conceptions of good science teaching: The role of classroom videos. *International Journal of Science Education, 28*(1), 1–24.

Zhu, Z. T. (2004). On the functioning modes of information technology in classroom instruction: Theoretical framework and case studies (信息技术在课堂教学中的基本模式:理论框架和案例研究) [EB/OL]. *Journal of Educational Technology & Communication*. Retrieved from http://www.etc.edu.cn/articledigest6/information.htm (in Chinese).

ERRATUM

Part VI
Science Teacher Education

Ling L. Liang

L.L. Liang et al. (eds.), *Chinese Science Education in the 21st Century: Policy, Practice, and Research*, Contemporary Trends and Issues in Science Education 45, DOI 10.1007/978-94-017-9864-8

DOI 10.1007/978-94-017-9864-8_20

In Table VI.1, the headings of the second and third columns were incorrectly interchanged in Part VI, page 374. The correct version as per Table VI.1 is left column being a China example and the right column being a US example.

The online version of the updated original Part can be found at
http://dx.doi.org/10.1007/978-94-017-9864-8

L.L. Liang et al. (eds.), *Chinese Science Education in the 21st Century: Policy, Practice, and Research*, Contemporary Trends and Issues in Science Education 45, DOI 10.1007/978-94-017-9864-8_20

Table VI.1 A comparison of sample preservice biology teacher training programs (4 years): China vs. United States

Category of course			Credits required (an example: China) On average, 1cr. = 16 class hr.	Credits required (an example: United States)
General courses, such as history and foreign language			44	27
Science content courses	Courses in related subject area, such as chemistry and physics		33.5	24
	Biology courses	Basic courses in biology	33.5	13
		Specialized or advanced courses in biology	25	12
Teacher education courses	*Required courses related to teaching		16	31
	*Education research courses		4	0
	Teaching practice/student Teaching		11	12
	Field experience		Incorporated into teaching practice (1)	4
	Professional integrity		2	1
*Thesis			6	0
Total			175	124

About the Editors

Ling L. Liang is an Associate Professor of Science Education at La Salle University in Philadelphia, PA, USA. Dr. Liang holds a PhD in Science Education from Indiana University at Bloomington, USA. She received both undergraduate and graduate training in physics in China and won a first-class teaching award while she was teaching physics in a Chinese secondary school (7–12) between 1984 and 1989. Since 1996, she has been engaged in teaching and research on implementation of inquiry-centered approaches involving pre-service and/or in-service science teachers in the United States. As a guest professor appointed by Nanjing Normal University, Dr. Liang has ongoing collaborations with science educators and researchers in China. She has also co-edited a special issue on Science Education in China for the International Journal of Science Education and a textbook on elementary science education adopted in China.

Xiufeng Liu is a Professor of Science Education, State University of New York at Buffalo. He received his teaching diploma in secondary school chemistry from Wuhan Teachers College at Xiaogan (now Hubei Engineering College), Master of Education in Chemical Education from East China Normal University, and PhD in Science Education from University of British Columbia. He was a high school chemistry teacher in China and a research associate at the China National Institute for Education Research before pursuing his doctoral degree in Canada. Dr. Liu maintains strong personal connections with science education researchers in China by regularly giving talks and workshops in China on a variety of topics in science education. He has been appointed as a guest professor by a number of Chinese universities and is on the editorial board of the Journal of Chemical Education of the Chinese Chemical Society and Science Popularization of the Chinese Research Institute for Science Popularization.

Gavin W. Fulmer is an Assistant Professor in the Department of Teaching and Learning at the University of Iowa, and formerly at the National Institute of

L.L. Liang et al. (eds.), *Chinese Science Education in the 21st Century: Policy, Practice, and Research*, Contemporary Trends and Issues in Science Education 45, DOI 10.1007/978-94-017-9864-8

Education (NIE), Singapore. Prior to joining NIE, Dr. Fulmer was an Associate Program Director in the Division of Research on Learning in Formal and Informal Settings of the US National Science Foundation (NSF), and a Research Analyst with Westat, an employee-owned contract research (Maryland, USA). He holds a BS in mathematics and physics from the University of Redlands (California, USA) and a PhD in science education from the State University of New York at Buffalo (New York, USA). His research interests are in the assessment of students' science knowledge and teachers' classroom instruction, and in the study of alignment among assessments, educational standards, curriculum, and instruction. His interest in science education in China is driven by his experience living and working in Asia, including previous and ongoing research collaborations with colleagues in mainland China.

About the Section Editors

Lingbiao Gao is a Professor of South China Normal University and the Deputy Head of the Centre of Curriculum Studies in Basic Education of the Ministry of Education/South China Normal University. He was one of the experts in the team for China's national reform from 2000 to 2010 and a member of the National Working Committee on Curriculum and Textbook Development in Basic Education since 2010. He is also an Honorary Professor of the Hong Kong Institute of Education (2007–2013). In 1967, Professor Gao received his undergraduate degree in Physics from Sun Yat-Sen University in Guangzhou China. Prof Gao studied in the Centre of Science Education in the University of Waikato in New Zealand in the 1980s and earned his PhD from the University of Hong Kong in the 1990s. His research covered two major areas: science/physics education and educational assessment. His present research projects focus on textbook analysis/evaluation, student learning assessment, and conceptions of teachers and students. He has published more than 200 journal papers and book chapters and 30 books in the past 30 years.

Enshan Liu is a Professor and the Director of the Department of Biology Education and the director of Science Education Research Center of Beijing Normal University. He is currently a member of National Academic Committee Board of General Education Curriculum, a member of National Academic Committee Board of Teacher Education Curriculum, an executive member (and the former vice president) of The East-Asian Association for Science Education, the deputy president of China National Association of Biology Teacher, the chairman of National Committee of High School Biology Olympiad, Editor-in-Chief of *Biology Bulletin (Chinese)*. In the most recent biology curriculum reform started in 1999 in mainland China, Prof. Liu, as a principal investigator, completed projects in developing new National Junior Secondary School (grade 7–9) Biology Curriculum Standards in 2001, National Senior High School (grade 10–12) Biology Curriculum Standards in 2003, and revised version of grade 7–9 National Biology Curriculum Standards in

L.L. Liang et al. (eds.), *Chinese Science Education in the 21st Century: Policy, Practice, and Research*, Contemporary Trends and Issues in Science Education 45, DOI 10.1007/978-94-017-9864-8

2011. He led projects in developing standard-based new biology textbooks and teaching materials that are being used by millions of 7–9 and 10–12 grades students in China. In the last 12 years, Prof. Liu played an active role in national projects for professional development of in-service biology teachers. As part of the tasks, he completed eight online teachers' training projects in the past six years with target population of more than 10,000 biology teachers. Prof. Liu published six books for teacher education in recent years. He has also been taking part in and organizing China National Biology Olympiad since 1993.

Winnie Wing-mui So is the Head and Professor of Department of Science and Environmental Studies of the Hong Kong Institute of Education, and the Director of the Centre for Education in Environmental Sustainability. Academically, Prof. So has rich and diverse experiences associated with learning and teaching of science in primary and secondary classrooms as well as teacher education. Prof. So is also active with professional contribution, being an executive member of the World Education Research Association, Asia-Pacific Education Research Association, and East-Asian Association for Science Education, respectively.

Bing Wei is an Associate Professor of Science Education in the Faculty of Education, University of Macau. His research interests include social contexts of science curriculum, scientific literacy, history and philosophy of science and science teaching, and science teacher development. Dr. Wei has published papers on science/chemistry curriculum in China in *Science Education*, *International Journal of Science Education*, *Research in Science Education*, and *School Science and Mathematics*. He has also published several book chapters in English.

Epilogue

Xiufeng Liu, Gavin W. Fulmer, and Ling L. Liang

As we stated in the introduction to this book, one primary purpose of the book is to "give science educators in China an opportunity to tell their own stories of a decade-long reform directly and collectively to the English-literate world." Although the chapters may not necessarily tell the whole story of recent Chinese science education reforms, nor do they necessarily represent the current state of the art of science education research and practices throughout China, they do provide a reasonable "snapshot" of Chinese science education policies, research efforts, and practices. In addition to the above purpose, as editors we have also written introductions to the chapters in order to connect the reported Chinese science education policies, research efforts, and practices with the English science education literature and to situate the chapters within specific Chinese contexts. Our goal for the introductions was to help readers better appreciate Chinese science education in both Chinese and international contexts. Ultimately, this book intends to serve as a mechanism to integrate Chinese science education into global science education. In this concluding section, we summarize major issues that connect Chinese science education with science education outside, particularly in the USA. We further discuss some broader implications about science education globalization and international collaboration, followed by recommendations for Chinese science education researchers and science education researchers in the developed world.

Science Curriculum and Instruction Reforms: From Visions to Classroom Practices

Much of what has been taking place in Chinese school science education over the past decade has been centered on the science curriculum reforms initiated by the Chinese Ministry of Education in 2001 (see Chaps. 1, 2, 3, 4, 5, and 6). Two central

L.L. Liang et al. (eds.), *Chinese Science Education in the 21st Century: Policy, Practice, and Research*, Contemporary Trends and Issues in Science Education 45, DOI 10.1007/978-94-017-9864-8

themes of Chinese science curriculum reforms are scientific literacy and science inquiry as both content and pedagogy. As discussed in the chapters, emphases on these two themes are due to the influence of science curriculum reform documents in developed countries, such as the *Benchmarks for Science Literacy* (Project 2061 1993) and the *National Science Education Standards* (NRC 1996). Curriculum reforms in China and the USA, like in all other countries, have been a constant occurrence over the past century (Liu 1996; Ravitch 2000). Often reforms come and go without necessarily leaving a satisfactory feeling among educators, policy makers, and the general public. As stated in the Part I Introduction, as an example, the recent wave of science education reform in the USA, marked by the *Benchmarks for Science Literacy*, the *National Science Standards*, and the NCLB legislation, was more a failure than success. No one would question the value and the good intention of the reforms; what failed was the implementation. Although systematic and comprehensive evaluation of the recent Chinese science curriculum reforms is not available, studies reported in Chaps. 3, 4, and 5 present some mixed results, i.e., elementary schools seem to have implemented the reformed curriculums to some extent, but the implementation of the reformed science curriculums in junior and senior high schools has been met with resistance. Junior and senior high schools face different constraints, particularly standardized testing, compared to elementary schools in China. As the USA has learned over recent efforts in educational reforms, in order for curriculum reforms to achieve their intended effects, all system components (e.g., teacher education, assessment, resources) must be aligned (Carnegie Corporation of New York 2009). Both China and developed countries including the USA face the same challenge of implementing curriculum reform visions. Addressing this challenge requires both policy innovation and research support; coordinating policy innovation with research will more likely result in desirable outcomes of science curriculum reforms.

The science curriculum reform initiated at the turn of the twenty-first century has brought visible changes to K-12 science education in China, from the curriculum structure and the textbooks to the classroom practices. However, at a deeper level, the knowledge-centered and exam-driven education system and people's beliefs remain largely unchanged in reality. In some teachers' words, the implementation of new curriculum standards is like "adding fresh water to the same Chinese herbal medicine." Given the perceived extremely high stakes of the college entrance examinations in a student's life, teachers, students, and their parents start the preparation as early as kindergarten. It appears that changes in the national college admission criteria would play a critical role in determining the success or failure of any serious curriculum reform efforts. In addition, even as "formative assessment" and "assessment for learning" were introduced as important aspects of the reformed curriculum and instruction, the external evaluation systems at middle and high school levels are largely unchanged. Such a mismatch between external evaluation and classroom instruction has hindered the progress of the reforms. Many Chinese teachers have described themselves as "dancing in shackles" (Wu 2004, cited in Wang and Zhao 2011).

In the Chinese context, direct teaching has been proven to be effective in developing student understanding of content knowledge or in helping students prepare for the standardized exams. However, Chinese "direct teaching" should not be simply interpreted as "rote learning." In fact, Chinese teachers focus on developing conceptual understanding and meaningful learning through teacher lecturing, questioning, explaining, and having students practice vigorous problem solving. One highly regarded teaching practice is called "举一反三," i.e., when a teacher demonstrates how to solve one problem in class, the students are encouraged to find and solve three or more problems in different contexts within or outside of the classroom. Chinese teachers also pay attention to individual differences in teaching (因才施教). However, inquiry-oriented instruction is something novel for most people. In order to help teachers incorporate multi-dimensional goals of science education in teaching (i.e., the development of knowledge and skills, process and methods, and attitudes and values) and to cultivate and develop students' scientific literacy for the twenty-first century, curriculum developers may consider providing necessary scaffolds for both teachers and students, by incorporating different instructional and curriculum design approaches such as the 5-E model, the project-based approach, and the model-based approach. Lecturing and teacher explanation, as part of the direct teaching method embraced by both students and teachers, do not have to be abandoned in an inquiry-oriented curriculum. In a 5-E model, for instance, teachers' lectures and explanations could be provided in the "explain" phase—after students have had a chance to explore and ask questions. Promoting inquiry-oriented teaching should not be simply interpreted as replacing direct instruction with inquiry entirely. Rather, inquiry learning should be integrated into the curriculum design, and inquiry-oriented teaching strategies should be added to teachers' pedagogical repertoires (Cobern et al. 2010; Saunders-Stewart et al. 2012).

Over the past 10 years, the basic education curriculum reform has made significant progress and promoted the development of rural education in China. Some promising practices based on the theoretical model of integration of urban and rural education have been summarized and published in Chinese journals (e.g., Tan et al. 2012). More empirical research studies in this area are needed. Such studies will be of great interest for other developing countries that are facing similar challenges.

Preservice and In-Service Science Teacher Education

Three chapters in the book address preservice and in-service science teacher education. As presented in the chapters as well as highlighted in the Part VI Introduction, both Chinese preservice and in-service teacher education tend to place more emphasis on developing pedagogical content knowledge (PCK). Chinese preservice teacher education and in-service teacher professional development models with a focus on topic-specific PCK are well supported by the existing literature. In these models, instructional development and professional development are often merged and become an integral part of the process of improving science learning and teaching in schools. In China, teachers seek professional development and proceed with a clear career path starting from a "junior rank" toward an "intermediate" and a "senior" rank and finally achieving the "master teacher" rank (or 特级教师). Only a small fraction of the teacher population reaches the final master level. This system appears to work well as most Chinese teachers are respected and treated as professionals by society, while they often consider themselves as lifelong learners. However, the understanding and development of teachers' beliefs about the nature of science and inquiry and beliefs about science learning and teaching seem to be largely ignored in both preservice and in-service teacher education programs in China. Teachers tend to teach science in the way that they were taught. In a recent study investigating science education as an undergraduate major in China (Zhang and He 2013), the students who experienced science education teacher preparation programs were strongly dissatisfied with their pedagogy courses. One cited example was that the teacher educator lectured on student-centered pedagogy by simply reading his PowerPoint slides to the class. Therefore, when teacher education programs are examined, it is not enough to simply check the amount of required coursework. It is equally important to examine how the science content and pedagogical courses are taught. To develop teachers' informed views on learning and teaching of science and to foster an informed understanding of nature of science and scientific inquiry, teachers and teacher candidates need to experience being active learners who construct their understanding through explicit, inquiry-oriented instruction. Research has shown that Chinese science curriculum reforms have particularly encountered a setback in implementing the integrated science curriculum in junior high schools (Zhang and He 2013), and the nature of science seems to be a weak aspect in Chinese science textbooks and possibly teacher knowledge (Chap. 6 of this book; Wei et al. 2013).

Overall, although Chinese science teachers are strong in PCK, they may not be well informed in cultural awareness and individual student learning differences. Chinese science teachers may be content experts, but they are typically weak in interdisciplinary science. The opposite is true for American science teachers. There are clear complementary strengths when comparing the Chinese preservice and in-service teachers with that in the USA. In a recent book chapter, Liu (2014) calls the above differences, along with differences in teacher education candidate selectivity, a "paradigm contrast."

Thus, for teacher education, the USA may learn from China about how to develop science teachers' PCK through carefully designed and coordinated subject-specific and topic-specific coursework and professional development activities. In the meantime, Chinese educators can learn from other countries about how to develop and cultivate teachers' teaching orientations or personal educational philosophy by broadening preservice teachers' knowledge and beliefs about educational context, how students learn and develop, as well as how to address student diversity and better meet the needs of individual students and society. To achieve this and break the exam-driven cycle, a coordinated change in the assessment system seems to be particularly critical for any educational reform efforts to be fully implemented in China.

Assessment Systems

The three chapters (Chaps. 9, 10, and 11) on assessment present some major changes as well as reform efforts related to science assessment in China. In a country that is known for its external, high-stakes, standardized testing-driven instruction, the recent science education reforms promoting multifaceted approaches to student assessment are encouraging. While high-stakes standardized testing will stay for at least some time in the future in China, how to incorporate other forms of assessment, particularly formative assessment and performance assessment, into science teaching is a major challenge for continuing science education reforms in China. This challenge is consistent with a similar emphasis in the USA for systems of assessment, but from a different direction—for the USA, the challenge was introducing standardized testing into science teaching (NRC 2005). A system of assessment consists of both teacher-made assessment and external assessment and includes diagnostic, formative, and summative assessments (Liu 2009). In the USA, traditionally more emphases were placed on diagnostic and teacher-made assessments; recent reforms intend to increase emphases on external standardized tests as well as formative assessments. While China and the USA are currently placing different emphases on different types of assessments, both countries are moving toward more diversity and multiplicity in approaches to student assessment.

Research on Informal Science Learning

As presented in Chaps. 12, 13, 14, 15, and 16 and the Part V Introduction, research on informal science education in China is still in a beginning stage. Although there is a hierarchical national infrastructure in China to carry out science popularization and there have been major national campaigns to increase Chinese citizens' scientific literacy, the theoretical foundation of those campaigns appears to be outdated from a Western point of view. In developed countries, scientific literacy was an emphasis in the 1950s and 1960s, followed by an emphasis on public understanding of science (i.e., beyond knowledge to include attitude toward science) in the 1970s and 1980s, and an emphasis on public engagement with science since the 1990s (Bauer et al. 2007; Wynne 1995). Further, there are efforts to incorporate public engagement with science into informal science education and integrate formal and informal science education (Center for Advancement of Informal Science Education 2009; Stocklmayer et al. 2010). Chinese informal science education research and practice are still guided by a science popularization framework, and the interaction between formal and informal science education is minimal. Since improving citizens' scientific literacy remains a task for all countries, the large-scale and nationally coordinated efforts to increase citizens' scientific literacy in China warrant attention by researchers in other countries. On the other hand, China can benefit from examining the historical evolution of informal science education in developed countries in order to develop a more comprehensive informal science education system that is guided by modern informal science learning theories.

Science Education Research as a Discipline

Although the Chinese National Association for Science Education was established in 2009, science education research as a discipline in China remains in a preliminary stage (Chap. 1). The chapters in this book are indications of the quality of science education research in China. The gap in rigor between Chinese science education research and that in the developed world is apparent.

According to Fensham (2004), science education research as a discipline needs to meet criteria related to structure, substance, and outcome. The structural criteria include academic recognition/ranks at universities, research journals, professional associations, research conferences, research centers, research training, and funding for research. While it is possible now at some Chinese universities to obtain a full-professor rank in science education and there are a few science education research centers at a few Chinese universities, currently there is no national journal devoted to science education, nor is there a regular national science education research conference in China. Doctoral training in science education remains primarily in disciplinary science departments (e.g., chemistry, physics), and there are no national science education research funding programs in China. Thus, Chinese science education as a discipline still needs to build its structural foundation.

In terms of the substance and outcome criteria for science education as a discipline, major domains of science education inquiry in China remain limited and are specifically confined within disciplinary science education (e.g., chemical education). Development of foundational science education theories is extremely rare. On the other hand, science education research in China is closely tied to practices and the central government's education policies, e.g., science curriculum reform documents (Wei 2012), which is clearly due to the pedagogical research tradition discussed in the Part I Introduction. As Jenkins (2001, 2004) defines, the empirical tradition is primarily concerned with developing and testing general science education theories, while the pedagogical tradition is primarily concerned with improvement of science curriculum and instruction in specific disciplines. Science education research in developed countries follows primarily the empirical tradition, while Chinese science education research follows primarily the pedagogical tradition. The advantage of the pedagogical tradition is that research is closely related to practice and policy; thus, the body of discipline-specific knowledge is solid. On the other hand, the empirical tradition has the advantage of developing a strong body of general science education knowledge; thus, transferability of knowledge across disciplines is strong. Once again, the complementarity between Chinese science education research and Western science education research is apparent.

The above themes and the complementarity between Chinese science education and Western science education have important implications. In an era of new waves of science education reforms in many countries, such as the *Next Generation Science Standards* in the USA, there are grand challenges facing science education in the USA and in many other countries (please refer to the April 19, 2013, issue of *Science*). Addressing those grand challenges, such as "identifying the underlying mechanisms that make some teacher professional development more effective than others," requires global solutions. As we have stated above, the relative strengths and weaknesses of Chinese science education and Western science education make international collaboration desirable. Unlike economic globalization—which implies intense competition against each other—science education globalization implies collaboration because science and engineering talents are no longer confined within country boundaries, but are world assets flowing freely between countries (Liu et al. 2012).

Based on the above premise that international collaboration in science education is beneficial to all countries, we make the following recommendations:

For Chinese Science Educators

1. Recognize the importance of and accept the reality that English is the medium of communication for international science education (Martin and Siry 2011). In order to participate fully in international science education (i.e., publishing in English-language science education journals, presenting at international science education conferences), Chinese science educators need to learn the use of not only English language in general but also technical writing conventions such as the AERA Standards for Reporting on Empirical Social Science Research in AERA Publications (AERA 2006) and the Publication Manual of the American Psychological Association (APA 2013). Such efforts will greatly increase the visibility and accessibility of findings and publications from mainland China to scholars around the world.
2. Make sincere efforts to develop general science education theories that are relevant both to Chinese science education across individual science disciplines and to science education in other countries. This may start with a critical review and testing of general science education theories developed in the West, followed by further development of alternative theories from within Chinese science education.
3. Expand the research tradition from pedagogical only to both pedagogical and empirical. There is a strong need to provide science education researchers, particularly doctoral students, with a solid foundation of both qualitative and quantitative research designs. There must be a shift in research approaches from the currently dominant form, which are often personal reflections and position

statements, toward scientific research characterized by theory-driven empirical studies (NRC 2002). This will greatly increase the perceived value and impact of research both within and outside of China.

For Science Educators in the Developed World

1. Expand the research tradition from an empirical tradition only to both an empirical and pedagogical traditions. While keeping the tradition of developing general science education theories, more emphasis needs to be placed on discipline-specific research and practice. To this end, enhancing subject matter knowledge of both science teachers and science education researchers is necessary.
2. Make science education research more directly relevant to both policy and practice. Research in the West has been criticized for lacking relevance to practice and policy. In order to address this perceived weakness of Western science education research, more attention must be paid to identifying research questions that are important in practice and policy. This may require a stronger connection with current educational practice within schools and more information about the import of context of schools and classrooms. This may also require a more flexible interpretation of the scope of proposed educational theories.
3. Continue seeking a balance between national standardization of curriculum and assessment and local educational autonomy and teacher professionalism. As some Western countries are moving toward more centralized standardized testing and accountability, they must be cognizant of the negative effects experienced in China and other counties with such standardized testing. In particular, the higher the stakes and the greater consequences for students, teachers, and schools, the greater the potential there is for the assessment system to be distorted or diverted. It is necessary to continue to advance theory and best practices for assessment systems for science teaching and learning.
4. Make an explicit effort to communicate and connect with researchers in mainland China, as well as other developing nations. These education research communities are engaged in a variety of research that can interest scholars in many areas. The relationships between policy and practice in these countries can provide useful points of comparison for Western counterparts on policy and the process of interpretation and implementation into practice. Furthermore, the differences and similarities in contexts and findings can be powerfully informative about the potential for general education theories and the limits to such general efforts.

In conclusion, we hope this book piques the interest of science educators both inside and outside China in seeking collaboration. We also hope that this book makes a contribution toward science education globalization.

References

American Education Research Association (AERA). (2006). AERA standards for reporting on empirical social science research in AERA publications. *Educational Researcher, 35*(6), 33–40.

American Psychological Association. (2013). *Publication manual of the American psychological association*. Washington, DC: American Psychological Association.

Bauer, M. W., Allum, N., & Millder, S. (2007). What can we learn from 25 years of PUS survey research? Liberating and expanding the agenda. *Public Understanding of Science, 16*, 79–95.

Carnegie Corporate of New York. (2009). *The opportunity equation: Transforming mathematics and science education for citizenship and global economy*. New York: Carnegie Corporate of New York.

Center for Advancement of Informal Science Education (CAISE). (2009). *Many experts, many audiences: Public engagement with science and informal science education: A CAISE inquiry group report*. Washington, DC: CAISE.

Cobern, W. W., Schuster, D., Adams, B., Applegate, B., Skjold, B., Undreiu, A., Loving, C. C., & Gobert, J. D. (2010). Experimental comparison of inquiry and direct instruction in science. *Research in Science & Technological Education, 28*(1), 81–96.

Fensham, P. J. (2004). *Defining an identify: The evolution of science education as a field of research*. Dordrecht: Kluwer Academic Publishers.

Jenkins, W. E., et al. (2001). Research in science education in Europe: Retrospect and prospect. In H. Behrend (Ed.), *Research in science education: Past, present, and future* (pp. 17–25). Hingham: Kluwer Academic Publishers.

Jenkins, W. E. (2004). Science education: Research, practice and policy. In E. Scanlon (Ed.), *Reconsidering science learning*. London: Routledge/Falmer.

Liu, X. (Ed.). (1996). *Mathematics and science curriculum change in the People Republic of China*. Lewiston: Edwin Mellon Press.

Liu, X. (2009). *Essentials of science classroom assessment*. Thousand Oaks: Sage Publications.

Liu, X. (2014). Paradigm contrast: A personal reflection of approaches to preservice science teacher education in China and North America. In M. Wen (Ed.), *Educational clashes and conformities of American tenets and Chinese wisdoms: In search of "Middle Ground" by teacher educators from China* (Chapter 10, pp. 127–142). New York: Teachers College Press.

Liu, X., Liang, L. L., & Liu, E. (2012). Science education research in China: Challenges and promises. *A Special Issue of the International Journal of Science Education, 34*(13), 1961–1970.

Martin, S. N., & Siry, C. (2011). Networks of practice in science education research: A global context. *Journal of Research in Science Teaching, 48*(6), 592–623.

National Research Council. (1996). *National science education standards*. Washington, DC: National Academies Press.

National Research Council. (2002). *Scientific research in education*. Washington, DC: National Academies Press.

National Research Council [NRC]. (2005). *Decision making for the environment: Social and behaviouralbehavioral science research priorities*. Washington, DC: National Academies Press.

Project 2061. (1993). *Benchmarks for science literacy*. New York: Oxford University Press.

Ravitch, D. (2000). *Left back: A century of battles over school reform*. New York: A Touchtone Book.

Saunders-Stewart, K. S., Gyles, P. D. T., & Shore, B. M. (2012). Student outcomes in inquiry instruction: A literature-derived inventory. *Journal of Advanced Academics, 23*(1), 5–31.

Stocklmayer, S. M., Rennie, L. J., & Gilbert, J. K. (2010). The role of the formal and informal sectors in the provision of effective science education. *Studies in Science Education, 46*(1), 1–44.

Tan, J., Wei, D., & Ran, L. (2012). On the integration process complexity of the urban and rural education practice. *Advances in Education (教育进展)*, *2*, 51–55 (in Chinese).

Wang, J., & Zhao, Z. (2011). Basic education curriculum reform in rural China: Achievements, problems, and solutions. *Chinese Education and Society, 44*(6), 36–46.

Wei, B. (2012). In pursuit of professionalism in the field of chemistry education in China: The story of Zhixin Liu. *International Journal of Science Education, 34*(13), 1971–1989.

Wei, B., Li, Y., & Chen, B. (2013). Representations of nature of science in selected histories of science in the integrated science textbooks in China. *School Science and Mathematics, 113*(4), 1–11.

Wynne, B. (1995). Public understanding of science. In S. Jasanoff, G. E. Markle, J. C. Petersen, & T. Pinch (Eds.), *Handbook of science and technology studies*. Thousand Oaks: Sage.

Zhang, H., & He, H. (2013). Effectiveness of undergraduate science education program in China: Perceptions of seniors and alumni. *International Journal of Science Education, 34*(13), 1991–2013.

Index

L.L. Liang et al. (eds.), *Chinese Science Education in the 21st Century: Policy, Practice, and Research*, Contemporary Trends and Issues in Science Education 45, DOI 10.1007/978-94-017-9864-8

D